4v193

NATURAL ENVIRONMENT RESEARCH COUNCIL
INSTITUTE OF GEOLOGICAL SCIENCES

MEMOIRS OF THE GEOLOGICAL SURVEY OF GREAT BRITAIN
SCOTLAND

The Geology of East Fife

(Explanation of the Fife portion of 'One-inch' Geological Sheet 41 and part of Sheet 49)

By

I. H. FORSYTH, B.Sc. AND J. I. CHISHOLM, M.A.

with contributions by
R. W. Elliot, B.Sc., E. H. Francis, D.Sc., F.R.S.E. and
R. B. Wilson, D.Sc., F.R.S.E.

EDINBURGH: HER MAJESTY'S STATIONERY OFFICE

The Institute of Geological Sciences was formed by the incorporation of the Geological Survey of Great Britain and the Museum of Practical Geology with Overseas Geological Surveys and is a constituent body of the Natural Environment Research Council.

© *Crown copyright 1977*
First published 1977

ISBN 0 11 880164 3

Geology of East Fife (*Mem. Geol. Surv.*) PLATE I (*Frontispiece*)

RAISED BEACHES, KINCRAIG, NEAR ELIE

EXPLANATION OF PLATE I (*Frontispiece*)

The beaches and their cliffs were cut in the soft tuffs and agglomerates of the Kincraig neck during short pauses in the late-Glacial marine regression (D 1766)

PREFACE

THE district of east Fife as described in this memoir includes most of the area covered by Sheet 41 of the 'One-inch' Geological Map of Scotland and the south-eastern part of Sheet 49. These sheets were originally surveyed by H. H. Howell, Sheet 41 being published in 1861 and Sheet 49 in 1884. A second edition of Sheet 41, incorporating the results of a partial revision by A. Geikie and J. S. G. Wilson, was produced in 1889. A resurvey of Sheet 41 was begun in 1932 and was almost complete when the Second World War intervened. The surveyors were J. K. Allan, Dr. F. W. Anderson, D. Haldane, Mr. J. Knox and Mr. T. R. M. Lawrie, under G. V. Wilson as District Geologist. Work was continued after the war under the guidance of J. B. Simpson, Dr. J. R. Earp and Mr. T. R. M. Lawrie as District Geologists. The resurvey was completed in 1947 by Mr. G. S. Johnstone, and between 1961 and 1963 Dr. E. H. Francis and Mr. I. H. Forsyth, with some assistance from Mr. J. I. Chisholm, re-examined all the important sections. Messrs. Forsyth and Chisholm supervised a programme of Geological Survey boreholes in 1963-64, the results of which have already been published. The east Fife portion of Sheet 49 was resurveyed between 1961 and 1963 by Mr. Chisholm, with some assistance from Mr. Forsyth. A second Geological Survey drilling programme, in the St Andrews area, was carried out in 1966 and supervised by Mr. Chisholm: the results are discussed in the present volume. Staff of the Geophysical Department provided valuable assistance in the investigation of specific field problems, mostly concerning the location and delimiting of igneous intrusions. Cores from the relatively few commercial boreholes that have been drilled for coal, oil-shale or water were examined by R. G. Carruthers, D. Haldane, W. Manson, G. A. Goodlet and Messrs. Lawrie, Forsyth and Chisholm. 'Solid' and 'Drift' editions of One-inch Geological Sheet 41 were published in 1970 and Sheet 49 is now in press. Geological maps of east Fife at a scale of 1:10 560 are available for public reference at the Edinburgh office of the Institute of Geological Sciences, from which copies can be obtained on application.

Neither of the original one-inch maps was accompanied by a sheet explanation, but the second edition of Sheet 41 was followed in 1902 by a memoir by A. Geikie on the geology of eastern Fife. This memoir covered the Fife portions of Sheets 41 and 49, including the tract of Old Red Sandstone rocks which lies north-west of the area reviewed in the present volume. A small part of the east Fife district, around Largo, falls within the area covered by the memoir entitled 'The economic geology of the Fife Coalfields, Area III' by Mr. J. Knox, which was published in 1954. Well sections and information on the hydrogeology of east Fife are given in a Geological Survey Water Supply Paper entitled 'Records of wells in the areas of Scottish one-inch geological sheets Kinross (40), North Berwick (41), Perth (48) and Arbroath (49)', by Mrs. N. P. D. Jackson (published 1967).

The present memoir has been written mainly by Mr. I. H. Forsyth and Mr. J. I. Chisholm, and edited by Mr. T. R. M. Lawrie. Dr. R. B. Wilson identified the Carboniferous fossils, most of which were collected by Messrs. P. J. Brand and D. K. Graham, and prepared the chapter on Carboniferous palaeontology and the lists of Carboniferous fossils. Dr. E. H. Francis wrote much of the chapter on the volcanic necks, to which Mr. R. W. Elliot contributed the account of the petrography of the associated igneous intrusions. Certain aspects of Carboniferous palynology were investigated by Dr. R. Neves, University of Sheffield, and Dr. J. R. Haynes and Dr. R. C. Whatley, University College of Wales, Aberystwyth, identified and commented on the environmental significance of the foraminifera and the ostracods found in the late-Glacial marine and estuarine sediments. Radiometric age-dating of some of the igneous rocks was carried out by the Isotope Geology Unit under Dr. N. J.

Snelling. Chemical analyses of igneous and carbonate rocks were made by staff of the Laboratory of the Government Chemist. The photographs were taken by W. D. Fisher and Messrs. A. Christie, R. Anderson and I. Bowler.

This memoir received final approval from Sir Kingsley Dunham, F.R.S., as Director before his retirement in December 1975.

A. W. WOODLAND
Director

Institute of Geological Sciences
Exhibition Road
South Kensington
London SW7 2DE

CONTENTS

(References are listed at the end of each chapter)

	PAGE
PREFACE BY THE DIRECTOR	iii
LIST OF ILLUSTRATIONS	vii

CHAPTER 1. INTRODUCTION 1
Area and physical features, 1. Geological sequence, 3. Outline of geological history, 4.

CHAPTER 2. UPPER OLD RED SANDSTONE 7

CHAPTER 3. CALCIFEROUS SANDSTONE MEASURES 9
General account: stratigraphy, 9; lithology and facies, 14; conditions of deposition, 18; trace-fossils, 19. Description of sections: Fife Ness, 21; Anstruther–Pathhead, 21; Anstruther–Fife Ness, 27; Fife Ness–Boarhills–Dunino, 30; St Andrews–Ceres, 36; Cameron–Lochty–Carnbee–Elie–St Monance, 51.

CHAPTER 4. LOWER LIMESTONE GROUP 60
Introduction, 60. Lithology, 61. Stratigraphy, 62.

CHAPTER 5. LIMESTONE COAL GROUP 82
Introduction, 82. Lithology, 84. Stratigraphy, 85. St Monance Coalfield, 86. Elie Coalfield, 86. Rires Coalfield, 89. Elie–Largoward district, 89. Ceres–Denhead district, 94.

CHAPTER 6. UPPER LIMESTONE GROUP 103
Introduction, 103. Lithology, 103. Stratigraphy, 103.

CHAPTER 7. PASSAGE GROUP 111

CHAPTER 8. COAL MEASURES 117

CHAPTER 9. CARBONIFEROUS PALAEONTOLOGY 122
Introduction, 122. Calciferous Sandstone Measures, 123. Lower Limestone Group, 130. Limestone Coal Group, 132. Upper Limestone Group, 133. Passage Group, 134. Coal Measures, 134.

CHAPTER 10. OLIVINE-DOLERITE SILL-COMPLEX 136
Introduction, 136. Field characters, 139. Petrography, 148.

CHAPTER 11. QUARTZ-DOLERITE AND THOLEIITE INTRUSIONS 162
Introduction, 162. Sills, 163. Dykes, 167. Petrography, 168.

CHAPTER 12. VOLCANIC NECKS 171
Introduction, 171. Structure and lithology, 172. Related intrusions and cryptovolcanic structures, 173. Mechanism of neck emplacement, 174. Age-relations, 177. Field relations, 180. Petrography of the neck intrusions, 206.

	PAGE
CHAPTER 13. STRUCTURE	221

Introduction, 221. Folds, 221. Easterly to south-easterly normal faults, 224. North-easterly zones of complex structure, 227. Minor wrench-faults, oblique faults and thrusts, 229.

CHAPTER 14. PLEISTOCENE AND RECENT 233

Glacial erosion features and till deposits, 233. Glacial meltwater deposits, 235. Late-Glacial marine deposits and shoreline features, 237. Deposits of the post-Glacial marine transgression and regression, 248. Freshwater alluvium and peat, 250. Blown sand, 250.

CHAPTER 15. ECONOMIC GEOLOGY 252

Coal, 252. Limestone, 252. Ironstone, 253. Oil-shale, 253. Building stone, 253. Roadstone, 253. Mineral veins, 254. Sand and gravel, 254. Brick-clay, 255. Semi-precious stones, 255. Water supply, 255.

APPENDIX 1. LIST OF CARBONIFEROUS FOSSILS.. 257

APPENDIX 2. LIST OF GEOLOGICAL SURVEY PHOTOGRAPHS 269

INDEX 271

ILLUSTRATIONS

TEXT-FIGURES

		PAGE
Fig. 1.	Geological sketch-map of east Fife	2
Fig. 2.	Sketch-map showing approximate outcrops of the subdivisions of the Calciferous Sandstone Measures	12
Fig. 3.	Vertical sections in the Calciferous Sandstone Measures between Pathhead and Anstruther, and near Fife Ness	20
Fig. 4.	Vertical sections showing lateral variation in parts of the Pittenweem Beds and Sandy Craig Beds	37
Fig. 5.	Borehole sections in the Pathhead Beds in the Denhead area	45
Fig. 6.	Vertical sections in the Pathhead Beds	47
Fig. 7.	Generalised vertical section of strata proved in boreholes around Lochty	53
Fig. 8.	Generalised vertical sections of the Lower Limestone Group in Fife and Midlothian	64
Fig. 9.	Comparative vertical sections in the Lower Limestone Group	67
Fig. 10.	Comparative vertical sections of the Limestone Coal Group in Fife	83
Fig. 11.	Comparative vertical sections in the lower part of the Limestone Coal Group in the Elie–Largoward area	90
Fig. 12.	Comparative vertical sections in the Limestone Coal Group in the Ceres area	95
Fig. 13.	Generalised vertical section of the Upper Limestone Group	104
Fig. 14.	Sketch-map of the shore section at Temple, Lower Largo	114
Fig. 15.	Generalised vertical section of the Coal Measures in the Largo district	118
Fig. 16.	Sketch-map of the olivine-dolerite sill-complex showing areal distribution of the various rock-types	140
Fig. 17.	Schematic diagram illustrating some of the processes involved in the emplacement of volcanic necks in a sedimentary pile and, particularly, the intermittent subsidence of bedded pyroclastic rocks from the inner flanks of superficial tuff-rings	176
Fig. 18.	Diagram showing the stratigraphical position of the volcanic necks, at present levels of erosion, exposed between Lundin Links and St Monance	178

		PAGE
FIG. 19.	Sketch-map showing distribution of volcanic necks in east Fife	181
FIG. 20.	(a) Map and horizontal section of the Lundin Links neck	182
	(b) Map and horizontal section of the Viewforth neck	182
FIG. 21.	(a) Map and horizontal section of the Ruddons Point neck	185
	(b) Map and horizontal section of the Kincraig neck	185
FIG. 22.	(a) Map and horizontal section of the Craigforth neck and structures between it and the Chapel Ness intrusion	187
	(b) Map and horizontal sections of the Elie Harbour and Elie Ness necks	187
FIG. 23.	Map of shore between Elie Ness and Ardross Farm, showing the Ardross Fault, and plans with horizontal sections of the Wadeslea and Ardross necks	192
FIG. 24.	(a) Map of shore between Ardross Farm and Newark Castle, showing the Ardross Fault, Coalyard Hill and Newark necks and associated structures	195
	(b) Map and horizontal section of shore between Newark Castle and St Monance, showing the Dovecot, Davie's Rock and St Monance cryptovolcanic structures and necks	195
FIG. 25.	Map and horizontal section of the Kinkell Ness neck	199
FIG. 26.	Sketch-map showing structural features in east Fife	222
FIG. 27.	Sketch-map showing distribution of till deposits and glacial erosion features	234
FIG. 28.	Sketch-map showing distribution of late-Glacial marine deposits	238
FIG. 29.	Drift map of St Andrews	243
FIG. 30.	Diagrammatic horizontal section along Canongate Road west of junction with Largo Road, St Andrews	246
FIG. 31.	Diagrammatic horizontal section through drift deposits at St Andrews	246

PLATES

PLATE	I.	Raised beaches, Kincraig, near Elie	*Frontispiece*
			FACING PAGE
PLATE	II.	Intrusion in Ruddons Point neck, showing transition from basalt through basaltic breccia to tuff	184
PLATE	III.	A Columnar jointing in basalt intrusion, Kincraig neck	186
		B Irregular intrusion of bleached basalt cutting shales with ironstone bands, near Davie's Rock, St Monance	186

			FACING PAGE
PLATE	IV.	A Basaltic bomb in the Elie Ness neck, showing impact effects in the underlying tuffs	190
		B Collapsed bedding in tuffs, eastern part of Elie Ness neck	190
PLATE	V.	A Cross-bedding and wedge-bedding in tuffs, Elie Ness neck	191
		B Centroclinally bedded tuffs in the Kinkell Ness neck	191
PLATE	VI.	A Flow-banding in tuffisite, north-eastern part of Coalyard Hill neck	194
		B Tuffisite at western margin of the St Monance neck	194
PLATE	VII.	The Rock and Spindle, Kinkell Ness neck	200
PLATE	VIII.	Largo Law, a complex volcanic centre	203

Chapter 1
INTRODUCTION

AREA AND PHYSICAL FEATURES

THE district of east Fife which is described in the following account comprises that part of the county of Fife which lies east of a line joining Lundin Links, Ceres, Kemback and Guard Bridge. It includes the area covered by Sheet 41 of the 'One-inch' Geological Map of Scotland[1] (with the exception of a small area south of the Firth of Forth) and the south-eastern part of Sheet 49. It is bounded to the east and south by the North Sea and the Forth estuary. Its western limit is taken at the western margin of Sheet 41 and to the north its limit has been drawn partly at the Dura Den Fault and partly at the junction between the Upper Old Red Sandstone and the Carboniferous north-east of Guard Bridge. The main features of the geology and the topography are shown in Fig. 1.

East Fife is an area of mainly low ground but rises to a height of 292 m at Largo Law (Plate VIII), a volcanic pile which makes a conspicuous feature especially when viewed from the west or south-west. Other prominent hills include Clatto Hill, Drumcarrow Craig, Dunicher Law, Kellie Law and Balcarres Craig. The higher ground is formed mainly by dolerite sills or volcanic necks. The extreme eastern part of the district is composed almost entirely of sedimentary rocks and the terrain is generally very low and featureless. The main stream draining the area is the Kenly Water, which rises near Largoward and flows east-north-eastwards past Dunino to the sea near Boarhills. Smaller sub-parallel streams include the Kinness and Cambo burns. The Ceres Burn flows northwards to join the River Eden. The Keil and Dreel burns are the largest of the southward-flowing streams.

Because of its peninsular isolation, east Fife has been little affected by industrialisation. Formerly, linen manufacture was an industry of some importance, particularly around Ceres, and for several centuries coal was mined from many small pits in the western part of the district and in the St Monance area. These industries have now disappeared from the district and today the most important occupations in east Fife are agriculture, fishing and tourism. The largest centre of population, and the most important holiday resort, is the ancient cathedral city of St Andrews, location of the oldest university in Scotland and also famous as the 'home of golf'. Other holiday resorts are Lundin Links, Largo and Elie. St Monance, Pittenweem, Anstruther and Crail combine tourism with inshore fishing; St Monance has in addition a boat-building industry. All these towns and villages are on the coast: inland villages include Ceres, Kemback, Strathkinness, Boarhills, Kingsbarns, Dunino, Arncroach, Largoward, Colinsburgh and Kilconquhar. Agriculture flourishes especially on the lower ground near the coast which the relatively dry and sunny climate makes one of the best areas in Scotland for arable farming. The higher ground is used to a greater extent for pasture land.

---

[1] Following the adoption by IGS of 1:50 000 as the scale for publication of geological maps in this series, new maps (including Sheet 49) will be published at this scale. Existing maps on the 1:63 360 scale (including Sheet 41) will be converted to the new scale when they become due for reprinting.

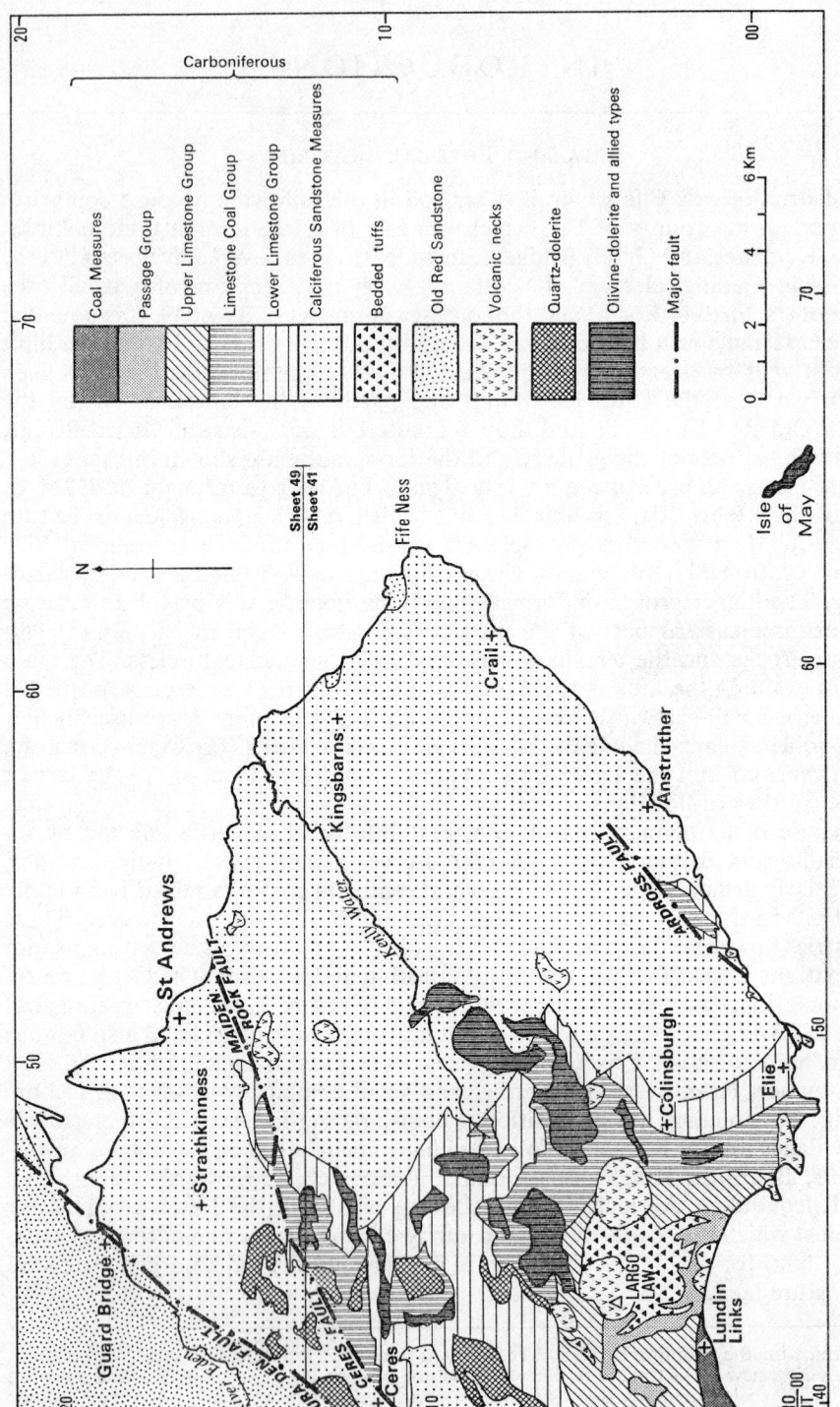

Fig. 1. *Geological sketch-map of east Fife*

Geological Sequence

The geological formations occurring within east Fife are summarised below.

SUPERFICIAL DEPOSITS (DRIFT)

RECENT AND PLEISTOCENE
  Blown sand
  Peat
  Freshwater alluvium
  Salt marsh deposits
  Present beach deposits
  Marine and estuarine alluvium (post-Glacial and late-Glacial)
  Glacial and fluvio-glacial sand and gravel
  Boulder clay

SOLID FORMATIONS

Thickness in metres

CARBONIFEROUS

UPPER CARBONIFEROUS

Upper Coal Measures:	sandstones, siltstones and mudstones	150
Middle Coal Measures:	sandstones, siltstones, mudstones, seatearths and coals	150
Lower Coal Measures:	sandstones, siltstones, mudstones, seatearths, coals and tuffaceous sediments	100+
Passage Group:	basaltic tuffs with subordinate basaltic lavas, sandstones, mudstones, shales, seatearths and thin coals	?300
Upper Limestone Group:	sandstones, siltstones, mudstones, marine limestones, coals, tuffs and volcanic detritus	400
Limestone Coal Group:	sandstones, siltstones, mudstones, seatearths and coals	240–275

LOWER CARBONIFEROUS

Lower Limestone Group:	sandstones, siltstones, mudstones, coals and marine limestones	120–230
Calciferous Sandstone Measures:	sandstones, siltstones, mudstones, dolomites, seatearths and coals	2000+

OLD RED SANDSTONE
UPPER

Balcomie Beds:	sandstones and mudstones, red, purple and yellow in colour, with lenticular bands of carbonate conglomerate	50+

INTRUSIONS

Quartz-dolerite sills and dykes of late Carboniferous or early Permian age.
Olivine-dolerite, teschenite and basanite sills of Upper Carboniferous or early Permian age.
Tuff, agglomerate and basanitic intrusions, mainly in necks, of Upper Carboniferous or Permian age.

Outline of Geological History

The Upper Old Red Sandstone rocks are confined to two small outcrops on the coast at Balcomie and Cambo. Consequently not a great deal can be said about the palaeogeography of east Fife at the time they were laid down. The absence of marine bands and the occurrence of red beds and lenticular bodies of carbonate conglomerate (probably erosion products of penecontemporaneous caliche-like accumulations) both suggest a semi-arid fluviatile environment with intermittently low water-table. The age of these beds is uncertain and at Balcomie they pass conformably upwards into strata which on facies grounds are assigned to the lowest subdivision of the Carboniferous.

This passage from Upper Old Red Sandstone to Calciferous Sandstone Measures (most if not all of which are Viséan in age) is marked by the disappearance of red beds and carbonate conglomerates and the appearance of non-marine faunal bands and seatearths. Gradually a rhythmic pattern of sedimentation, including the accumulation of coal-forming material, became established and marine incursions, almost certainly from the east or south-east, began to occur. As time went by the resultant marine bands tended to become more widespread and persistent and to have more abundant and varied faunas. There are however certain differences between the sedimentary cycles in the Calciferous Sandstone Measures and those that occur higher in the Carboniferous. The former show a much greater amount of lateral variation. They commonly include argillaceous non-marine bedded dolomite bands, and ostracods are abundant both in these and in the carbonaceous shales. This suggests that within the coastal or deltaic framework, to which coal-bearing cyclic sediments are usually ascribed, a non-marine environment in which ostracods (and sometimes bivalves such as *Naiadites*) flourished and dolomite sometimes accumulated was much more frequent and widespread in east Fife during most of the Viséan than it was anywhere in Scotland later in the Carboniferous. While the Calciferous Sandstone Measures were being deposited there were occasional episodes when red beds and carbonate conglomerates reappeared to the exclusion of coals and marine bands. This suggests a fall in the water-table, disappearance of swamps and withdrawal of the sea from the area at such times. Thick sandstones are present throughout the Calciferous Sandstone Measures sequence, and commonly occur in association with the red beds; they appear to have been laid down under fluviatile conditions, probably in channels crossing the raised delta-tops.

Marine conditions were more widespread and persistent in east Fife during the period of deposition of the Lower Limestone Group (which corresponds almost exactly with the P_2 stage of the Viséan) than in any other part of the Carboniferous. At times the water was sufficiently free from land-derived sediment to allow limestones to form, particularly in the lower half of the group. The cyclic pattern of sedimentation continued, however, and at intervals swamp vegetation flourished on the delta-tops, sometimes for long periods, leading to the formation of thick coal seams. During the succeeding Limestone Coal Group times (approximately the E_1 stage of the Namurian) the cyclic pattern continued but the marine incursions were rare and short whereas swamp vegetation frequently colonised the area for long periods.

There appears to have been some local volcanic activity in east Fife early in Calciferous Sandstone Measures times, after which volcanism appears to have died out until the period of deposition of the Limestone Coal Group. During Upper Limestone Group times (most of the E_2 stage of the Namurian) there was a major volcanic episode which in the western part of the district resulted in the deposition of thick accumulations of tuff and volcanic detritus, probably derived from a number of vents. Elsewhere cyclic sedimentation continued, with a change of emphasis from mainly swamp vegetation and fluviatile sands in the earlier part to persistent marine incursions in the later part. The history of Passage Group times remains rather obscure but outbursts of volcanic activity took place, particularly around Largo Law; they may well have been both widespread and prolonged. In places, however, sediments were deposited. They were mostly argillaceous and some were laid down under marine conditions, but they include sandstones probably of fluviatile origin. Swamp vegetation colonised parts at least of the area for several, mostly short, periods. The Lower Coal Measures (Westphalian A) are known only from one old borehole, which recorded a mainly variegated sequence probably of volcanic origin, and from coal-bearing sediments interbedded with tuffs that have subsided into a volcanic neck. This suggests either a continuance or a recrudescence of active volcanism during Lower Coal Measures times. The Middle Coal Measures (Westphalian B) show a return to cyclic sedimentation, with frequent prolonged periods when swamp vegetation flourished, at least in the Lundin Links area to which they are now confined. Similar conditions probably continued into Upper Coal Measures times (Westphalian C–D) but any coals formed were later oxidised.

The Carboniferous strata have been folded, for the most part quite gently, and they have also been subjected to major faulting along north-easterly, north-westerly and west–east lines. Some of the north-easterly faults show evidence of major horizontal movement and near these the strata have been thrown into complex tight folds. Some of the folding and faulting may be Namurian in age but some is certainly later than Westphalian B. Uplift and erosion accompanied the later folding and faulting, and during the subsequent period of arid or semi-arid conditions the Upper Coal Measures and, in places, older strata were reddened and oxidised. A sill-complex of olivine-dolerite and allied types was intruded into Carboniferous strata up to and including the Upper Limestone Group. Its relationship to the faulting has not been determined and its age is uncertain, but it seems likely that it is related to the volcanism that extended from Upper Limestone Group to Coal Measures times. Sills and dykes of quartz-dolerite and tholeiite were emplaced some 290 to 300 million years ago, probably during the Stephanian. They appear to be later than most of the faulting but their age-relationship to the olivine-dolerite sill-complex is not known. Many volcanic necks cut Carboniferous strata from the Calciferous Sandstone Measures up to the Passage Group; some contain intrusions ranging from olivine-basalt to monchiquite in character, which are probably Stephanian in age. One small neck, with intrusions of a different type, cuts Middle Coal Measures strata. Some of the necks are definitely younger than some of the olivine-dolerite sills but the evidence as to their relationship to the quartz-dolerite intrusions is inconclusive and conflicting. In the Elie–St Monance coast section the necks appear to be younger than the folding but older than important faulting. The precise order of events in east Fife during the Upper Carboniferous has therefore not been fully elucidated. In particular it remains uncertain whether or not there was a major volcanic episode in late Westphalian, Stephanian or Lower Permian times.

The geological history of east Fife in Mesozoic and Tertiary times is obscure, but it is probable that the physical features of the district had been largely carved out by the end of the Tertiary. Like the rest of Scotland, east Fife was glaciated at least once during the Pleistocene period. The ice, in its eastward passage towards the North Sea, smoothed and moulded the landscape into its present form, leaving behind an extensive sheet of ground-moraine or till which covers most of the lower ground. During the retreat stages of the ice-sheet meltwaters deposited spreads of sand and gravel near the ice margins; in coastal areas these spreads are closely associated with deposits, ranging from clay to gravel, which were laid down in the late-Glacial sea. Sea-level at this time was considerably higher than it is today, reaching a maximum of about 36 m above O.D. In places old shoreline features still survive, notably at Kincraig (Plate I, frontispiece), where three conspicuous benches have been cut in volcanic agglomerate. By Boreal times the sea had fallen below its present level and a peat layer accumulated on the newly emerged land. Patches of this peat survived erosion and were buried beneath the deposits of the post-Glacial (or Flandrian) marine transgression, when the sea again encroached on the land, reaching a maximum height of almost 9 m above O.D. Subsequently the sea gradually withdrew to its present position, leaving behind a well-marked wave-cut platform covered mainly by sand and gravel and backed in many places by old sea-cliffs. I.H.F.

Chapter 2

UPPER OLD RED SANDSTONE

THE Upper Old Red Sandstone rocks on the north-west side of the Dura Den Fault lie just beyond the north-western limit of the district under review: they will be described in the forthcoming memoir on the geology of the country around Perth and Dundee.

Two small inliers of red beds among the Carboniferous rocks exposed on the coast at Cambo Sands near Kingsbarns and at Balcomie near Fife Ness have been assigned to the Upper Old Red Sandstone (MacGregor 1968, p. 23; cf. Geikie 1902, pp. 75–76). The rocks consist of grey, red and purplish sandstones and reddish mottled mudstones with bands and nodules of concretionary limestone and dolomite ('concretionary cornstone'). Conglomeratic bands consisting mainly of mudstone and cornstone pebbles ('conglomeratic cornstone') are also present, and some of these contain pebbles of lava (MacGregor 1968, pp. 185–186). The latter resemble Lower Old Red Sandstone rather than Carboniferous lava types, suggesting that the Lower Old Red Sandstone may lie not far beneath. No fossils have been found.

At Balcomie there is a conformable upward passage into grey strata assigned to the Fife Ness Beds at the base of the Carboniferous (p. 21) but at Cambo Sands the rocks are less well exposed, and the extent of Upper Old Red Sandstone strata is not clearly defined. The base of the sequence is not exposed in either of the inliers.

In facies these rocks are similar to those of the Upper Old Red Sandstone in some other parts of central Scotland (Geikie 1902, pp. 75–76), but they do not closely resemble any part of the Upper Old Red Sandstone sequence in the nearest large outcrop, that of north Fife and Kinross (Chisholm and Dean 1974). They are, however, similar to the rocks of the 'red facies' (p. 17) in the Calciferous Sandstone Measures of east Fife. Their age is not known with certainty; they contain lava pebbles of Lower Old Red Sandstone type, and they lie in the lowest exposed part of a conformable sequence in which the oldest datable strata are of Viséan age.

DETAILS

Balcomie. Strata of Upper Old Red Sandstone facies, here termed the Balcomie Beds, crop out on the shore around the golf course at Balcomie between a point [NO 62281061][1] about 800 m north of Balcomie Farm and a point [NO 36141015] north-west of Constantine's Cave. In the western part of the outcrop, around Fluke Dub [NO 627107], the strata consist of red and purple mudstone with bands of grey, red and purple mottled sandstone, yellow and grey concretionary limestone and dolomite and lenses of conglomerate containing mudstone and carbonate pebbles. Pebbles of lava of Lower Old Red Sandstone type (S 54300–1[2], 56641) have been found in conglomerates at two localities [NO 62611073, NO 62551073]. The strata pass upwards without any sharp break into rocks of Carboniferous aspect; the dividing line has been

---

[1] National Grid References are given in this form throughout the Memoir.
[2] Numbers preceded by S refer to rock slices in the Geological Survey Scottish collection.

drawn at the top of the highest carbonate conglomerate, which is well exposed at high water mark [NO 62281061]. The strata above the carbonate conglomerate belong to the Fife Ness Beds at the base of the Calciferous Sandstone Measures. They are described as the Fluke Dub section (p. 21).

The central and eastern parts of the outcrop are disturbed by cryptovolcanic structures (p. 198) containing sediments of Carboniferous appearance. The rocks are partly concealed by loose boulders and shingle, but red, purple and yellow mottled mudstone with lenticular bands of purplish-grey sandstone appears to make up most of the sequence, with nodules of concretionary dolomite and beds of dolomite conglomerate in places, especially around McGowan's Harbour [NO 631105].

Cambo Sands. Poorly exposed strata containing red beds are seen on the shore between the Kingsbarns and Cambo faults at Cambo Sands. They are believed to correlate with the Balcomie Beds. The lowest part of the sequence crops out at the north end of the sands, close against the Kingsbarns Fault; here greenish and purplish sandstones in a small structural dome [NO 60481238] are overlain by about 2 m of red, purple and yellow mottled mudstone with a large polygonal system of cracks at the top. Above this lies a thin lenticular bed of soft sandstone and conglomerate containing pebbles of mudstone, limestone and lava (S 56637, 56639; MacGregor 1968, p. 186). Material from this bed has filtered down and filled the cracks in the mudstone below. The lavas comprise feldsparphyric and aphyric varieties similar to the basic andesites of the Lower Old Red Sandstone of the Ochil and Sidlaw Hills, but are very altered. Red mudstone overlies the bed with lava pebbles and this in turn is overlain, near high water mark, by a massive bed of cream and purplish mottled concretionary limestone and dolomite containing veins and irregular masses of black chert. Exposures of sandstone, red mudstone and concretionary dolomite around The Humbie [NO 606121] probably lie higher in the sequence than the beds just described, and further south, at Cambo Ness [NO 609118], greenish and purplish cross-stratified sandstones are exposed, dipping to the south. These last resemble beds near the base of the Fife Ness Beds at Fluke Dub (p. 21) and the dividing line between the Upper Old Red Sandstone and Calciferous Sandstone Measures has therefore been drawn, rather arbitrarily, between the sandstones at Cambo Ness and the strata exposed around The Humbie.

I.H.F., J.I.C.

References

CHISHOLM, J. I. and DEAN, J. M. 1974. The Upper Old Red Sandstone of Fife and Kinross: a fluviatile sequence with evidence of marine incursion. *Scott. Jnl Geol.*, **10**, 1–29.

GEIKIE, A. 1902. The geology of eastern Fife. *Mem. geol. Surv. Gt Br.*

MACGREGOR, A. R. 1968. *Fife and Angus geology: an excursion guide*. Edinburgh and London: Blackwood.

Chapter 3

CALCIFEROUS SANDSTONE MEASURES

GENERAL ACCOUNT

STRATA belonging to the Calciferous Sandstone Measures crop out over most of east Fife east of a line from Elie to St Andrews, and also in the Ceres district (Fig. 1). Around Largo and Largoward they are covered by younger beds and little is known about them. Exposures on the coast are good but inland they are largely confined to streams and disused quarries. The information from these surface sections is supplemented by a few boreholes, the most important of which were put down in 1964 and 1966 by the Geological Survey at Anstruther and near St Andrews.

STRATIGRAPHY

The lowest beds of the Calciferous Sandstone Measures in east Fife are known only from the coast section north-west of Fife Ness where they rest conformably on red beds which on facies grounds are assigned to the Upper Old Red Sandstone. There is a passage by alternation between the latter and the strata of Carboniferous facies, and the base of the Calciferous Sandstone Measures is taken in the middle of this transition zone, at the top of a prominent bed of carbonate conglomerate. The top of the division is drawn at the base of the St Monance Brecciated Limestone, the presumed local equivalent of the Hurlet Limestone. Currie (1954, pp. 531–533) tentatively placed the upper limit of the Lower Bollandian (P_1) Stage of the Viséan at the base of the Hurlet and drew its lower limit at the Fordell (? = Burntisland) Marine Band. The latter horizon has not however been recognised in east Fife. Currie (1954, pp. 531–532) also suggested that the 'Encrinite-bed' of east Fife (now known as the Pittenweem Marine Band), which lies at least 1200 m above the base of the Calciferous Sandstone Measures in the Pittenweem area, might be of Cracoean (B) age. This was based on the suggested equivalence (Kirkby *in* Geikie 1902, p. 104; Wilson 1952, p. 317) of this band with the Lower Cove Marine Band of Berwickshire. Wilson (1974, p. 47) indicated that there is good faunal evidence for the correlation of the 'Encrinite-bed' and other associated marine bands in east Fife with a group of such bands in the Lothians (named by him the Macgregor Marine Bands) which includes the Cove Marine Bands. This correlation is also supported by the evidence of the miospore assemblages (Neves and others 1973, p. 51) and is accepted here. The great thickness of the Calciferous Sandstone Measures in east Fife (see below) suggests the possibility that the lowest part of the succession may be of Tournaisian age, but palynological evidence (Neves and others 1973, pp. 44–46) suggests that the sequence at least down to the base of the Anstruther Borehole is of Viséan age and belongs to their TC zone, the base of which is drawn some distance above the Tournaisian–Viséan boundary. Present indications are, therefore, that most if not all of the Calciferous Sandstone Measures in east Fife belong to the Viséan.

The area of maximum development almost certainly lies in the south-eastern part of the outcrop where a thickness probably exceeding 2000 m is attained

TABLE 1

Subdivisions of the Calciferous Sandstone Measures in east Fife
(details of thickness and facies relate to coast sections shown in Fig. 3)

Stratigraphical Subdivisions		Thickness (m)	Facies
Lower Limestone Group		—	—
Calciferous Sandstone Measures	Pathhead Beds	311	Grey facies, many coal horizons, several fully marine bands
	Sandy Craig Beds	557+	Grey and red facies, few coal horizons, rare marine bands, thick sandstones common
	Pittenweem Beds	220+	Grey facies, many coal horizons, several fully marine bands
	Anstruther Beds	813+	Grey facies, many coal horizons, several restricted marine bands, dolomite bands common
	Fife Ness Beds	229+	Grey and red facies, few coal horizons, no marine bands known, thick sandstones common
Upper Old Red Sandstone		—	—

in the well-exposed coast section between Pathhead and Anstruther. This figure is comparable with those found in West Lothian (Mitchell and Mykura 1962), hitherto regarded as the area of thickest development of the Calciferous Sandstone Measures in Scotland. Neither the Burdiehouse Limestone nor the cementstone facies can be recognised in east Fife, so that the threefold subdivision of the succession used in the Lothians (e.g. Mitchell and Mykura 1962) cannot be applied and the Calciferous Sandstone Measures are mapped as a single unit. Four lithostratigraphical subdivisions unique to east Fife have, however, been recognised in the Anstruther–Pathhead coast section and the Anstruther Borehole section. A fifth east Fife subdivision, thought to lie below the other four, has been pieced together from coastal sections around Fife Ness, where it is regarded as passing down into the Upper Old Red Sandstone. These subdivisions are listed in Table 1 together with their characteristic lithologies. Table 2 gives lists of the marine bands in each subdivision, with a provisional correlation.

It is believed that these lithostratigraphical subdivisions can be traced in a general way over most of east Fife (see Fig. 2) but it must be emphasised that there is a considerable reduction in the thickness of the strata between certain

STRATIGRAPHY

TABLE 2

Correlation of marine bands in the Calciferous Sandstone Measures
(all marine band and *Lingula* band names except those marked with an asterisk are new)

	Coast Section ANSTRUTHER–PATHHEAD	Coast Section ANSTRUTHER–FIFE NESS	Coast Section FIFE NESS–ST ANDREWS	Inland Sections ST ANDREWS–CERES AREA	Inland Sections DUNINO–CARNBEE AREA
PATHHEAD BEDS	St Monance White Limestone* Pathhead Upper* Pathhead Lower* Ardross *Lingula* band Upper Ardross Limestone* Lower Ardross Limestone* West Braes			St Monance White Limestone* Pathhead Upper* Pathhead Lower* Upper Ardross Limestone* Lower Ardross Limestone* West Braes	Newbigging Claremont Blebo Cottage Hole Denbrae Farm Carnbee? Gordonshall?
SANDY CRAIG BEDS	Boat Harbour		St Nicholas?	New Mill? Denbrae House?	Lochty?
PITTENWEEM BEDS	Pittenweem Harbour *Lingula* band Pittenweem Kirklatch Cuniger Rock	Crail Harbour Westland Skelly Pans	St Andrews Castle Witch Lake West Sands	Knockhill Nydie Kincaple?	Brigton? Ovenstone *Lingula* band? Chesters Ovenstone?
ANSTRUTHER BEDS	Chain Road Billow Ness Anstruther Wester Anstruther Borehole marine and *Lingula* bands	Barns Mill Caiplie *Lingula* band? Innergellie? Goats?	Kenly Mouth? Craig Hartle North Craig Hartle South Craig Hartle Lower Randerston marine and *Lingula* bands Wormistone Upper Wormistone Lower		

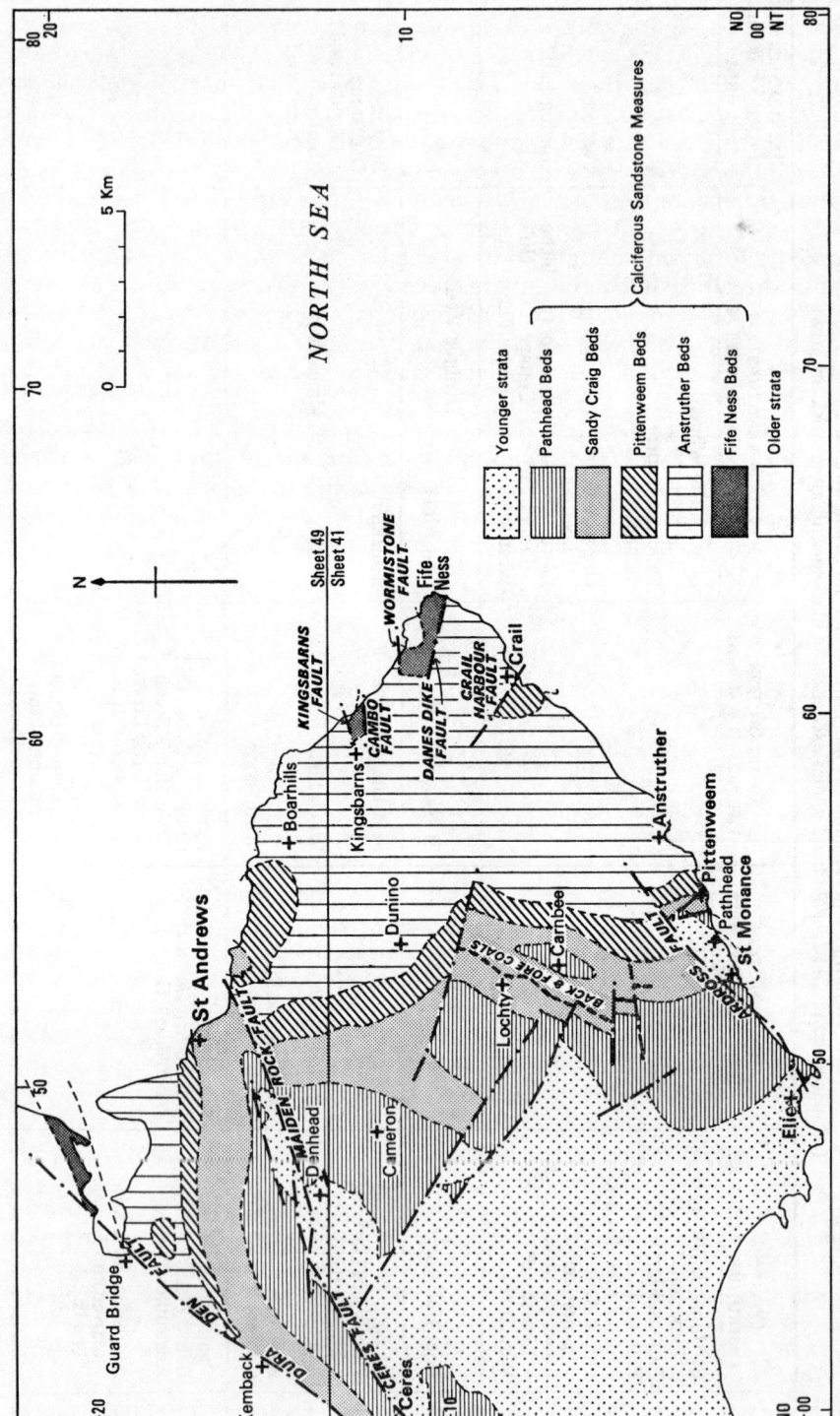

Fig. 2. Sketch-map showing approximate outcrops of the subdivisions of the Calciferous Sandstone Measures

STRATIGRAPHY

recognisable marine bands northwards and westwards from the Anstruther–Pathhead coast section (Fig. 4), and there is no doubt that in the ground west of Ceres the thickness of the whole of the Calciferous Sandstone Measures is greatly reduced. The thickness variations are accompanied by considerable lateral facies variation, and this makes correlation difficult. In addition, the geological structure is largely unknown in detail away from the coast, where it can be seen to be complex, with numerous folds and faults. Consequently neither the stratigraphy nor the outcrop distribution of the Calciferous Sandstone Measures is sufficiently well known for the various subdivisions to be accurately mapped or for a full stratigraphical succession to be worked out. Nevertheless, despite correlation difficulties and gaps in the exposed sequences, a composite section can be compiled from coast sections around Fife Ness and between Anstruther and Pathhead which provides a general stratigraphic framework to which other sections can be related. This composite section is described on pp. 21–26.

Marked lateral variation is one of the features of the Calciferous Sandstone Measures in east Fife that distinguishes them from the rest of the Carboniferous, which also consists mainly of coal-bearing cyclic sediments. The other main features of difference are the thinness of most of the coal seams and the abundance of non-marine ostracods and of dolomite bands.

PREVIOUS RESEARCH

Geikie (1902) provided a comprehensive list of 19th century publications on the area, and brief summaries of only the more important of these contributions are included here. Landale's study of the east Fife coalfield is devoted mainly to higher strata, but includes a valuable section (Landale 1837, pp. 295–299) on the Back and Fore coals. Brown (1861) made a detailed study of the coast section from Elie to St Andrews, in which he drew particular attention to the presence of a crinoidal marine band far below the limestones of the Lower Limestone Group. He called this bed the 'line of lower encrinites', from which the widely used term 'Encrinite-bed' has been derived. Brown reported its presence at Pittenweem, south of Crail, near the Rock and Spindle and at St Andrews. The first edition of one-inch Geological Sheet 41, which was published in 1861, showed many of the structural features and faunal bands exposed on the coast, including the Encrinite-bed. Geological Sheet 49, which followed in 1884, provided similar information about the coast sections in the St Andrews area, and showed the inland outcrops of the Encrinite-bed and locally of the '*Myalina* Bed'. This appears to be the first published used of the latter term, later altered to '*Myalina* Limestone' in Geikie (1902) and now replaced by the name 'St Andrews Castle Marine Band'.

By this time Kirkby had begun his detailed investigation of the coast section, especially in the part between Pathhead and Anstruther. His first stratigraphical paper (1880) dealt only with the marine bands, but it is clear from the text that he had already measured the Pathhead–Anstruther sequence. His section, however, was not published until much later, by Geikie (1902, pp. 77–99). Kirkby recognised several faunal bands in addition to those noted by Brown, and drew attention to the frequent close association of marine bands and coal seams, of which he detected over 50 in the Anstruther–Pathhead section.

He also published a detailed account (Kirkby 1901) of the Randerston and Wormistone sections. Geikie (1902) used Kirkby's section as the basis of his account of the Calciferous Sandstone Measures and also quoted observations by Kirkby on the correlation of the sections on the two limbs of the Anstruther Anticline.

Craig and Balsillie (1912) described the occurrence of marine bands near Pitscottie but placed them in the Lower Limestone Group. Tait and Wright (1923) revised Kirkby's account of the highest strata at Pathhead and provided a detailed description of the type section of what is now called the Pathhead Upper Marine Band. Kirk (1925a, 1925b) produced detailed maps and descriptions of the coast section between Kinkell and Kingask. Cumming's (1936) account of the Elie–St Monance sector of the coast is concerned primarily with the volcanic and structural aspects of the geology, but also contains a brief description of the stratigraphy. More recently Greensmith (1965) has suggested that the Calciferous Sandstone Measures of east Fife were deposited in a deltaic environment under the influence of rivers flowing in southerly and southwesterly directions. Bennison (1960, 1961) dealt mainly with the non-marine bivalves, but also included some discussion of the conditions of deposition of the strata in which these shells occur.

Forsyth and Chisholm (1968) described the sequence in the Anstruther Borehole and suggested a correlation between the lower half of this section and the coast section at Randerston. They also recorded sections at the top of the Calciferous Sandstone Measures from the Drumcarro and Higham boreholes. MacGregor's (1968) guide to the geology of Fife and Angus devotes several excursion itineraries wholly or partly to the Calciferous Sandstone Measures, dealing mainly with the most interesting parts of the coast section. Neves and others (1973) described the miospore assemblages in the Calciferous Sandstone Measures in east Fife and indicated the limits on the coast section of the three of their five miospore concurrent range zones which they found there. The correlation of the Randerston section indicated by the miospore assemblages is at variance with the one put forward on stratigraphical grounds by Forsyth and Chisholm (1968): this matter is discussed further below (pp. 22, 31).

LITHOLOGY AND FACIES

CLASTIC ROCKS

Sandstone forms a large proportion of the Calciferous Sandstone Measures sequence throughout east Fife. Detailed petrological accounts of sandstones from this area, and of similar sandstones of the same age further west, have been published by Greensmith (1961; 1965, pp. 229–230). Individual sandstone beds vary widely in thickness, ranging up to a maximum of 36 m. For stratigraphical reasons and for ease of description, it has been found convenient in this account to distinguish beds over about nine metres thick as 'thick sandstones'. These are typically soft, white, cream or pink in colour, fine to medium in grain, and cross-stratified, with, in many instances, a marked convolution of the upper layers (cf. Greensmith 1965, p. 230). Thick sandstones often rest on erosion surfaces and many lie in upward-fining clastic sequences; they are most common in the Fife Ness Beds and the Sandy Craig Beds (see below and Fig. 3).

Thinner sandstones are a constant feature at all levels in the sequence and in all areas.

Argillaceous beds are generally silty to some degree and vary in character from black fissile shales transitional to oil-shale, through grey, bedded mudstones (often fossiliferous), to poorly-bedded rocks with listric surfaces and irregular fracture. Rocks of this last type are termed seatearths if, as is common, they contain traces of rootlets. Poorly-bedded mudstones vary from grey through various mottled colours to purple and red; the reddish colours are the most characteristic feature of the red facies (see below).

Alternating sequences of sandstone, siltstone and mudstone are widespread, forming zones of transition from mudstone to sandstone and vice versa. Trace-fossils (see below) are often well displayed in strata of this type.

CARBONATE ROCKS

Carbonate rocks are made up of one or more of the minerals siderite, iron-rich magnesite, ankerite, dolomite and calcite. When fresh, rocks of this type are usually grey in colour but in the intertidal zone they develop surface crusts whose colour may be used as a guide to composition. Iron-rich carbonates weather to red colours, dolomite (which is generally slightly ferriferous) weathers orange or yellow and calcite weathers grey or white.

For descriptive purposes the carbonate rocks are divided into three types based on field occurrence. These are bedded carbonates, nodular carbonates and carbonate conglomerates.

Bedded carbonates

Bands and flat nodules of clay ironstone up to about 0·3 m thick are commonly present in bedded mudstones at all levels in the sequence. They may be finely laminated or massive and may contain fossils. Usually they consist of siderite, but iron-rich magnesite ($\omega = 1\cdot750$ to $1\cdot782$) has also been detected.

Dolomite beds, usually less than a metre in thickness but ranging up to three metres, are found near the bases of many bedded mudstones, especially in the Anstruther Beds; they are also found, but less commonly, in the Fife Ness Beds, Pittenweem Beds and Sandy Craig Beds. The dolomite is often ferriferous (see Muir and others 1956, pp. 101–102; cf. Greensmith 1965, p. 232), and calcite, ankerite and siderite are also present in some. They have often been described as 'limestones' or 'cementstones' (e.g. Geikie 1902, pp. 70–146; Kirkby *in* Geikie 1902, pp. 126–130) but in the present account the term 'dolomite' is normally used except where a named 'limestone' is referred to, or in discussion of old borehole records in which the term 'limestone' may have been used for both limestones and dolomites. A great diversity of textures can be recognised. Varieties consisting chiefly of bioclastic material can be distinguished from types containing no recognisable organic remains, and mixed varieties are also common. Beds and layers made up largely of ostracod remains are common, as are 'musselbands' composed of squashed bivalve shells, generally of the non-marine genera *Naiadites* and *Carbonicola*. Marine or quasi-marine bivalves such as *Schizodus*, *Sanguinolites* and *Myalina* have also been recorded from some beds, and algal bodies are an important constituent at some horizons.

Shelly crinoidal limestones, often partly or wholly dolomitised and similar to those found in the overlying Lower Limestone Group, are found at intervals in the upper parts of the sequence, particularly in the Pathhead Beds. The thickest is the St Monance White Limestone.

Nodular carbonates

Nodular carbonates are as common as bedded carbonates and like them are found at all levels in the succession. They tend to occur in poorly-bedded rather than well-bedded mudstones however and are also found in siltstones and sandstones, especially where these are penetrated by rootlets. Nodular ironstones, consisting of microcrystalline siderite in spherulitic, brecciated or more homogeneous masses, are a common feature of rooty horizons in the grey facies wherever this is found. Nodular dolomites are less common and are associated chiefly, but not exclusively, with the red facies in the Sandy Craig Beds and the Fife Ness Beds. They are also a feature of the Upper Old Red Sandstone beds at Balcomie (p. 7). The nodules generally show 'catsbrain' mottling (cf. Greensmith 1965, p. 236) and under the microscope commonly show evidence of repeated brecciation and recementation, a textural feature characteristic of the calcite nodules called 'cornstones' in other areas (Muir and others 1956, plate 2, fig. 3; Burgess 1960). It is likely that they have a similar origin, in soil profiles.

Carbonate conglomerates

The Sandy Craig Beds at Pittenweem contain several lenticular bands of conglomerate composed of pebbles of dolomite and siderite up to about two centimetres in diameter. Similar bands are found in the Upper Old Red Sandstone at Balcomie (p. 8; Greensmith 1965, pp. 256–257); thus, like the nodular dolomites, the dolomite conglomerate lithology is associated with the red facies.

COAL

Seams of coal are present at intervals throughout much of the sequence, but are absent from the Fife Ness Beds and rare in the Sandy Craig Beds. The thickest seams are the Back and Fore coals (p. 52) which, according to Landale (1837, p. 295), reach thicknesses of 2·7 m and 2·1 m respectively. Some other seams exceed 0·6 m in thickness but the majority are less than 0·3 m thick.

VOLCANICLASTIC ROCKS

Thin bands of pale-coloured kaolinitic mudstone and graded siltstone, probably of volcanic origin (cf. Francis 1961), have been detected at a few horizons; they may have been derived from the Garleton Hills or Burntisland lava fields, the nearest known contemporaneous volcanic centres. In the Anstruther Beds a band of ironstone near the base of the Anstruther Borehole was regarded by Francis (1968, pp. 123–124) as possibly tuffaceous. Higher up the sequence several bands of kaolinitic mudstone and graded siltstone have been detected at outcrop in the Sandy Craig Beds, and similar bands have been encountered in boreholes in the Pathhead Beds in north-east Fife.

NATURE OF SEQUENCE

Large parts of the succession consist of thin, often rhythmic, alternations of sandstone, siltstone and mudstone interspersed with subordinate coals and carbonate rocks. Upward-coarsening and upward-fining clastic units can both be recognised but indeterminate sequences are also widespread. Where coal seams and rooty horizons are well developed the interval between two successive rooty horizons may be termed a cyclothem or sedimentary cycle, and each such cycle may consist of an upward-coarsening unit, an upward-fining unit, or, commonly, a combination of the two.

Upward-coarsening units are recognisable throughout large parts of the sequence, particularly in the Anstruther Beds, Pittenweem Beds and Pathhead Beds, and vary up to about 15 m in thickness. In general the lowest part of each unit consists of mudstone or siltstone, with body-fossils and bedded carbonates. Trace-fossils are sometimes found in the zone of transition from mudstone to sandstone.

Upward-fining units are less common than upward-coarsening units and range up to about 45 m in thickness in those examples where a thick sandstone (see below) forms part of the unit. Typically each unit rests on an erosion surface; the lowest part of the unit may consist of coarse, medium-grained or fine sandstone and this passes upwards into siltstone or mudstone which may be poorly bedded, with rootlets and carbonate nodules. Both body-fossils and trace-fossils are generally rare. Some upward-fining units lie in clearly-defined channel-shaped erosional hollows cut into the underlying strata but in other units the basal erosion surface appears to be nearly plane, at least in the limited exposures available.

Indeterminate sequences, in which neither upward-coarsening nor upward-fining characteristics can be recognised, are common. Sequences of irregularly alternating mudstone and sandstone fall into this category, as do some of the thick sandstones.

SEDIMENTARY FACIES

Three main facies types are recognised: a grey facies, a red facies and a thick sandstone facies. The grey facies accounts for by far the largest part of the succession, consisting of upward-coarsening and upward-fining units, and indeterminate sequences, in which silty and argillaceous beds are grey in colour. Coals and rooty horizons are widespread, as are bedded dolomites, and both bedded and nodular ironstones are common. The red facies is of much more restricted occurrence. It consists mainly of upward-fining units and indeterminate sequences in which argillaceous beds are red or purple in colour, although sometimes mottled with yellow and grey. Mudstones may have a seatearth-like texture but rootlets and coals are rare and nodular carbonates tend to be dolomitic rather than sideritic. Dolomite conglomerates are also found. Reddening affects sandstones in both the grey and the red facies but is generally considered to be a secondary feature. The red facies is characteristic of the Upper Old Red Sandstone at Balcomie (p. 7) but in the Carboniferous is very subordinate to the grey facies; it is known only at the base, in the Fife Ness Beds, and near the top, in the Sandy Craig Beds (Fig. 3).

The thick sandstone facies, as its name implies, consists of sandstones between about 9 and 36 m in thickness. These cannot be assigned confidently to either grey or red facies, although they seem to occur mainly in those parts of the sequence where the red facies is best developed (Fig. 3), namely in the Fife Ness Beds and Sandy Craig Beds. Examples also occur, however, in the Anstruther Beds, Pittenweem Beds and Pathhead Beds.

Conditions of Deposition

The rapid alternation of coal seams with marine strata has long been taken to indicate a coastal or deltaic environment for the deposition of the Calciferous Sandstone Measures (Geikie 1902, pp. 70–71), but it is still not possible to assign each lithology encountered in the sequence to a specific deltaic environment (Greensmith 1965). In general, however, the predominating grey facies probably represents near-coastal environments in which the water-table at most times lay near or above ground level, whereas the less common red facies may represent environments more distant from the coast, in which the water-table at times fell below ground level. Comparison with the Rhône delta (Oomkens 1967) suggests that upward-coarsening and upward-fining units may represent the fillings of stretches of open water and channels respectively. The thick sandstones, some of which lie in upward-fining sequences, may have been deposited in large fluviatile channels, and although found at intervals throughout the sequence they are most abundant at levels where the red facies is developed, suggesting that deposition in fluviatile environments was relatively more important in areas farther from the coastline.

Although the Calciferous Sandstone Measures in east Fife—like much of the Carboniferous succession throughout the Midland Valley of Scotland—consist largely of coal-bearing, cyclically-deposited sediments, with both marine and non-marine faunal bands, it must be emphasised that they differ in certain important respects. Firstly, lateral variation appears to be much greater and only major marine bands can be correlated with any confidence for more than a few kilometres. For example, the Cuniger Rock and Pittenweem marine bands can almost certainly be traced from Pittenweem to Crail, a distance of seven or eight kilometres, and the thickness of the strata between them alters only slightly (from 183 m to 165 m), but the pattern of sedimentation in these two sections is quite different.

Secondly, ostracods are very abundant at many horizons, both in black shales and in bedded dolomites. Only in the Lower Limestone Group of Renfrewshire and west Glasgow (Forsyth and Wilson 1965, plate iv) is there any comparable abundance of ostracods and even there the thickness of strata involved is much less and the coal bearing cyclic nature of the sequence is largely lost. Thirdly, bedded, usually non-marine, dolomites are present in many cycles whereas in the rest of the Scottish Carboniferous they are absent or very rare, the non-marine lower part of the Blackhall Limestone in the Lower Limestone Group of Renfrewshire and west Glasgow (Forsyth and Wilson 1965, p. 71) being the most notable of the few known exceptions. Abundant non-marine ostracods occur in some of these bedded dolomites but they also occur in black shales; some of the bedded dolomites contain non-marine bivalves, as well as or instead of ostracods, others have no recognisable organic remains and a few contain restricted marine faunas.

I.H.F., J.I.C.

Trace-Fossils

Trace-fossils are found at intervals throughout the sequence and can be divided into three groups on the basis of the type of body-fossil fauna with which they are most commonly associated.

The first group consists of the septate trace-fossils *Teichichnus*, *Diplocraterion* and *Rhizocorallium* (Chisholm 1970b), together with *Chondrites* and a small form of *Planolites* (Chisholm 1968, 1970a). Members of the group are most often found with or near marine body-fossil faunas and, as might be expected, they are common in the Pathhead Beds but are only rarely encountered at lower horizons. The best-preserved examples of *Teichichnus* and *Rhizocorallium* are found in sandstones beneath the St Monance White Limestone at Pathhead (p. 26), and good examples of *Diplocraterion* can be seen in the sandstone cliffs at the west end of the Witch Lake, St Andrews (p. 38). *Chondrites* is well developed in striped beds above the West Braes Marine Band west of Pittenweem (p. 26).

The second group consists of *Monocraterion* and 'small simple pipes' (Chisholm 1968). These two trace-fossils are typically found in association with non-marine bivalves and ostracods, although about 15 per cent of their recorded occurrences lie close to marine bands. The assemblage is by far the most abundant of the three and is particularly characteristic of the Anstruther Beds. A typical and well-preserved development of *Monocraterion* is exposed east of the bathing pool at Pittenweem (p. 25), and good specimens of 'small simple pipes' are to be found in striped beds at the western side of the Billow Ness Bathing Pool (p. 23).

The third group consists of the Duloch form of *Planolites* (Chisholm 1970a) and two forms previously undescribed from the Scottish Carboniferous. These are a small horizontal bilobate burrow resembling *Aulichnites* and a large horizontal backfilled burrow similar to *Beaconites antarcticus* Vialov (Gevers and others 1971, plate 18). The members of the group are generally found in unfossiliferous, diffusely-bedded sandstone–siltstone sequences and although far less common than the members of the other two groups, they are absent only from the Pathhead Beds. The Duloch form of *Planolites* is more commonly encountered than the other two members of the group; it is best seen east of the bathing pool at Pittenweem (p. 25) where it occurs at two horizons, above and below an intercalation of red beds.

The first and second groups of trace-fossils have probably both developed in paralic environments, although the first group favoured fully marine conditions while the second group appears to have inhabited water of reduced ('non-marine') salinity. The third group may have developed in a fluvial environment.

J.I.C.

DESCRIPTION OF SECTIONS

The coast sections around Fife Ness, the Anstruther Borehole and the Anstruther–Pathhead coast section together cover the known range of Calciferous Sandstone Measures in east Fife, and constitute a useful composite section (Fig. 3) for reference and comparison. These sections are therefore described first, in ascending stratigraphical order (pp. 21–26). Then follow the remaining sections, both coastal and inland, which are arranged geographically in an

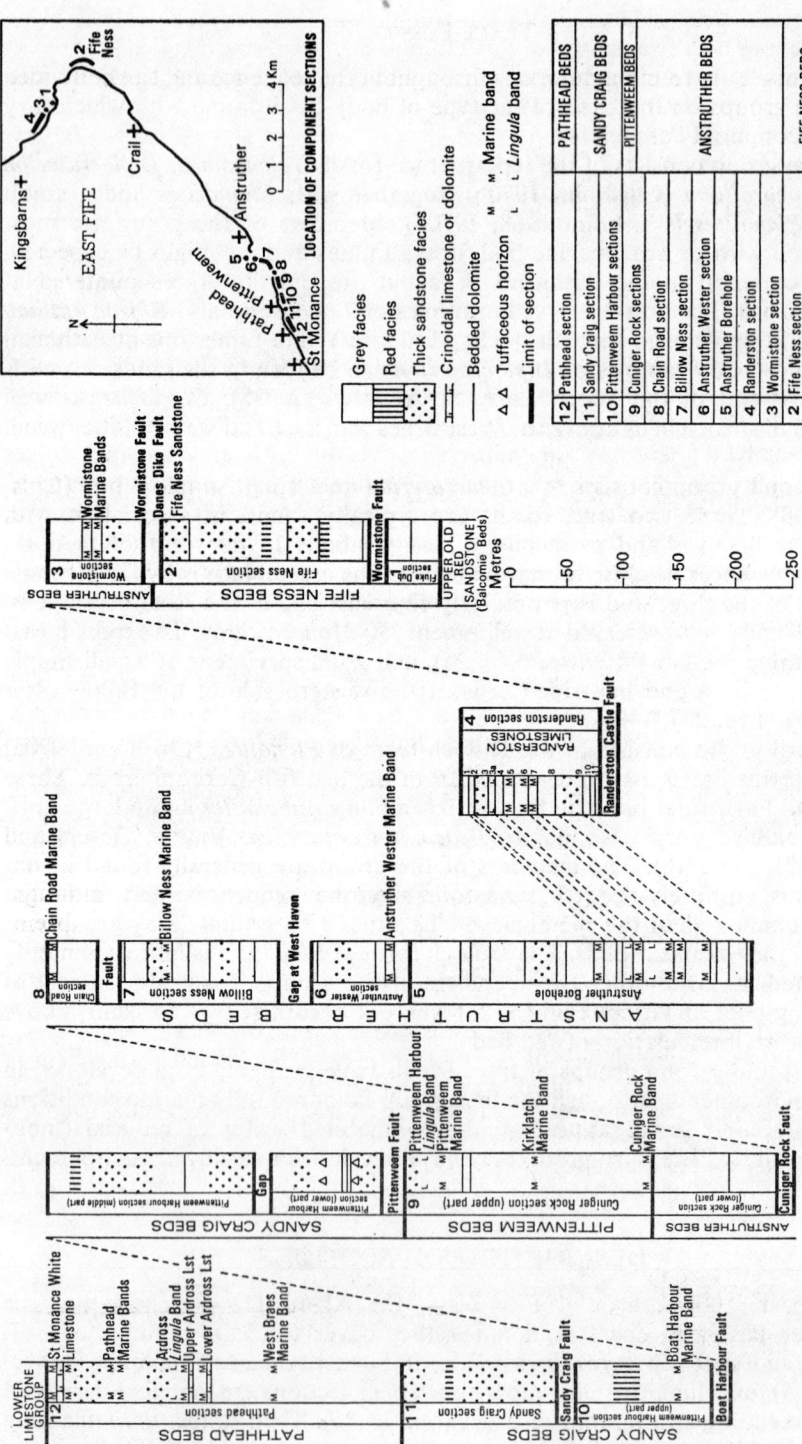

FIG. 3. *Vertical sections in the Calciferous Sandstone Measures between Pathhead and Anstruther, and near Fife Ness*

anticlockwise manner, starting with the area between Anstruther and Fife Ness and proceeding northwards by way of Dunino and Kingsbarns to St Andrews, then westwards to Kemback and Ceres, and thence southwards across the central part of the district to St Monance.

Coast Section around Fife Ness

FIFE NESS BEDS

The Fife Ness Beds (new name) are the lowest subdivision of the Calciferous Sandstone Measures in east Fife. They are characterised by the presence of thick white sandstones, the absence of coals and marine bands and the rarity of seatbeds, bedded dolomites and non-marine faunal bands. Argillaceous beds are mainly poorly-bedded grey or red mudstones. The thick sandstone facies is well developed, therefore, and representatives of both grey and red facies are also present (Fig. 3). The base of the subdivision is drawn at the bottom of the Fluke Dub section, where there is a conformable passage into the Balcomie Beds of the Upper Old Red Sandstone, but the top of the division is faulted, so that the highest beds are not exposed. Faulting also interrupts the continuity of sequence within the subdivision and consequently its full thickness is not known, but the thickness of strata exposed is about 229 m. The only other known outcrop of the subdivision is at Cambo Ness (p. 32).

DETAILS

Fluke Dub section. The section runs from the top of a band of carbonate conglomerate [NO 62281061] at the top of the Balcomie Beds westwards for about 250 m, as far as the point [NO 62051071] where the Wormistone Fault crosses the high water mark. The strata dip north-westwards at moderate angles, and consist of some 60 to 90 m of alternating sandstone and grey mudstone, with a few thin bands of bedded carbonate, one of which contains ostracods.

Fife Ness section. The Fife Ness section runs from the eastern margin [NO 63281014] of a cryptovolcanic disturbance near Constantine's Cave round Fife Ness to the Dane's Dike Fault. There appears to be no overlap with the Fluke Dub section. About 180 m of strata, mainly thick white sandstones with grey and red mudstones and some siltstones, are seen dipping eastwards at angles up to 40°. A bed of black shale with thin bands of dolomite lies near the base. There are also four other bands of bedded dolomite, with faunas of ostracods, *Spirorbis* and fish remains, and algal material is present in one band [NO 63581015]. The Fife Ness Sandstone, at the top of the section, is over 23 m thick and forms Fife Ness itself.

Anstruther Borehole and Anstruther–Pathhead Coast Section

On the coast between Anstruther and Pathhead an ascending sequence 1600 m thick is exposed, and the Anstruther Borehole [NO 56530350] adds another 300 m of strata to the bottom end of this sequence without reaching the base of the Calciferous Sandstone Measures. Faults cause several gaps in the sequence and the total thickness of the section must therefore be well over 1900 m. The succession comprises strata belonging to the four highest subdivisions of the Calciferous Sandstone Measures (see Table 1), but the facies changes between these subdivisions are gradational and there is therefore some difficulty in selecting boundary horizons: the actual horizons chosen are discussed in the accounts of the individual subdivisions. The Anstruther–Pathhead section was measured in detail by Kirkby (*in* Geikie 1902, pp. 77–99) but the account given here is based on a re-measurement carried out by the present authors.

ANSTRUTHER BEDS

The Anstruther Beds (new name) are over 800 m thick and consist mainly of the grey facies—alternations, usually cyclic, of mudstone, siltstone and sandstone with rootbeds and coal, and bands of bedded dolomite. There are no red beds and few thick sandstones (Fig. 3). Many of the mudstones and bedded dolomites contain faunas of ostracods, fish debris and non-marine bivalves among which *Naiadites obesus* is very common: the dominance of this species is a characteristic feature of this subdivision. *Carbonicola* is also present, but *Curvirimula* has not been recorded. A trace-fossil assemblage consisting of a form of *Monocraterion* and a small burrow known as 'small simple pipes' (p. 19) is associated with these non-marine faunas and is another feature of the subdivision. Marine bands are quite common in the lower part but are rather rare in the upper part. Their faunas are usually restricted to molluscs and *Lingula*: crinoid debris and articulate brachiopods are rare, while corals and goniatites are unknown. The base of this subdivision was not reached in the Anstruther Borehole. The top is drawn at the base of the Cuniger Rock Marine Band.

DETAILS

Anstruther Borehole. The Anstruther Borehole was drilled from the old pier at Anstruther Wester in 1964 (Forsyth and Chisholm 1968, pp. 74–76, plate iv). The bore started at a horizon near the base of the Anstruther Wester section and penetrated 300 m of strata, the lowest 136·5 m of which are correlated, on stratigraphical grounds, with the part of the Randerston section that includes limestones nos. 1–9 (but see p. 31). The sequence is typical of the Anstruther Beds, consisting of cyclic alternations of mudstone, siltstone and sandstone with bands of bedded dolomite and coal seams. Most of the mudstones and bedded dolomites yielded non-marine faunas of bivalves (mostly *Naiadites obesus*), ostracods and fish remains but the horizons regarded as the equivalents of Randerston limestones nos. 1, 4, 6, 7 and 8 contain in addition marine molluscan faunas, with or without *Lingula*. *Lingula* was also found at the horizon of Limestone No. 3A and not far above the horizon of Limestone No. 5. Only the equivalent of Limestone No. 7 contains articulate brachiopods.

Anstruther Wester section. This section extends from the axis of the Anstruther Anticline [NO 567033] south-westwards as far as the sands of West Haven [NO 56350300]. As the attitude of the beds on either side of West Haven is very similar and there is no evidence of faulting, it is likely that the sands conceal about 30 m of strata between the top of the Anstruther Wester section and the base of the Billow Ness section. The lowest 26 m of the sequence overlap with the top of the Anstruther Borehole section and are poorly exposed. The strata belong to the grey facies except for one thick sandstone but the cyclic pattern is imperfectly developed and the coals are few and thin. The section is otherwise similar to that in the Anstruther Borehole. The Anstruther Wester Marine Band (new name) [NO 56530330] contains a restricted molluscan fauna, and Kirkby (*in* Geikie 1902, p. 98, bed 718) mentions another marine band six metres higher in the sequence, but the fossil material subsequently collected from this band, though possibly marine, is too poorly preserved to be determined.

Billow Ness section. The Billow Ness section extends from the west side of West Haven to a small north-westerly fault [NO 55900269], which separates it from the Chain Road section to the west. The gap in sequence caused by the fault may be small, however; it was ignored by Kirkby (*in* Geikie 1902, p. 94). The dip is to the south-west, at angles up to 25°. The sequence consists of about 145 m of strata similar to those found at lower levels in the Anstruther Beds, with bedded dolomites in a mainly cyclic sequence of sandstone, siltstone and mudstone in which seatbeds are well-developed, especially in the lower part. Faunas consist mainly of non-marine bivalves, ostracods and fish debris but the Billow Ness Marine Band (new name) [NO 56060272] contains marine bivalves and *Lingula*; *Lingula* has also been found between a coal and a bedded dolomite two cycles higher in the sequence. Good examples of the trace-fossil 'small simple pipes' (p. 19) are preserved in siltstones at the west side of the Billow Ness bathing pool [NO 56100277]. There are two thick sandstones in the section; one of them forms a prominent crag called Johnny Dow's Pulpit [NO 56180280].

Chain Road section. A small section, comprising about 45 m of strata of grey facies, extends westwards along the shore near high water mark in a fault-bounded outcrop below Chain Road. A sandstone at the top of the section resembles one at the base of the Cuniger Rock section but the correlation is uncertain, and there may equally well be a small break between the two sections. The only noteworthy feature of the section is the presence, near the top, of the Chain Road Marine Band (new name) [NO 55860270] with a molluscan fauna.

Cuniger Rock section (lower part). The section extends along the seaward side of the Cuniger Rock Fault from low water mark [NO 55980265] as far as the base of the Cuniger Rock Marine Band, near Cuniger Rock itself [NO 55660271]. The sequence dips to the north-west, and consists of 110 m of irregular alternations of sandstone, siltstone and mudstone, with a single band of bedded dolomite. I.H.F.

PITTENWEEM BEDS

In the Anstruther–Pathhead coast section the Pittenweem Beds (new name) belong mainly to the grey facies (Fig. 3) and are made up of alternations, cyclic in places, of mudstone, siltstone and sandstone with rootbeds and thin coals. Ironstone bands are common in the mudstones but bedded dolomites are rare, a point of contrast with the Anstruther Beds. Non-marine faunas are similar to those at lower levels, though of less frequent occurrence, but some of the marine band faunas are richer, containing crinoids and articulate brachiopods in addition to molluscs and *Lingula*. The Pittenweem Beds, with the Cuniger Rock Marine Band at their base, succeed the Anstruther Beds without a break and 220 m of strata are exposed; these, however, are separated from the overlying Sandy Craig Beds by the Pittenweem Fault, so that the top of the division is not seen here. In the St Andrews area (p. 38) this part of the sequence is more completely exposed and there the base and top of the division are drawn at the West Sands Marine Band and the St Andrews Castle Marine Band respectively. In the Anstruther–Pathhead section the equivalent of the latter band is believed to lie among the strata cut out by the Pittenweem Fault, or possibly in the gap in exposures at the mouth of Pittenweem Harbour.

DETAILS

Cuniger Rock section (upper part). The section extends from the base of the Cuniger Rock Marine Band (new name) at Cuniger Rock [NO 55660271] to the point [NO 55100249] where the Pittenweem Fault runs into the cliffs by Pittenweem Harbour. The strata dip north-westwards at angles up to 45°. The section contains all 220 m of the Pittenweem Beds that are exposed on the Anstruther–Pathhead coast. There is one thick sandstone at the top, but the rest of the sequence consists of alternations of mudstone, siltstone and sandstone with rootbeds and thin coals; a cyclic pattern is apparent in places, especially in the upper part. A few of the mudstones are barren of fauna or yield only fish debris. Others contain in addition *Spirorbis*, *Euestheria*, *Naticopsis* or non-marine bivalves. There are three marine bands. The Cuniger Rock Marine Band, at the base of the section, lies in a 3-m bed of mudstone with two thin carbonate bands and has a fauna of molluscs, bryozoa and crinoid debris. The Kirklatch Marine Band (new name) [NO 55430265] contains a restricted fauna of *Sanguinolites clavatus* and indeterminate bivalves. The Pittenweem Marine Band (new name), in a 6-m bed of mudstone with carbonate bands [NO 55250258], has the richest and most varied fauna in the Pittenweem Beds, containing abundant crinoid debris, ribbed brachiopods and various molluscs. It is the 'line of lower encrinites' of Brown (1861) and the 'Encrinite-bed' of Geikie (1902) and later authors. About 11 m above the marine band lies the Pittenweem Harbour *Lingula* Band (new name), just above a coal 60 cm thick which Kirkby (*in* Geikie 1902, p. 87, bed 368) thought had been worked.

SANDY CRAIG BEDS

The Sandy Craig Beds (new name) are exposed on the coast between the harbour at the east end of Pittenweem and the bathing pool at the west end. The sequence, at least 550 m thick, is of mixed facies and the red, grey and thick sandstone facies are all represented (Fig. 3). The base of the section [NO 55180236] is cut off by the Pittenweem Fault, and two other faults cut out strata within the section. The top of the division is drawn at the base of a black shale, now rather poorly exposed, at the bathing pool [NO 54350224]. The sandstones range up to 36 m in thickness and the highest one is unusually coarse and pebbly. Strata of the red facies include poorly-bedded red or mottled mudstones and lenticular beds of concretionary dolomite and of dolomite conglomerate. Coals and rooty beds are rare, as are bedded carbonates, and there is only one marine band (the Boat Harbour Marine Band); it has a fauna consisting largely of brachiopods. Non-marine faunas, of ostracods and fish remains, are also rare.

DETAILS

Pittenweem Harbour section (lower part). The base of the section lies at the point [NO 55180236] where the Pittenweem Fault meets the low water mark, and there is a gap between this section and the top of the Cuniger Rock section. The sequence, 99 m thick, dips north-westwards and consists of irregular alternations of sandstone, siltstone and mudstone, with a group of thin bedded dolomites [NO 55100236] not far above the base. Some thin layers of detrital material, probably of volcanic origin, are also present in the mudstone; one such band lies just below the group of bedded dolomites and another lies about 23 m higher in the sequence. Rare fish fragments are the only fossils. The section includes 20 m of strata poorly exposed on the floor of Pittenweem East Harbour and ends at a gap in the exposures at the harbour entrance

[NO 54930234]. This section is rather different from the rest of the Sandy Craig Beds in that it belongs mainly to the grey facies (though it is scarcely typical of the latter) and it is possible that it belongs to the Pittenweem Beds.

Pittenweem Harbour section (middle part). About 15 m of unexposed strata including, possibly, the St Andrews Castle Marine Band are thought to lie in the gap at the harbour mouth. The middle part of the Pittenweem Harbour section starts on the west side. The strata dip north-westwards and are 184 m thick. The sequence consists mainly of thick sandstones, some of which are coarse-grained, with some generally poorly-bedded and locally rooty mudstones. The only fossiliferous horizon is a thin band of dolomite [NO 54810228] with ostracods. Near the top of the section is a zone, six metres thick, containing red and yellow mottled mudstone, nodules of concretionary dolomite and lenses of cross-stratified carbonate conglomerate up to a metre in thickness [NO 54650224]. The section ends near low water mark at the Boat Harbour Fault [NO 54600218].

Pittenweem Harbour section (upper part). The Boat Harbour Fault trends north-north-eastwards across the tidal zone at a small angle to the strike of the rocks. There is no overlap between the sections on either side of it, but the amount of strata cut out may not be great. The upper part of the Pittenweem Harbour section starts near high water mark on the west side of the fault [NO 54700236] and continues to the Sandy Craig Fault [NO 54520238]. The sequence is 122 m thick, with north-westerly dip. It is unusually argillaceous and consists largely of poorly-bedded mudstones, some of which are grey and rooty, whereas others are mottled red and yellow, with dolomite and sideritic concretions. Near the base lies the Boat Harbour Marine Band (new name) [NO 54670234], with a fauna of bivalves and brachiopods. The fauna has affinities with those in the Pittenweem Beds and on the palaeontological evidence the band could be correlated with the St Andrews Castle Marine Band. This correlation is not compatible with the other lithological and stratigraphical evidence, however, and is considered to be unlikely (p. 127). A lenticular sandstone and a lenticular bed of carbonate conglomerate, both up to four metres in thickness, occur in the middle part of the sequence [NO 54560236].

Sandy Craig section. The highest 137 m of the Sandy Craig Beds are exposed, dipping north-westwards at 50° to 60°, between low water mark on the west side of the Sandy Craig Fault [NO 54530215] and Pittenweem bathing pool, where the top of the subdivision is drawn just above the top of a thick sandstone [NO 54340222]. The sequence consists mainly of sandstone, in beds 10 to 36 m thick. The thickest of these beds is at the base and forms Sandy Craig [NO 54470211]. The topmost sandstone, at the east side of the bathing pool, is very coarse, with small quartz pebbles. Below it lies a sequence, about 27 m thick, of varied strata including red mudstone, concretionary dolomite and a cross-stratified lens of carbonate conglomerate up to nearly two metres thick [NO 54390222]. Kirkby (*in* Geikie 1902, p. 82) recorded ostracods and fish remains in the mudstone above the conglomerate, and *Curvirimula scotica* and ostracods in another band about 3·6 m below it. The sandstone above the last-mentioned band contains well-preserved examples of the trace-fossil *Monocraterion* (p. 19) and the sandstone below it contains the Duloch form of *Planolites* (p. 19; Chisholm 1970a, plate i).

PATHHEAD BEDS

The Pathhead Beds (new name) are exposed in unbroken sequence on the shore west of Pittenweem, and are 311 m thick. The base of the subdivision is defined by a marked change of facies at the top of the highest thick sandstone of the Sandy Craig Beds, at Pittenweem bathing pool. The top of the subdivision, and of the Calciferous Sandstone Measures, is drawn at the base of the St Monance Brecciated Limestone, the local correlative of the Hurlet Limestone of the Glasgow area (Forsyth and Chisholm 1968, p. 76; Forsyth 1970). The Pathhead Beds are made up largely of strata belonging to the grey facies but include a few thick sandstones in the upper part (Fig. 3). Rootbeds and thin coals are present throughout. Kirkby (*in* Geikie 1902, pp. 81–82) placed the Back and Fore coals (see p. 54) in the lower part of the Pathhead Beds but it now seems more likely that they lie in the Sandy Craig Beds. There is only one marine band in the lower part of the subdivision but the upper contains a *Lingula* band and five marine bands with rich and varied faunas which resemble those found in the Lower Limestone Group rather than those in the rest of the Calciferous Sandstone Measures.

DETAILS

Pathhead section. The section extends along the coast from Pittenweem bathing pool [NO 54350224], where it follows on from the top of the Sandy Craig section, to Pathhead [NO 53810212] where the base of the St Monance Brecciated Limestone is exposed. The strata dip north-westwards at 50° to 60°.

The lowest 161 m consist of alternations of mudstone, siltstone and sandstone with many rootbeds and a few thin coals. At the base, Kirkby (*in* Geikie 1902, p. 82, beds 188–190) recorded a non-marine fauna of *Curvirimula scotica*, ostracods and fish remains in a blackband ironstone but the bed is now obscured by the bathing pool. About 110 m higher in the sequence lies the West Braes Marine Band (new name) [NO 54190226], with a rather sparse fauna of molluscs and *Lingula* in mudstone; the overlying siltstones contain a well-preserved development of the trace-fossil *Chondrites* (p. 19).

The highest 150 m of the section consist mainly of mudstones containing marine bands, and of thick sandstones. The Ardross limestones, so named from their occurrence near Ardross Farm (p. 56; Cumming 1936, pp. 346–348), lie among marine mudstones in the lower part of the section [NO 54120226]. They are both dolomitised crinoidal beds about 30 cm thick, and are about 17 m apart. The upper one is now largely obscured by sand and by seawalls. The Ardross *Lingula* Band (new name) lies 14 m above the Upper Ardross Limestone, and is separated from the overlying Pathhead Marine Bands [NO 53900217] by about 48 m of sandstone with some mudstone and coal. The Pathhead Lower Marine Band (Kirkby *in* Geikie 1902, p. 77, beds 12–16) is thin and contains a variable band of coral limestone (Tait and Wright 1923, pp. 176–178). The Pathhead Upper Marine Band was not recorded by Kirkby, perhaps because the strata among which it occurs are disturbed. It lies in a bed of mudstone 5·5 m thick. About 30 m of sandstone separate the Pathhead marine bands from the highest marine band, which includes the St Monance White Limestone. At high water mark [NO 53810212] the limestone is about five metres thick. It contains corals and is white or pale-grey with a prominent orange-coloured dolomitised band near the centre. The top has a nodular texture, with patches of greenish clay. Below the limestone lie about five metres of varied strata with a prominent development of *Teichichnus* and related trace-fossils. Lateral variations in the limestone and overlying strata were traced across the foreshore by Tait and Wright (1923, pp. 168–170), who found that near low water mark the limestone, here only 2·4 m thick, is crinoidal, with corals abundant only near the top; it is overlain by very fossiliferous mudstones which appear to be the lateral equivalents of the upper part of the limestone as developed at high water mark. Above these variable beds lie a pale-grey mudstone and a thin coal, which are separated from the base of the St Monance Brecciated Limestone by about four metres of varied strata with a marine fauna.   I.H.F.

Area between Anstruther and Fife Ness

Coastal and inland sections in this area lie mainly in the Anstruther Beds (Fig. 2). The coastal exposures are almost continuous, from the axis of the Anstruther Anticline at Anstruther to the Dane's Dike Fault, near Fife Ness, but inland exposures are few. They include sections in small streams and old quarries, and there are a few borehole records.

Details

COAST FROM ANSTRUTHER TO FIFE NESS

Cellardyke section. Between the east side of Anstruther Harbour [NO 56980345] and the Peatlow Rock Fault [NO 57770377] lies a sequence of strata 46 m thick, which is believed to correlate with the top part of the Anstruther Borehole section. The strata are well exposed and dip south-eastwards at about 10°. The probable equivalent of the Kilrenny Mill Musselband [NO 57370352], in a sandstone near the middle of the sequence, is full of *Carbonicola*. It was first noticed by Brown (1861, p. 399) and was also described by Kirkby (1880, p. 578) and by Greensmith (1965, pp. 235–236); the latter regarded a series of seven sharp localised downfolds in this sandstone as 'collapse-erosion' structures.

Kilrenny section. Between the upthrow side of the Peatlow Rock Fault and a gap in the exposures at Innergellie Haven [NO 58700489] lies a south-easterly-dipping sequence of strata about 87 m thick. It is well exposed, except for a 10-m gap at the mouth of Cellardyke Harbour, and several of the sandstones form prominent scarps. The lowest 67 m are believed to correlate with an equivalent thickness at the top of the Anstruther Borehole, so that there is an overlap with the Cellardyke section. The Kilrenny Mill Musselband (new name) [NO 58130442] lies about 44 m above the base, in a sandstone 0·2 m to 1·2 m thick. Just below the top of the section [NO 58680495] lies the Innergellie Marine Band (new name), with a restricted molluscan fauna, in a 7-cm dolomite band; it is believed to correlate with the Anstruther Wester Marine Band of the Anstruther–Pathhead section (p. 23). The bedded dolomites immediately west of the small gap in exposures at Innergellie Haven were correlated by Geikie (1902, p. 129) with other dolomites at Billow Ness, but detailed comparison of sections indicates that the latter are some 150 m higher in the succession. They also contain abundant *Naiadites obesus* whereas those at Innergellie do not.

Caiplie section. Between the east side of Innergellie Haven and a fault [NO 60360608] near Barns Mill lies a sequence of easterly- and south-easterly-dipping strata nearly 300 m thick. The section is believed to correlate with part of the Anstruther–Pathhead section included in the Anstruther Wester and Billow Ness sections (as first suggested by Brown 1861, p. 400) and may also include some strata equivalent to parts of the Chain Road and Cuniger Rock sections. In the lowest 112 m of the section the sandstones are generally well exposed but the strata between them are mostly obscured by sand and shingle. A 'petrified forest' consisting of casts of *Lepidodendron* in sandstone (Brown 1861; Kirkby *in* Geikie 1902, p. 107) is exposed near Caiplie [NO 59050518].

In the higher part of the section, east of Caiplie, exposures are rather better but there are still numerous small gaps. Geikie (1902, p. 109) stated that Kirkby found marine fossils in a bedded dolomite and shale [NO 59200517] just east of Caiplie but only non-marine bivalves and ostracods have subsequently been obtained from these beds. A *Lingula* band, however, has been found in the shale immediately above a band of dolomite 45 m higher in the succession [NO 59520541]. This band is thought to be the one reported near Caiplie by Brown (1861, p. 400) and is now named the Caiplie *Lingula* Band: it may correlate with the Billow Ness Marine Band or the *Lingula* band just above it. There are several thick sandstones in this part of the sequence. The thickest, 24 m thick, lies just above the Caiplie *Lingula* Band. About 28 m higher in the sequence, just below the thick sandstone that forms Crail Coves [NO 59910570], lies a bed, 76 cm thick, composed partly of coal and partly of cone-in-cone limestone. The associated strata are somewhat reddened and there is no doubt that this is an example of replacement of coal by limestone, as described in the Coal Measures of Ayrshire by Mykura (1960).

Pans section. About 60 m of strata, dipping east at 10°–20°, crop out on the shore between the fault at the top of the Caiplie section and the Pans Fault [NO 60650654]. The sandstones are well exposed but not all the intervening strata are seen. About 35 m above the base lies the Barns Mill Marine Band (new name) [NO 60550642], with a fauna of molluscs. The band is known only from loose blocks, which lie strewn along a gap in the exposures. There can be little doubt that this is the band correlated by Geikie (1902, p. 110) with the bed now known as the Chain Road Marine Band of the Anstruther–Pathhead section, and this correlation is accepted here.

Crail section. Between the Pans Fault and the Crail Harbour Fault [NO 61350742] lies the Crail section, a sequence of easterly- and north-easterly-dipping strata about 180 m thick. The Pans Marine Band (new name) [NO 60770666], just above the base, was recorded by Geikie (1902, p. 110) in a bed of shale which is no longer exposed. However, a marine fauna has been obtained from fossiliferous lenses at the top of the underlying sandstone. The nature of the fauna suggests that the correlation put forward by Geikie, with the bed now known as the Cuniger Rock Marine Band, is correct. The band was probably discovered first by Brown (1861, p. 387), although he regarded it as the equivalent of his 'line of lower encrinites'—the Pittenweem Marine Band. The Westland Skelly Marine Band (new name) [NO 61090718] lies about 93 m above the base of the section. It contains a sparse fauna of molluscs and probably correlates with the Kirklatch Marine Band of the Anstruther–Pathhead section. In the overlying sandstone large tree stumps are preserved in their position of growth.

At the top of the section, in the cliff [NO 61050739] by the harbour, lies the Crail Harbour Marine Band (new name); it was not mentioned by Geikie (1902), probably because at that time it lay concealed under a grassy slope. Recent erosion on the slope has left the strata exposed. The band consists of about three metres of mudstone, with bands of ironstone and decalcified limestone, and a 30-cm bed of crinoidal limestone at the base. The fauna is rich, with brachiopods, molluscs and crinoids, and closely resembles the faunas of the Pittenweem and Witch Lake marine bands. If the marine bands are correlated rightly, it follows that all but the lowest seven metres of the section, the part below the Pans Marine Band, belong to the Pittenweem Beds. A comparison of the sequence with the equivalent strata in the Pathhead–Anstruther coast section (Fig. 4), seven kilometres distant, shows that the Crail section is somewhat thinner, with more sandstone and more bedded dolomites than the Anstruther–Pathhead section. Some of the strata in the Crail section are reddened, and a few coal seams are partly altered to limestone.

Roome Bay section. Between the Crail Harbour Fault and the Roome Bay Fault [NO 62000784] 64 m of strata crop out on the shore in Roome Bay. They certainly lie below the level of the Crail Harbour Marine Band, and are probably below the level of the Pans Marine Band. The beds dip gently seawards and are affected by gentle flexuring. Most of the sequence is sandstone, but there is one bedded dolomite and at least one coal, which is only 15 cm thick in the tidal zone but thickens to about 60 cm north-north-eastwards and was formerly worked near high water mark [NO 617078]. Strata at a slightly higher stratigraphical level than those described crop out in a small structural dome [NO 621077] by the Roome Bay Fault at Roome Rocks: they are mostly sandstones but include a dolomite bed 30 cm thick.

Kilminning Castle section. Between the north-east side of the Roome Bay Fault and a pair of faults [NO 634090] 450 m north-east of Kilminning Castle lies a sequence of strata about 207 m thick. The lowest strata are seen at the north-east end of the section. At the Roome Bay end of the section the strata are folded into a shallow syncline, so that the top of the sequence is exposed both in the cliffs at Roome Harbour [NO 62050785] and on the shore east of Roome Rocks. These highest beds include a coal which has been largely altered to cone-in-cone limestone and another, 38 cm thick, which appears to have been worked at crop. As the section is followed past Kilminning Castle the direction of dip swings from south-west through south to south-east. The strata between the sandstones are not everywhere exposed but the number of bedded dolomites seen suggests that the section belongs to some part of the Anstruther Beds, as would be expected on structural grounds.

The Goats Marine Band (new name), first recorded by Brown (1861, p. 400) and also mentioned by Kirkby (*in* Geikie 1902, p. 111), lies in a dolomite band 18 cm thick [NO 63000844]. It contains coiled nautiloids and orthocones and is tentatively correlated with the Anstruther Wester Marine Band. The only other notable bed is the Kilminning Castle Musselband, 15 m lower in the sequence than the Goats Marine Band. It is a thin ferruginous sandstone full of *Carbonicola antiqua* (Kirkby 1880, p. 578; Bennison 1960), and is exposed on both sides of the promontory [NO 63190861] on which stand the remains of Kilminning Castle. If, as suggested above, the Goats Marine Band correlates with the Anstruther Wester Marine Band, then the musselband probably lies at about the horizon of the Kilrenny Mill Musselband; this accords with Kirkby's interpretation which, however, was later disputed by Bennison (1960, p. 137).

Dane's Dike section. Between the end of the Kilminning Castle section and the Dane's Dike Fault [NO 636094] the continuity of sequence is broken by several faults, but these are believed to be small, and the strata exposed probably lie within the Anstruther Beds, not far below the level of the Kilminning Castle section.

INLAND AREA BETWEEN ANSTRUTHER AND FIFE NESS

Pitcorthie. According to Geikie (1902, p. 119) oil-shale was worked at one time from a pit near West Pitcorthie [NO 571070]. This shale yielded a large fish fauna (Walker 1872; Traquair 1901). Several, mostly shallow, boreholes were put down in search of oil-shale in 1913–14 along the Kilrenny Burn south and west of West Pitcorthie. Pitcorthie No. 6 Borehole [NO 56920632] reached a depth of 183 m and recorded a sequence of alternating mudstone, siltstone and sandstone with bands of 'limestone' and faunas of ostracods and non-marine bivalves—a lithological assemblage typical of the Anstruther Beds. In quarries at Muiredge [NO 563069], sandstone, black shale and limestone were formerly exposed, but only loose blocks are now visible. Whinnyhall Quarry at Spalefield [NO 556066] is now obscured, but sandstones and mudstones are exposed in a streamlet 180 to 700 m north-west of West Pitkierie [NO 556056].

Thirdpart. Temporary sections near Thirdpart [NO 590058] were recorded by Wattison (1962). One section [NO 586069] includes shales and limestone with a fish fauna which he compared with those of the Pitcorthie oil-shale and the Burdiehouse Limestone of Midlothian.

Cornceres. The outcrop of a limestone formerly exposed in quarries at Cornceres [NO 580053] (Geikie 1902, p. 120; Walker 1872) was shown on the original geological maps but its course cannot now be traced. It probably lies in the same part of the sequence as the strata at Pitcorthie, but in view of the large number of bedded dolomites ('limestones') in the Anstruther Beds, no exact correlation can be established. Several other quarries in the vicinity are now obscured or show only sandstones. Exposures in the Kilrenny Burn consist mainly of sandstones dipping to the south-east at about 25°.

Ribbonfield area. A limestone three metres thick, recorded in quarries [NO 596080] north-east of Troustrie (Geikie 1902, p. 120) was shown on the original geological maps but can no longer be seen though the sandstone above it is still visible. A borehole [NO 59340829] at Sypsies Plantation, not far down-dip from the quarries, reached a depth of 97 m, proving an alternating sequence typical of the Anstruther beds, with faunas of ostracods, fish remains and non-marine bivalves. There are several bands of limestone including one, 3·5 m thick at 75 m depth, which is probably the band formerly worked at the quarries. About a kilometre farther west, Toldrie No. 9 Borehole (1914) [NO 58460855] was put down to a depth of 42·7 m and cut an alternating sequence with a few thin bands of limestone and two thin coals. Ostracods and non-marine bivalves were recorded. All but one of the quarries in the vicinity are now completely obscured: in the one exception sandstone is visible.

Another limestone, said to have been shelly (Geikie 1902, p. 120), was formerly quarried about 400 m north of Ribbonfield. A borehole [NO 59430895] sited just down-dip from the quarries recorded a 2-m bed of broken shelly limestone, perhaps the quarried bed, at a depth of 8·7 m. The bore continued to a depth of 95·7 m, proving an alternating sequence with thin limestones, reminiscent of the Anstruther Beds. Limestone was also formerly quarried some 550 m west of Grassmiston [NO 602097] with a northerly dip of 18° to 20° but these quarries also are now obscured.

Strata formerly exposed in the railway cutting [NO 606084] north-west of Crail consist mainly of sandstones and shales with several thin coals, including one 76 cm thick which is overlain by a shell-limestone. The dip is mainly to the east and north-east at angles of 30°–50°. These strata lie on the upthrow side of the Crail Harbour Fault and probably belong to the Anstruther Beds.

Crail. Crail No. 1 Borehole [NO 62750879] proved a sequence 152 m thick, of which the top 55 m are mainly thick red sandstones; below lies an alternating sequence of sandstone, siltstone and mudstone with rooty horizons, coals and bands of bedded ironstone and dolomite. The faunas consist of ostracods, fish remains and non-marine bivalves with, in the lowest band, *Lingula* and *Sanguinolites?*. The section undoubtedly belongs to the Anstruther Beds, and may well be the equivalent of the upper part of the Anstruther Borehole section. Near Balcomie a bore [NO 62070994] encountered fully three metres of olivine-basalt of Markle type at a depth of 23·3 m. It probably lies in the Fife Ness Beds and if, as is believed, it is a lava, it provides the only indication of volcanicity in east Fife during the Lower Carboniferous.

Area between Fife Ness, Boarhills and Dunino

Most of the inland area is underlain by strata belonging to the Anstruther Beds, although higher subdivisions may be present in the west (Fig. 2). The rocks, however, are too poorly exposed for the subdivisions to be accurately delimited. In the east, strata belonging to lower subdivisions are exposed on the coast but again the exact limits of outcrop cannot be defined inland. The coast sections extend with few gaps from the Wormistone Fault, near Fife Ness, to Kittock's Den; except for those at Cambo they are all thought to lie in the Anstruther Beds, although correlation with specific parts of the Anstruther–Pathhead section cannot, in general, be established. The rocks on the coast are cut by numerous faults and folded into small shallow domes and basins, many of which are slightly elongated in a north–south direction. It is presumed that similar structures occur inland but little is known about them.

DETAILS

COAST SECTION FROM WORMISTONE TO KITTOCK'S DEN

Wormistone section. The Wormistone section was measured on the west limb of the Wormistone Syncline, which lies between the Randerston Castle Fault [NO 61721102] and the Wormistone Fault [NO 62101072]. (The strata on the eastern limb are less well exposed and are complicated by minor folds, but appear to be similar.) About 85 m of strata are exposed, dipping to the south-east. Stratigraphically the section is believed to lie near the base of the Anstruther Beds, below the lowest strata in the Anstruther Borehole, but this inference is largely based on the lack of any resemblance to the pattern of sedimentation in any higher part of the Anstruther Beds and on the apparent northerly downthrow of the Randerston Castle Fault. The sequence in the Wormistone section consists of ten sedimentary cycles of sandstone, siltstone and mudstone. Most of these cycles have well-developed seatearths. There are a few bands of bedded dolomite and a few thin coals. The Wormistone Marine Bands, near the centre of the section [NO 61921083, NO 61971078], contain molluscan faunas. The part of the section measured by Kirkby (1901, p. 68; *in* Geikie 1902, p. 127) is noteworthy for a fauna rich in gastropods (Kirkby 1880, p. 573; 1901, pp. 68, 75; *in* Geikie 1902, p. 127) obtained from the bed now known as the Wormistone Lower Marine Band.

Randerston section. The Randerston section lies on the eastern limb of the Randerston Syncline, between the axis of a small anticline [NO 61621116] next to the Randerston Fault and low water mark on the synclinal axis [NO 61131161]. There appears to be no stratigraphical overlap with the apparently underlying Wormistone section. Part of the sequence, down to Limestone No. 8, also crops out on the west limb of the Randerston Syncline [NO 609117], which is truncated by the Cambo Fault. The section has long been famous. Brown (1861, p. 399) was probably referring to some of the Randerston limestones when he mentioned beds 'four or five feet thick, consisting of consolidated shells piled above each other in countless myriads' between Fife Ness and Kingsbarns. Kirkby (1880, pp. 572–577) recognised eleven limestones and listed their faunas; later (1901) he published a complete measured section which was reproduced by Geikie (1902, pp. 123–131) together with his faunal lists and some of his comments.

Kirkby thought that the Randerston section was to be correlated with the strata exposed around Billow Ness in the Anstruther–Pathhead section (see Fig. 3) because of similarities in both the lithological sequence and the molluscan and ostracod faunas. Such similarities as do exist, however, do not appear sufficient to justify this correlation. Forsyth and Chisholm (1968, pp. 74–76, plate iv) correlated most of the Randerston sequence with the lower part of the Anstruther Borehole and there can be no doubt that the Randerston section has a much greater resemblance, in both lithological sequence and faunas, to that part of the borehole than it has to any part of the Anstruther–Pathhead coast section. Neves and others (1973), however, have questioned this correlation on palynological grounds and suggested that the equivalents of the Randerston section lie some 600 m higher in the sequence, between the Chain Road Marine Band and a horizon about 40 m above the Cuniger Rock Marine Band. This conflict between palynological and general stratigraphical evidence remains unresolved.

The sequence consists of about 128 m of alternating sandstone, siltstone and mudstone with bands of bedded dolomite ('limestones'), several shell-beds (including limestones nos. 3, 5, 7, 8 and 10), and two coals. The limestones are numbered 1 to 11, from the top of the section downwards, following Kirkby (1880, pp. 572–579). Limestones nos. 7, 6 and 5 carry marine faunas, mainly of molluscs but with some brachiopods, and Limestone No. 3A contains *Lingula* as does No. 2 on the western limb. Limestone No. 8 is a shell-bed full of *Carbonicola*, which is seen only near low water mark and is probably lenticular. The marine fauna recorded from it by Kirkby (1880, p. 576) has not been confirmed and may have been obtained from loose blocks of Limestone No. 7, which appears to have been the one formerly quarried. Limestone No. 9 is a distinctive bed containing algal bodies. Between the Randerston and Randerston Castle faults about 30 m of strata are exposed; they consist mainly of sandstone in beds up to five metres thick but include a shell-bed with *Naiadites obesus* and a thin dolomite bed with *Lingula* in the overlying shale.

Cambo. Between the Cambo and Kingsbarns faults there is an uplifted block of strata which are seen in exposures scattered over Cambo Sands and on the north side of Cambo Ness [NO 609118]. Rocks at the northern end of the sands, near the Kingsbarns Fault, are of Upper Old Red Sandstone facies, and are correlated with the Balcomie Beds (p. 8) but to the south, near Cambo Ness, these appear to pass below sandstones and mottled grey, red and yellow mudstones which are assigned to the Fife Ness Beds.

Kingsbarns section. The section lies between the Kingsbarns Fault and the Babbet Ness Fault [NO 594140] and includes about 75 m of strata. Except near the bounding faults, where there are small folds, the rocks dip gently south-westwards. The lowest 30 m or so, from the Babbet Ness Fault to near Airbow Point [NO 59651370], are poorly exposed but include sandstone, mudstone and dolomite. The top 45 m are well exposed and consist of alternations of sandstone, siltstone and mudstone with bedded dolomites and faunas of ostracods, fish debris, non-marine bivalves and *Spirorbis*. The marine fauna recorded by Kirkby (*in* Geikie 1902, p. 131) has not been found, although the band of bedded dolomite from which it was obtained is exposed at high water mark [NO 60071266] just north-west of Kingsbarns Harbour; it contains *Teichichnus*, *Diplocraterion* and *Chondrites*, an assemblage of trace-fossils which is known to be associated with marine faunas elsewhere (p. 19). As this band is extensively recrystallised the fossils present are poorly preserved; such material as could be identified appears to be all referable to *Naiadites*. Near the top of the section is a bed of dolomite, 1·4 m thick, which is full of *Naiadites obesus*; it is well exposed around The Lecks [NO 605125]. MacGregor (1968, p. 188) has drawn attention to lateral variation and channelling in some of the beds in the Kingsbarns section. The sequence clearly belongs to the Anstruther Beds, probably to the lower part, but it is doubtful if the detailed correlation with the Randerston section put forward by Geikie (1902, p. 131) can be sustained.

Babbet Ness Fault to mouth of Kenly Water. The rocks in this stretch of coast lie on the flanks of the Babbet Ness Anticline, a north-easterly-trending structure whose axis [NO 590142] lies just west of Babbet Ness. The same 55 m of sequence are seen on both limbs of the fold and consist of alternations of seatearths, sandstones, siltstones and mudstones with thin beds of dolomite and ironstone. They contain faunas of ostracods, fish remains, non-marine bivalves and *Spirorbis* and resemble sequences in the Anstruther Beds. On the western limb a thick sandstone at the top of the sequence rests on about 1·7 m of interbedded sandstone, mudstone and dolomite with the trace-fossils *Monocraterion*, *Chondrites* and *Diplocraterion*, together with the non-marine fossils '*Naticopsis*' and *Naiadites*; these strata are cut out by the overlying sandstone farther east, at Salt Lake [NO 586144], and are also absent where the same thick sandstone crops out on the eastern limb at Babbet Ness [NO 594141].

Mouth of Kenly Water to Craig Hartle Fault. The strata between the mouth of the Kenly Water [NO 581144] and the eastern end of the Craig Hartle Fault [NO 580148] lie on the limbs of a northerly-trending asymmetrical fold, the Craig Hartle Syncline. The strata on the east limb dip to the north-west at low angles. Stratigraphically they probably follow on, after a small gap, above the sandstone at the top of the section to the east. On the west limb of the syncline the strata are better exposed, dipping east at angles of up to 40°, and a section about 53 m thick can be made out. It consists of alternations of sandstone, siltstone and mudstone with thin coals and several bands of bedded dolomite. *Naiadites obesus* and *Carbonicola antiqua* are abundant in two of these bands. A sandstone near the base contains fossil tree trunks. The Craig Hartle Lower Marine Band (new name) is a bed of flat-laminated sandstone with *Sanguinolites*, and lies just above a band of silty dolomite with coprolites [NO 57901476] near the base of the section. On the east limb of the syncline there is a band of mudstone at the equivalent horizon, but no marine fauna has been detected. The Craig Hartle South Marine Band (new name) [NO 57971476] lies near the top of the 53-m section on the west limb of the syncline. It contains crinoid ossicles and molluscs in a 20-cm bed of dolomite and is probably the band from which Kirkby (*in* Geikie 1902, p. 132) recorded the occurrence of an aviculoid bivalve and other fossils. The general nature of the sequence in these sections suggests that they belong to the Anstruther Beds and that they may have correlatives in the Billow Ness section (p. 23), but a detailed correlation cannot be established.

The strata on the shore are separated from those exposed in the Kenly Water itself by the Kenly Mouth Fault [NO 58081438], an east–west fracture of unknown throw. The Kenly Mouth Marine Band (new name), with a fauna of molluscs and ostracods, is exposed [NO 58081431] on the west bank of the stream not far south of the fault. It presumably lies in the Anstruther Beds. The strata farther upstream are described elsewhere (pp. 35–36).

Craig Hartle north section. About 69 m of strata are exposed on the limbs of the Craig Hartle Anticline; they are cut off from the adjoining sections to the west and south by the Craig Hartle Fault. They are seen best on the east limb of the structure, just west of Craig Hartle [NO 58111487], where they dip east at about 25°. The sequence resembles those found in the Anstruther Beds and contains the Craig Hartle North Marine Band (new name) [NO 57891485] near the centre. The marine band is probably the one from which Kirkby (*in* Geikie 1902, p. 132) obtained *Schizodus* and *Sanguinolites*; it has also yielded *Naiadites crassus*, *Archaeocidaris* and crinoid columnals. Kirkby recorded abundant *Naiadites obesus* above the marine shells. The section does not appear to overlap with the 53-m section south of the Craig Hartle Fault (described above), and probably lies beneath it.

Craig Hartle Fault to Kittock's Den. Between the western end of the Craig Hartle Fault [NO 574151] and the Buddo Ness Fault [NO 562152] the strata dip to the west or north-west but are rather poorly exposed except for the sandstones and two thin bedded dolomites. Between the Buddo Ness Fault and the centre of a small syncline [NO 55231515] not far west of the mouth of Kittock's Den, they are well exposed and form the Buddo Ness section, at the east end of which, around Buddo Ness [NO 559154], the strata lie in four well-defined folds and contain two thin bands of dolomite with faunas of *Spirorbis*, ostracods and fish remains. West of Buddo Ness higher strata appear. They include [at NO 55801531] 3·6 m of mudstone with ironstone nodules and a marine fauna of crinoids, bivalves, *Lingula* and *Orbiculoidea*. The fauna is distinctive enough to identify the band as the West Sands Marine Band of St Andrews (p. 125), which marks the boundary between the Anstruther Beds and the Pittenweem Beds. Above the marine band, around the mouth of Kittock's Den, lies a sequence of thick sandstones and at a higher level, in the cliffs behind [NO 55271513], lies a bed correlated with the Witch Lake Marine Band of St Andrews (p. 38). A comparison of the upper part of the Buddo Ness section with the same sequence elsewhere (Fig. 4) shows that there is much more sandstone here than in the Pathhead–Anstruther coast section, and that the sequence as a whole thins to the north and north-west.

West of the synclinal axis at Kittock's Den the strata are described as part of the Maiden Rock–Kittock's Den section (pp. 39–40). In Kittock's Den itself, white sandstones are exposed, dipping to the south and south-west at angles between 13° and 37°. In a few places black shales with ironstone bands, containing ostracods and fish debris, are also visible. The section is almost continuous with the Buddo Ness section.

INLAND AREA AROUND DUNINO, KINGSBARNS AND BOARHILLS

Cameron Burn. There are many small exposures along the Cameron Burn, from the eastern end of Cameron Reservoir [NO 480113] to the confluence [NO 54301174] with the Wakefield Burn and also locally in its north-bank tributaries. They are mainly of sandstone, but other lithologies appear locally, including mudstones, bedded dolomites, ironstones, seatrocks and thin coals. Some of the mudstones and dolomites have yielded *Spirorbis*, ostracods and fish debris, but no shell-bed, marine or non-marine, has been detected. In the upper reaches of the stream the rocks lie on the western limb of a syncline whose axial trace crosses the Cameron Burn about locality NO 49941195. To the east for about 400 m the dips are north-westerly and vary in amount up to 55°: farther east both amount and direction vary, suggesting that the stream crosses a number of shallow folds. Several different parts of the Calciferous Sandstone Measures are probably represented.

Gilmerton. Five bores put down around Gilmerton in 1892 (Geikie 1902, p. 118) all show irregular alternations of sandstone, siltstone, mudstone and seatrock with a few coals and limestones. One reportedly encountered old workings 11 m below the surface. Exposures in Bridgeton Den and the old quarry 150 m south of Gilmerton House [NO 511115] are almost all of sandstone with some shale and mudstone. A 'cherry coal' was formerly visible (Geikie 1902, p. 118). The dip is to the west at 10° to 18°. A coal seam 1·2 m thick is said (Geikie 1902, p. 118) to crop out on the east side of the den, and to have been worked west of the den. A seam of the same thickness was reported at a depth of 25·6 m in a shaft [NO 50661127] 400 m west of the south end of the den, but a bore (No. 4 of the series mentioned above) put down only 120 m west of the shaft in 1892 encountered no seam thicker than 35 cm. The sequence reported includes only four coals and one 'limestone' in a sequence of sandstone, mudstone and seatearth 128 m thick. Farther north-west, near the ruins of Gilmerton Cottage [NO 504116], a coal overlain by black shale is said to crop out, dipping to the south (Geikie 1902, p. 118). Landale (1837, p. 298 and map), on the other hand, believed the dip to be northerly in that vicinity. In a gully 500 m east of Gilmerton House black shales with ostracods are exposed on the west side, with sandstone and sandy mudstone on the east side. A coal 90 cm thick is said to crop out in the bottom, where there are traces of old workings.

A small exposure [NO 51661161] 75 m north of the gully shows black shale on seven centimetres of decalcified limestone with abundant marine shells. The fauna of this band, here named the Brigton Marine Band, suggests a correlation with the St Andrews Castle Marine Band (p. 126), or alternatively with the Craig Hartle North Marine Band (p. 33). Unless a fault intervenes between the exposures, either correlation would suggest that the strata around Gilmerton lie lower in the sequence than the Sandy Craig Beds, and would thus throw doubt on Landale's (1837, p. 298) and Geikie's (1902, p. 118) identification of the Back and Fore coals in this area, an identification which in any case is inadequately based (see p. 52).

Kinaldy (Wakefield) Burn. The stratigraphical range of the rocks exposed in this stream, which joins the Dunino Burn at Dunino [NO 53771120], is probably quite large.

For about a kilometre downstream from Lathockar Mill [NO 495095] there are frequent exposures, mainly of sandstone, but also including seatearth, siltstone and mudstone with a thin dolomite [NO 49910961] containing *Spirorbis* and fish remains. An adjacent exposure of mudstone has yielded *Euestheria*. Downstream the burn flows over alluvium for some 600 m and then enters Kinaldy Den where sill and vent rocks are exposed and a coal was formerly visible (Geikie 1902, p. 118). Farther downstream, as far as Tosh [NO 524105], exposures are poor but westerly dips seem to prevail.

Between Tosh and Dunino, where the stream is known as the Wakefield Burn, exposures become progressively better downstream. At first they consist mainly of sandstone, dipping to the east at 18° to 35°, but the dips decrease downstream to zero and then become generally westerly and rather variable in amount. Siltstones, mudstones, dolomites, thin coals and seatbeds are exposed in addition to sandstones, and several continuous lengths of section are recorded. Faunas are restricted to ostracods and fish remains. At one exposure [NO 53621119] on the south bank, there is an adit of an old mine, but nothing is known about the workings concerned.

Dunino Burn. There are frequent exposures in the Dunino Burn from a point [NO 53190939] near Chesters down to its confluence [NO 54301174] with the Cameron Burn to form the Kenly Water. The direction of dip is generally between north-west and north-east: the amount of dip ranges up to 27°. The strata exposed are mainly sandstones, especially in the lower reaches of the stream: one forms the Bell Craig [NO 540109]. A marine band, which from its locality [NO 53650968] has been named the Chesters Marine Band, was found in a bed of mudstone, 1·2 m thick, containing ironstone nodules. It has a fauna resembling that of the Witch Lake Marine Band, with which it is provisionally equated. The following section, which on structural evidence lies not far above the marine band, is exposed about 50 m downstream.

	m
Mudstone, silty .. .. (seen)	0·5
Sandstone with shale laminae	0·6
Siltstone with sandstone bands	0·8
Mudstone, silty, with sandy laminae	0·5
Mudstone with ironstone bands: full of *Carbonicola? antiqua* and *Naiadites obesus*	2·1
Mudstone, dark-grey	0·2
Dolomite, argillaceous, with *Lingula squamiformis*, *Naiadites obesus* and *Sanguinolites? striatus*	0.6
Mudstone, dark-grey	0·05
Dolomite, argillaceous	0·05
Mudstone, dark-grey	0·8

This section strongly suggests that a marine or quasi-marine band is here present a short distance above the Witch Lake Marine Band; this is analogous to the occurrence of the Pittenweem Harbour *Lingula* Band in the coast section. Some of the strata exposed in the Dunino Burn clearly belong to the Pittenweem Beds but in the eastern part of its course the strata exposed probably lie in the Anstruther Beds.

Kenly Water. Exposures of rock are numerous all along the Kenly Water, from Dunino to the sea, and in the lower reaches of its tributary, the Pitmilly Burn, but they are generally discontinuous and consist mainly of sandstone. In a few places argillaceous beds, with associated coals, dolomites and ironstones, are visible but usually these lie unexposed in the gaps between the sandstones. Dips are rather variable in both amount and direction, but tend to be westerly in the western part of the river's course and easterly in the eastern part, the change taking place near Lower Kenly [NO 562128]. Most of these strata are assigned to the Anstruther Beds.

Kilduncan Burn. There are scattered exposures, mainly of sandstone, in the Kilduncan Burn in a stretch north and west of North Quarter [NO 573113]. They include also a few exposures of mudstone and shale, locally with dolomite bands and ironstones. Some of the shales contain fish scales and ostracods. The dips are generally between south-westerly and north-westerly, at angles up to 23°. A bore [NO 57411180] beside the stream cut 46 m of sandstone, siltstone and mudstone, with a thin coal near the top.

Carhurly, Kippo and Cambo burns. In the Carhurly Burn beside Carhurly [NO 564097] there are exposures of sandstone and shale with dolomite and ironstone. One of the shales is black, irony and cannely; it contains fish remains and *Naiadites obesus*. The dips are north-westerly throughout.

In the Kippo and Cambo Burns, for about a kilometre east of Kippo [NO 577104], there are exposures of sandstone, with some argillaceous beds, two thin coals and a recrystallised shelly dolomite [NO 58401017]. The beds dip north-north-west at 20° to 25° in the stretch south-west of the railway and lie horizontal on the north-east side. Exposures also extend downstream almost continuously from some 200 m west of the bridge [NO 59901054]. As the dips are low and variable in direction the thickness of rock exposed is not great. The strata include alternations of sandstone, siltstone and mudstone with rooty beds and thin bands of dolomite, and contain non-marine faunas of bivalves, ostracods and fish remains. They probably lie in the Anstruther Beds, like the nearby Randerston shore section (pp. 31–32), from which the strata south of the grounds of Cambo House are probably separated by faulting. I.H.F.

St Andrews–Ceres Area

The structure of the area between St Andrews and Ceres is dominated by the Ceres–Maiden Rock Fault Zone (Fig. 2) In the fault zone the rocks are affected by strong local folding but to the south gentle domes and basins are superimposed on a westerly regional dip, while to the north gentle southerly or south-easterly dips prevail. There are good exposures on the coast, and inland there are sections in streams and old quarries, as well as borehole records. Representatives of all the stratigraphical divisions, except the lowest, have been recognised. The Anstruther Beds are poorly exposed but the Pittenweem Beds and the Sandy Craig Beds are well seen. The latter two subdivisions are thinner than in the Anstruther–Pathhead section, and appear to contain a higher proportion of sandstone. They also thin westwards within the St Andrews–Ceres area (Fig. 4). The Pathhead Beds are only slightly thinner than in the Anstruther–Pathhead section, however, and are more argillaceous (Fig. 6).

The thinning continues into areas farther west so that at Cults and in the Lomond Hills only about 30 m to 90 m of Calciferous Sandstone Measures intervene between the base of the Lower Limestone Group and the top of the Upper Old Red Sandstone (Geikie 1902, pp. 72–73). From the evidence in the St Andrews–Ceres area, and from the high proportion of marine strata in the Calciferous Sandstone Measures present in the Lomond Hills (MacGregor 1968, p. 245), it seems likely that it is the four lowest divisions which have been overlapped in these more westerly areas, and that it is an attenuated representative of the Pathhead Beds which is preserved there.

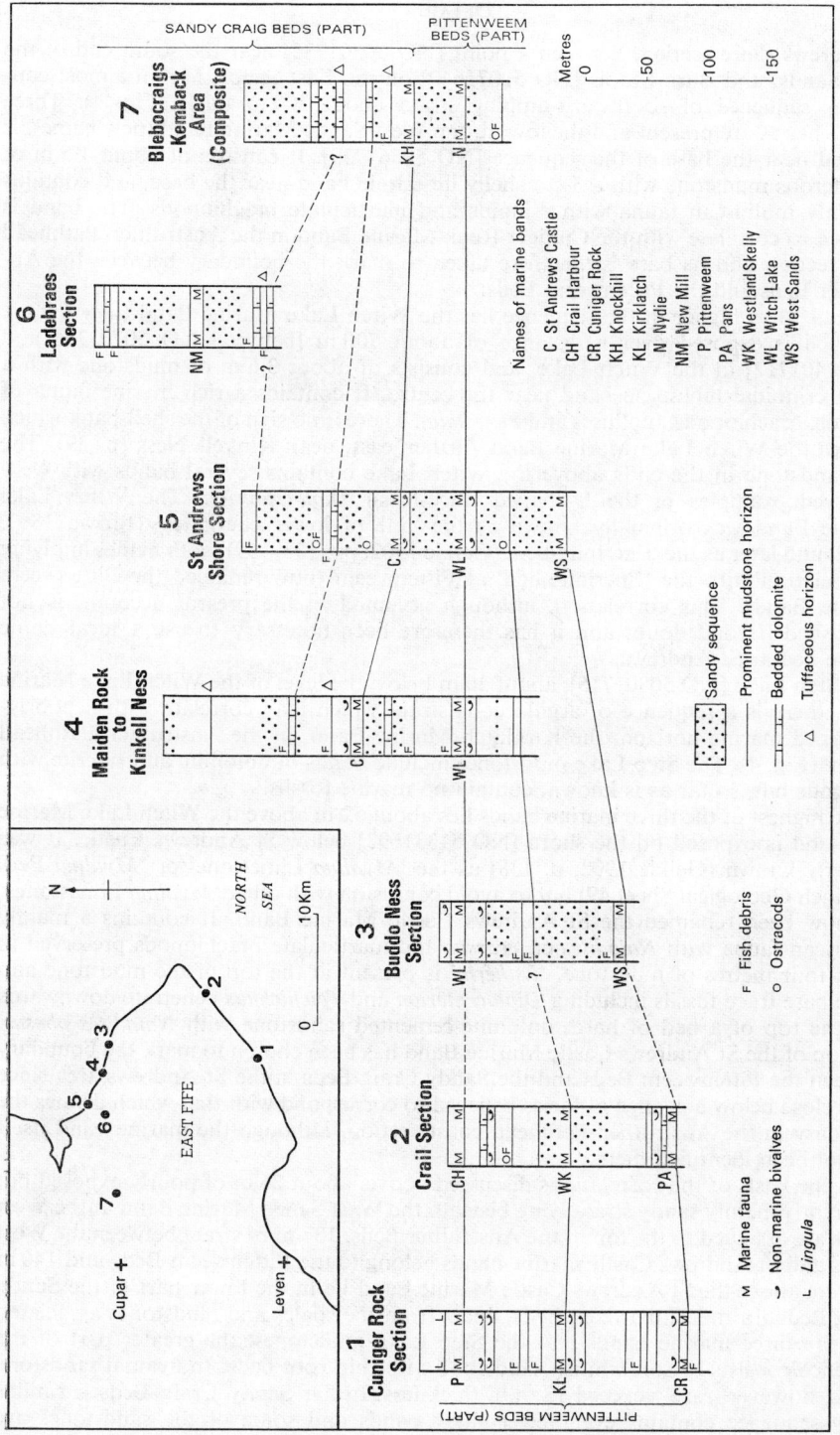

FIG. 4. Vertical sections showing lateral variation in parts of the Pittenweem Beds and Sandy Craig Beds

DETAILS

St Andrews shore section. Between a point [NO 50501735] near the south end of the West Sands, and Burn Stools [NO 52071640] on the East Sands, lies an almost continuous sequence of southward-dipping strata about 300 m thick (Fig. 4). Three marine bands are present and the lowest, the West Sands Marine Band (new name), is exposed near the base of the sequence [NO 50861729]. It consists of about 1·5 m of fossiliferous mudstone with a 5-cm shelly limestone band near the base, and contains a mainly molluscan fauna with crinoids and inarticulate brachiopods. The band is believed to correlate with the Cuniger Rock Marine Band in the Anstruther–Pathhead coast section and its base is therefore taken to mark the boundary between the Anstruther Beds and the Pittenweem Beds.

About 81 m higher in the sequence lies the Witch Lake Marine Band (new name). This bed is exposed over a distance of about 700 m [between NO 50701705 and NO 51401712] in the Witch Lake, and consists of about 3·6 m of mudstone with a 15-cm crinoidal limestone band near the centre. It contains a rich marine fauna of crinoids, brachiopods, molluscs and *Fenestella*. There is no sign of the shell-bank which overlies the Witch Lake Marine Band further east, near Kinkell Ness (p. 39). The 6-m sandstone in the cliffs above the Witch Lake contains several bands with well-preserved examples of the U-shaped trace-fossil *Diplocraterion*. The Witch Lake Marine Band was originally known as the 'line of lower encrinites' (Brown 1861, p. 389) and later as the 'Encrinite-bed' (Geikie 1902, pp. 138–139), both names implying a correlation with the 'Encrinite-bed' at Pittenweem (now renamed the Pittenweem Marine Band). This correlation, although accepted in the present account, is not established beyond doubt and it has therefore been necessary to use a local name for the bed at St Andrews.

At Step Lake [NO 50801715], about 30 m below the level of the Witch Lake Marine Band, there is a sequence of argillaceous strata which may correlate with a poorly-developed marine horizon, the Kirklatch Marine Band, in the Anstruther–Pathhead section (Fig. 4). The Step Lake mudstones include bands of dolomite and siderite with ostracods but, so far as is known, contain no marine fossils.

The highest of the three marine bands lies about 52 m above the Witch Lake Marine Band and is exposed on the shore [NO 51331692] below St Andrews Castle. It was formerly known (Geikie 1902, p. 138) as the '*Myalina* Limestone' or '*Myalina* Bed' (one-inch Geological Sheet 49) but to avoid confusion with other '*Myalina* Limestones' has now been renamed the St Andrews Castle Marine Band. It contains a mainly molluscan fauna with *Naiadites* cf. *crassus* and inarticulate brachiopods preserved in about four metres of mudstone. '*Estheria*' is present at the top of the mudstone and at the base trace-fossils including *Diplocraterion* and *Teichichnus* penetrate downwards into the top of a bed of hard, dolomite-cemented sandstone with *Naiadites obesus*. The top of the St Andrews Castle Marine Band has been chosen to mark the boundary between the Pittenweem Beds and the Sandy Craig Beds in the St Andrews area since it lies close below a change of facies believed to correspond with that which defines the boundary in the Anstruther–Pathhead coast section, although the marine band itself has not been identified there.

On the basis of the correlations discussed above, about 24 m of poorly-exposed but apparently mainly sandy strata lying beneath the West Sands Marine Band at the West Sands are assigned to the top of the Anstruther Beds, 137 m of strata between the West Sands and St Andrews Castle marine bands belong to the Pittenweem Beds and 140 m of strata above the St Andrews Castle Marine Band lie in the lower part of the Sandy Craig Beds. In the Pittenweem Beds, apart from the shales and mudstones associated with the three marine bands and the Step Lake mudstones, the greater part of the sequence consists of soft whitish sandstone with thin root beds. Individual sandstone bands, however, rarely exceed 12 m in thickness. In the Sandy Craig Beds a similar sandy sequence contains even fewer shaly bands and some of the sandstones are thicker, one bed reaching 30 m in thickness. About 52 m above the St Andrews Castle Marine Band a bed of dolomite-siltstone 1·4 m thick contains fish remains and has polygonal cracks on its upper surface. It may correlate (Fig. 4) with sandy dolomitic horizons at Kemback (p. 42), Kinkell Cave and Kinkell Ness (p. 40).

Shore section, Maiden Rock to Kittock's Den. In the three kilometres of coast section between Maiden Rock [NO 52641579] and the mouth of Kittock's Den [NO 55421509] lies a well-exposed sequence belonging to parts of the Pittenweem Beds and Sandy Craig Beds. An excursion guide to this part of the coast has been published by MacGregor (1968). The strata are thrown into a series of folds which decrease in intensity eastwards away from the Maiden Rock Fault (Geikie 1902, fig. 15). Individual horizons can be traced round the folds and it is clear that a relatively short stratigraphical interval, not more than about 300 m thick, is involved in these structures. A generalised sequence for the part of the section between Maiden Rock and Kinkell Ness is shown in Fig. 4, column 4. The lowest strata are seen in the first anticline east of Maiden Rock [at NO 52831578] and again in the structural domes north of Kingask [at NO 54341539, NO 54791536], while the highest strata are preserved in a syncline [NO 53431582] about 400 m west of Kinkell Ness.

That this sequence can be correlated with the section exposed at St Andrews, on the far side of the Maiden Rock Fault, was first realised by Brown (1861, pp. 387, 389), who recognised his 'line of lower encrinites' (now the Witch Lake Marine Band) both at the Witch Lake at St Andrews and near the Rock and Spindle in the present section. The general equivalence of the St Andrews and the Maiden Rock–Kittock's Den sections was accepted by Geikie (1902, pp. 134–139) but the correlation with the St Andrews section was not firmly established until Kirk (1925a, 1925b) determined the stratigraphical relationships between the individual outcrops of marine and other faunal bands within the Maiden Rock–Kittock's Den section itself.

Two distinct marine horizons are now recognised in this sequence. The higher band contains a fauna of crinoids, inarticulate brachiopods and molluscs with *Naiadites* cf. *crassus*, and is correlated with the St Andrews Castle Marine Band; the lower bed contains a richer fauna of crinoids, brachiopods, molluscs and *Fenestella* and is correlated with the Witch Lake Marine Band.

The level of the West Sands Marine Band is nowhere reached in the Maiden Rock–Kittock's Den section, but the equivalents of the Step Lake mudstones of the St Andrews section crop out in the centres of two structural domes below Kingask and in the first anticline east of Maiden Rock.

The equivalent of the Witch Lake Marine Band, originally described at a locality [probably NO 542154] near the Rock and Spindle, as the 'line of lower encrinites' (Brown 1861, p. 389) and later as the 'Encrinite-bed' (Kirkby 1880, p. 563; Geikie 1902, p. 135) crops out also at three localities [NO 52651580, NO 52751579, NO 52891578] east of Maiden Rock and at several points on the shore and in the cliffs between Kinkell Ness and Kittock's Den (Kirk 1925a, plate 48).

Mudstones immediately above and below the Witch Lake Marine Band east of the Rock and Spindle [at NO 542154] were noted by Kirk (1925a, pp. 367–369) as being 'full of marine fossils', but they contain only *Naiadites obesus* and *Carbonicola?* and would not now, therefore, be regarded as marine horizons. The higher of the two bands is a cross-stratified shell-bed varying from 0·6 to 1·8 m in thickness. It was first noted by Kirkby (1880, pp. 572–573) and was subsequently mistaken by Geikie, perhaps on the basis of a wrongly-located fossil collection (Geikie 1902, fig. 14, pp. 135–136), for the equivalent of the St Andrews Castle Marine Band; as Kirk (1925a, p. 370) pointed out, however, the bed is a local shell-bank, not present in adjoining exposures.

The St Andrews Castle Marine Band, formerly known as the '*Myalina* Limestone' and lying 52 m above the Witch Lake Marine Band at St Andrews, is exposed [NO 52991577] near Kinkell Cave and at the margin of the Kinkell Ness neck [NO 53981565]. It was first correctly identified in the Maiden Rock–Kittock's Den section by Kirk

(1925a, p. 370), for although Geikie had realised that the bed should be present there his identification of it was in fact mistaken (see above). At the exposure near the Kinkell Ness neck a 20-cm shell-bed made up mainly of *Naiadites obesus* lies at the top of the marine mudstone, and the same bivalve is present in mudstones both above and below the marine band at its Kinkell Cave outcrop. At both exposures a yellow-weathering bed of fine-grained dolomite containing fish remains is present between 14 and 20 m below the marine band, but this bed has died out before the St Andrews section is reached.

As previously stated, the top of the St Andrews Castle Marine Band has been selected as a convenient marker horizon to define the top of the Pittenweem Beds in the St Andrews area, so that the lower half of the Maiden Rock–Kittock's Den section falls into this subdivision, and the upper half falls into the Sandy Craig Beds. Apart from the Step Lake mudstones and the argillaceous strata associated with the Witch Lake and St Andrews Castle marine bands, the sequence in the Pittenweem Beds consists mainly of sandstone with thin seatearth bands, and contrasts strongly with the much more argillaceous facies of the equivalent beds farther south (Fig. 4). Within the St Andrews area alone, variations in both thickness and facies can be observed as the strata are traced from west to east. The thickness of the interval between the Witch Lake and St Andrews Castle marine bands, for example, increases from 52 m at St Andrews to 82 m near the Rock and Spindle, while the strata associated with the Witch Lake Marine Band change markedly in facies as well as in thickness when traced from west to east.

In the Sandy Craig Beds sandstone is predominant and it alternates with bands of seatearth and sandy mottled mudstone, some of which are red in colour; a few thin bands of shale also occur. A zone of flat-bedded dolomitic sandstones containing fish debris is present about 60 m above the St Andrews Castle Marine Band and is exposed conspicuously around the minor anticline and syncline at Kinkell Cave [NO 531158] as well as at Kinkell Ness [NO 538158], where the lowest bed, 20 cm thick, is a dolomite-siltstone containing a small proportion of quartz grains. The zone as a whole is believed to correlate with a 1·4-m bed of dolomite-siltstone 52 m above the St Andrews Castle Marine Band at St Andrews (Fig. 4). Dolomitic sandstones are present at various higher levels in the sequence also, and beneath one of them, lying about 120 m above the St Andrews Castle Marine Band near Kinkell Ness, a distinctive 10-cm band of pale-coloured laminated shale and siltstone, possibly of tuffaceous origin, is present [NO 53481575]. A band of similar colour and composition, but with a more fragmental texture, is present about 27 m above the same marine band near Kinkell Cave [NO 531158].

As suggested in Fig. 4, these zones of dolomitic material and horizons of possibly tuffaceous origin may correlate with similar rock types in the Ladebraes section and in the Blebocraigs–Kemback sequence, and it may be noted that dolomitic bands (although in an argillaceous sequence) and tuffaceous horizons are a feature of the strata assigned to the lower part of the Sandy Craig Beds in the coast section at Pittenweem also (Fig. 3).

Coast section, East Sands to Maiden Rock. Between the south end of the East Sands [NO 52101600] and the Maiden Rock [NO 52641579] lies an area of disturbed strata adjacent to the Maiden Rock Fault. The rocks generally dip towards the south except in the immediate vicinity of the fault, where they are folded sharply about north-easterly axes. The general disposition of the rocks in this stretch of coast is shown in Fig. 2 (see also Balsillie 1920, fig. 4; MacGregor 1968, p. 161). A prominent 3-m band of unbedded mudstone and sandstone containing nodular dolomitic concretions with 'catsbrain' mottling can be traced eastwards along the cliffs as far as a small fault [NO 52221582], to the east of which lies a sequence of sandstones with bands of shale and mudstone and the St Nicholas Marine Band (new name). The strata west of the fault probably lie lower in the sequence than those to the east. The marine band is exposed low on the foreshore [e.g. at NO 52431600 and NO 52581591] and also in a small faulted block [NO 52311590] nearer the high water mark. The last-mentioned exposure is the most easily accessible, and here the section is:

	m
Sandstone	3·1
Mudstone with sandstone laminae	1·1
Mudstone with ironstone nodules, some bioturbated; brachiopods, molluscs and *Fenestella*	1·5
Mudstone with *Lingula*	0·05
Shale, barren	0·15
Sandstone	7·6

A prominent band of mudstone about 30 m above the St Nicholas Marine Band forms an open gully among the sandstone reefs on the foreshore [NO 52431589] and from this band *Curvirimula* has been collected. The presence of this non-marine bivalve may indicate a horizon relatively high in the Calciferous Sandstone Measures (p. 129). On facies grounds the strata in this stretch of coast are all assigned to the Sandy Craig Beds; it is possible that the St Nicholas Marine Band correlates with the New Mill Marine Band (Table 2).

Ladebraes section. A sequence of strata dipping south at 10° to 20° is discontinuously exposed in the Kinness Burn between Carron Bridge [NO 48891585] and Cockshaugh Park [NO 50331631] in St Andrews. Two bands of bedded dolomite near the base of the section contain ostracods and fish debris, and the upper one [at NO 50121624] is underlain by a 10-cm bed of banded, possibly tuffaceous, siltstone. Above lies a very sandy sequence about 45 m thick, at the top of which there is a bed of mudstone, 1·5 m thick, containing a rich marine fauna of brachiopods, bryozoa and bivalves—the New Mill Marine Band (new name). This bed is exposed in the Kinness Burn [NO 49841602] 90 m downstream from New Mill but is usually below water level. Its fauna resembles those found at higher levels in the succession, in the Pathhead Beds and above, rather than those found at lower levels. Just above New Mill a 4·6-m bed of grey shale with sparse *Curvirimula*, fish debris and ostracods is overlain by sandstones and a 38-cm shaly coal [at NO 49671586]; the seatbed below the coal contains spherulitic nodules of an unusually magnesium-rich siderite ($\omega = 1.782$). Between this point and Carron Bridge there are several gaps in the section, and the sedimentary sequence is interrupted by the tuffs of the Bogward neck. At a point [NO 49471571] 400 m upstream from New Mill a band of bedded dolomite 0·6 m thick is underlain by a 5-cm shaly coal, but otherwise sandstone predominates among the rocks exposed.

Stratigraphically the Ladebraes section probably lies immediately above the St Andrews shore section; there may be either a small gap or a small overlap between them (Fig. 4). The facies suggests that the whole section lies in the Sandy Craig Beds.

Cairns Den. A southward-dipping sequence of strata is exposed in Cairns Den between the Bogward neck [NO 49301557] and the Cairnsmill neck [NO 49411503]. The strata are poorly exposed except for a stretch towards the northern end of the section, where some 24 m of soft sandstone and massive sandy mudstone containing zones of dolomitic concretions are overlain by shales with fish debris [NO 49481545]. On facies grounds this part of the section is assigned to the Sandy Craig Beds but at the southern end of the section limestones belonging to the Lower Limestone Group were formerly exposed (p. 69); the apparent absence of the Pathhead Beds (cf. Craig and Balsillie 1912, p. 23) is now attributed to the effect of a large fault which separates the two parts of the section.

Lumbo Den. Isolated southward-dipping exposures of sandstone, siltstone, shale and ironstone are visible in Lumbo Den between the margin of the Bogward neck [NO 49221556] and Lumbo Bridge [NO 48801481]. Immediately to the north of Lumbo Bridge the sequence includes limestones belonging to the Lower Limestone Group (p. 76) but northwards from a point 175 m downstream from the bridge the strata probably belong to the Calciferous Sandstone Measures. As in Cairns Den nearby the lowest part of the section, at least, is believed to lie in the Sandy Craig Beds. Both the Lumbo Den and Cairns Den sections are probably higher in the sequence than the Ladebraes section.

Strathkinness. White sandstones dipping southwards at low angles were formerly worked in a number of quarries lying immediately west of Strathkinness. Most of the quarries are now overgrown or filled up but at Bonfield Quarry [NO 45351600] about 1·5 m of highly dolomitic siltstone and sandstone with fish debris are exposed beneath the boulder clay. A *Lingula* band was reported in shales at a depth of 19·8 m in a water bore [NO 45731610] drilled in 1948 a few hundred metres east of Bonfield Quarry, but no specimens are available and the record cannot be confirmed. These strata, like those exposed in the quarries and mines around Blebocraigs (pp. 43–44), appear from their geographical position to lie higher in the succession than the St Andrews Castle Marine Band, and are therefore assigned to the Sandy Craig Beds.

Kincaple. Rocks probably belonging to parts of the Anstruther Beds and Pittenweem Beds occupy a large tract of poorly exposed ground south of the Eden estuary between Kincaple and St Andrews. The Kincaple Marine Band (new name) was formerly exposed in shales in the Den Quarry [NO 45491772] at the south end of Kincaple Den, but marine fossils can now be obtained only from the spoil tips. The fauna has affinities with those of the Witch Lake and St Andrews Castle marine bands, suggesting that the band lies in the Pittenweem Beds. In the den itself discontinuously exposed soft white sandstones with occasional rooty beds and a thin coal are arranged in a shallow synclinal structure. Soft white sandstones, in general dipping southwards at moderate angles, are or were exposed in a number of old quarries between Kincaple Den and Strathtyrum [NO 490172].

Knock Hill to Kemback. Soft white sandstones dipping generally south-eastwards at low angles are exposed over a wide area on Knock Hill and Kemback Hill from Longmuir [NO 45121686] to near Kemback [NO 42131521]. The region appears to be relatively free of faults so that the sections exposed in old freestone quarries and mines can be pieced together to form a single discontinuously exposed sequence about 240 m thick (Fig. 4, column 7). The Nydie and Knockhill marine bands, correlated on faunal grounds with the Witch Lake and St Andrews Castle marine bands respectively, are present in the lower half of the sequence, which therefore belongs mainly to the Pittenweem Beds, although some strata at the extreme base of the section may fall into the Anstruther Beds. The upper half of the sequence is consistently sandy, with a number of dolomitic horizons, and is assigned to the Sandy Craig Beds.

The lowest strata are exposed in Nydie Quarry [NO 43981671] where eight metres of interbedded shale, siltstone and sandstone with dolomitic bands overlie a worked sandstone of which only the top part is now visible. The finer beds contain ostracods, fish debris and poorly-preserved bivalves. Higher strata comprising sandstones, shales, rooty beds and coal are patchily exposed in various quarries in Nydie Wood [NO 444167], in Millstone Wood [NO 435163] and in the northern part [NO 427160] of Kemback Wood. The Nydie Marine Band (new name) is seen in one of the quarries in Nydie Wood [NO 44461679] where the section is:

	m
Limestone, shelly, ferruginous	0·2
Shale, grey, with crinoidal bands and rich brachiopod–mollusc fauna	1·3

	m
Coal, shaly	0·25
Seatearth	0·4
Sandstone	2·1
Siltstone, laminated	0·5
Sandstone, soft, white, massive (quarried bed) .. .. .. (seen)	2·0

The Nydie Marine Band was also formerly exposed farther west [NO 43521633] but this section is now obscured. The Knockhill Marine Band (new name), formerly exposed over a distance of 600 m in Knockhill Quarry, is not now visible but a marine fauna composed mainly of molluscs, including *Naiadites* cf. *crassus*, can still be obtained from the spoil tips. Soft white sandstones lying between the Nydie and Knockhill marine bands are exposed in a number of quarries. At Kemback Quarry [NO 42911611] 15 m of sandstone are seen and thicknesses of up to 9 m are exposed at several points [NO 42741602, 43351628, 43701628] in Kemback and Millstone Woods. At Knockhill Quarry, where sandstone has been worked underground, the section is:

	m
Horizon of Knockhill Marine Band (not seen)	–
Sandstone, fine-grained, soft, white, massive-weathering but cross-stratification visible in places: worked in quarry face	15·0
Sandstone, fine-grained, flaggy-weathering; rare *Stigmaria*	2·3
Sandstone, fine-grained, soft, white, massive-weathering but with traces of cross-stratification and convolution visible in places: worked underground from adits .. .. (seen)	9·0

Strata probably lying not far above the Knockhill Marine Band are exposed in a quarry [NO 42801580] 400 m west of South Flisk. The section is:

	m
Sandstone, fine-grained, soft, white	3·6
Siltstone, sandy, grey	0·05
Sandstone, as above	1·4
Siltstone, dark-grey, fissile: plant remains	0·3
Seatearth, light-grey	0·3
Sandstone, fine- to medium-grained, white: erosive base	1·4
Siltstone, sandy, dark-grey, massive; laminated porcellanous band (? tuffaceous) at top	0·07
Sandstone, fine-grained, white; argillaceous lenses at top .. (seen)	2·4

Strata still higher in the succession are exposed in a line of quarries and underground workings extending from South Flisk [NO 43131583] to a point [NO 42351539] near the western end of Blebocraigs. At Blebocraigs Quarry [NO 42661560] the section is:

	m
Sandstone, fine-grained, soft, white, bedded	3·6
Sandstone as above, rooty	1·4
Seatclay, grey	0·1
Sandstone, fine-grained, soft, white, cross-stratified	1·7
Sandstone, fine-grained, yellow and white, flat-bedded; many hard dolomitic bands	2·2
Sandstone, fine-grained, hard, yellow, flat-bedded, dolomitic	1·6
Siltstone, grey, laminated, sandy at base; fish debris	0·5
Sandstone, fine-grained, soft, white, massive; worked underground from adits .. .. (seen)	2·4

A somewhat similar, but less complete, section is exposed in quarries and adits at Flisk Quarry [NO 42921572], a few hundred metres to the east.

Stratigraphically higher beds outcrop between Blebocraigs Quarry and Broomside [NO 43611522]; soft white sandstones appear to make up much of the sequence and form drift-free ridges between the village of Blebocraigs [NO 430152] and Clatto Hill [NO 436158]. The sandstones are exposed in a few quarries, and are seen to be interbedded with harder, yellow, orange and brown dolomitic sandstones south of South Flisk [at NO 43221564], at Blebocraigs [NO 43011525] and between Blebocraigs and Kemback [at NO 42391530]. A 4·3-m bed of dolomitic sandstone with ostracods, exposed [NO 44031601] north of Easter Clatto, probably also lies in this part of the sequence.

In Dura Den white sandstone with thin bands of siltstone and mudstone is exposed at a number of points south of the Dura Den Fault between Kemback and Pitscottie [NO 417131]. These rocks are believed to lie in the Sandy Craig Beds since they are predominantly sandy and lie along the strike from the similar beds exposed at Blebocraigs.

Boreholes in the Denhead area. Four Geological Survey boreholes in the Denhead area south-west of St Andrews penetrated strata lying in the upper part of the Calciferous Sandstone Measures. They are: Drumcarro Borehole (1963) [NO 45911290], Denork Borehole (1966) [NO 45401409], Claremont Borehole (1966) [NO 45181419] and Mount Melville Cottages Borehole (1966) [NO 48281496]. The lithological sequences encountered in these boreholes are shown in graphical form in Fig. 5. Faunas collected from the marine bands in the boreholes have affinities with those known from the upper part of the Pathhead Beds in the Anstruther–Pathhead coast section, and the borehole sequences are therefore believed to lie in the same part of the succession. Comparison of the composite section of the boreholes with the coastal section of the Pathhead Beds (Fig. 6) suggests that all six marine bands recognised on the coast can be identified in north-east Fife, and it appears that the upper part of the Pathhead Beds is slightly thinner but much more argillaceous there than in the Anstruther–Pathhead section. The belief held by Craig and Balsillie (1912, p. 23), that the strata associated with the Ardross Limestones are thin or absent in north-east Fife, is now seen to have been erroneous.

The Drumcarro Borehole cut the St Monance Brecciated Limestone at a depth of 118·2 m and thereafter penetrated 123·7 m into the Calciferous Sandstone Measures, encountering two marine bands which are correlated tentatively with the Pathhead marine bands of the Anstruther–Pathhead coast section (Forsyth and Chisholm 1968, pp. 71–72).

The Denork Borehole appears to have encountered a zone of faulting between 47·8 and 49·4 m from the surface. Above this level lies a mainly argillaceous sequence, containing *Curvirimula* and *Naiadites?*, with two distinct marine horizons. The upper, at a depth of 24·7 m, contains a fauna of molluscs and brachiopods, and a crinoidal limestone 0·3 m thick, in about 2·7 m of mudstone. The lower band, at a depth of 37·4 m, contains only *Lingula* and *Schizophoria?* scattered through about 0·3 m of mudstone. The nine metres of mudstone below the lower marine band contain numerous dolomitic siltstone beds and several thin tuffaceous horizons. The marine bands clearly correlate with the equivalents of the Pathhead marine bands in the Drumcarro Borehole (Fig. 5).

A fauna of marine molluscs is present in mudstones in the fault zone itself but the stratigraphical position of this marine horizon remains in doubt. Below the level of the fault zone lies a mainly argillaceous sequence containing scattered marine molluscs, with a single band containing productoids at a depth of 54·8 m, and *Lingula* near the base, at a depth of 75·5 m. The marine mudstones are underlain by a prominent bed of black fissile shale 0·4 m thick. The sequence in this part of the bore clearly correlates (Fig. 5) with the marine mudstones in the lower half of the Mount Melville Cottages

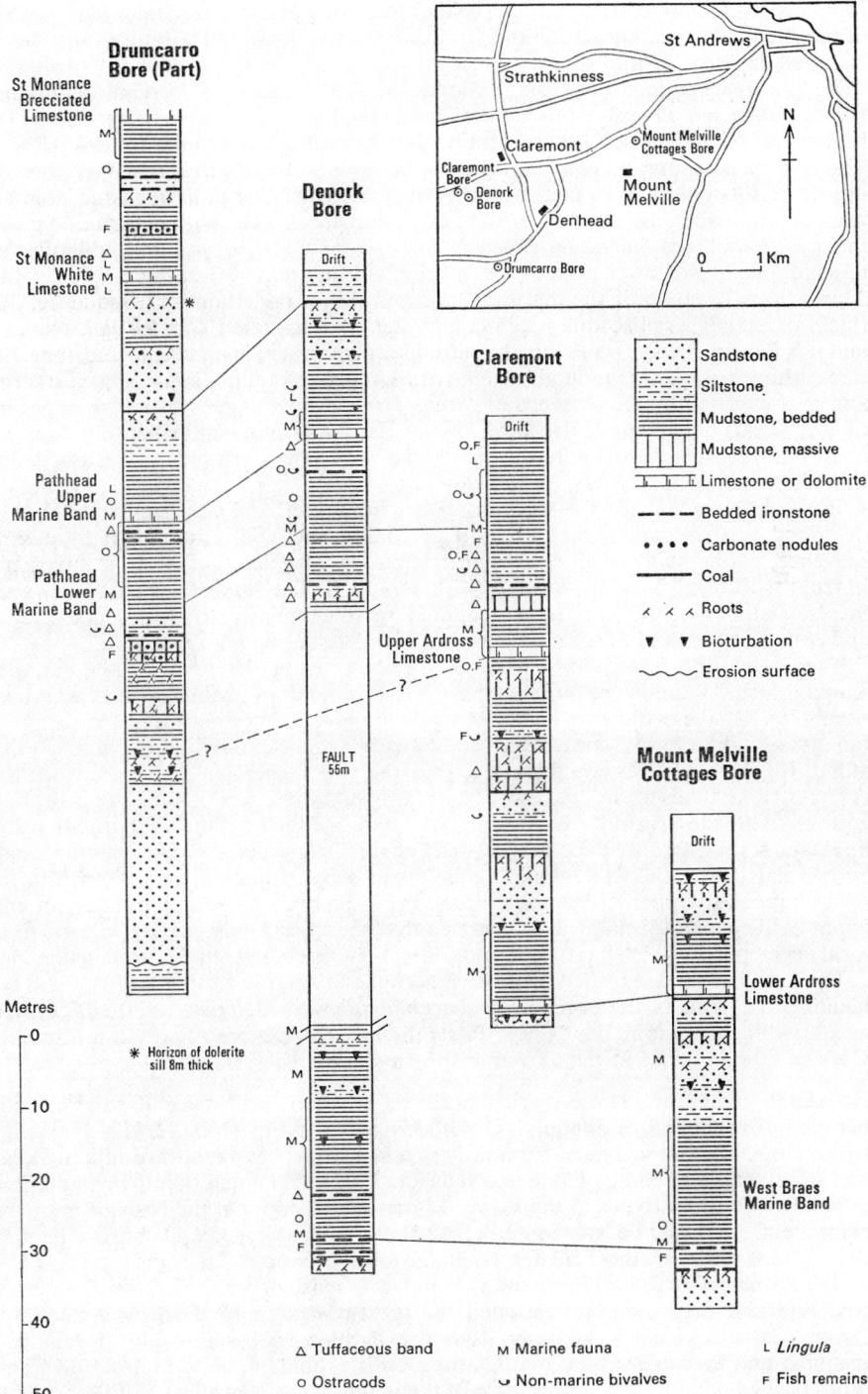

FIG. 5. *Borehole sections in the Pathhead Beds in the Denhead area*

Borehole, which are considered (see below) to represent the West Braes Marine Band of the Anstruther–Pathhead section. On this basis the throw of the fault in the Denork Borehole is about 55 m.

A thin marine band at 15·5 m depth in the Claremont Borehole contains a sparse fauna of *Lingula*, productoids and molluscs, and clearly represents the horizon of the Pathhead Lower Marine Band at Drumcarro and Denork (Fig. 5). The band lies among mudstones with *Curvirimula* and *Naiadites*, and there are thin dolomitic siltstones and tuffaceous horizons just below. The tuffaceous and dolomitic beds are underlain by six metres of mudstone with a rich marine fauna consisting mainly of molluscs including *Streblopteria ornata* with, in addition, crinoids and brachiopods in the lower part. A shelly crinoidal limestone 0·3 m thick, at a depth of 32 m, forms the base of the marine band, and this is underlain by a sequence, 36 m thick, of mainly argillaceous rocks containing only scattered *Curvirimula*, ostracods and fish debris. Below this relatively barren sequence there is a marine mudstone 12 m thick whose base lies at a depth of 80·2 m. A restricted fauna, consisting of molluscs and characterised by the presence of *Streblopteria ornata*, is present in the upper part of this band, but below a depth of 72·2 m the fauna is much richer, containing in addition brachiopods, trilobite remains and *Fenestella*. A crinoidal limestone 0·38 m thick is present near the base of the band.

In the Mount Melville Cottages Borehole, a marine band very similar to that just described was encountered at a depth of 23·8 m and this is underlain by a sandy sequence, cut by some small faults and containing, at a depth of 32 m, a thin marine horizon with bivalves and a productoid brachiopod. The sandstones pass downwards into mudstones, which contain a scattered marine fauna of molluscs and *Lingula*, and rest, at a depth of 56·1 m, on a bed of black fissile shale 0·7 m thick.

There are thus three major marine horizons below the level of the Pathhead Lower Marine Band in the strata cut by the Claremont and Mount Melville Cottages boreholes. Of these the two highest contain crinoidal limestones and relatively rich faunas, while the lowest contains a very restricted and scattered fauna, with no limestone. It seems most likely, therefore, that the two higher bands correlate with the Ardross Limestones of the coast section at Pathhead while the lowest band is probably equivalent to the sparsely fossiliferous West Braes Marine Band (Fig. 6). It will be apparent from Fig. 5 that the deepest part of the Drumcarro Borehole cannot readily be correlated with the Claremont Borehole section, and that in particular the horizon of the supposed Upper Ardross Limestone is not present at Drumcarro. This anomaly is not easily explained unless it is assumed that the thick sandstone at Drumcarro is a local development which has caused rapid lateral facies changes in the strata overlying it.

Four marine bands exposed in the area between St Andrews and Pitscottie have been identified in terms of the borehole sequence just described. These are the Blebo Hole and Newbigging marine bands near Pitscottie and the Denbrae Farm and Claremont Cottage marine bands in the Claremont Burn section (see below).

Newbigging of Blebo. Three marine horizons are known in a small stream section between Newbigging of Blebo [NO 430135] and a point [NO 42341315] south of Blebo Hole. They lie in a discontinuously exposed sequence of eastward-dipping strata, and were first described by Craig and Balsillie (1912) who noted that their faunas have affinities with those found in the Lower Limestone Group. On the basis of indifferent evidence from Cairns Den they believed that the limestones and marine bands in the highest part of the Calciferous Sandstone Measures (the part now known as the Pathhead Beds), which contain similar faunas, are not developed in northeast Fife and they therefore assigned the present section to the Lower Limestone Group. It is now clear, however, that the Pathhead Beds are fully developed in this area and that their absence at Cairns Den is probably due to faulting. Moreover it has proved possible to identify some of the bands in the Newbigging of Blebo stream section with bands in the local borehole sequence of the Pathhead Beds, to which stratigraphical subdivision the section is now assigned.

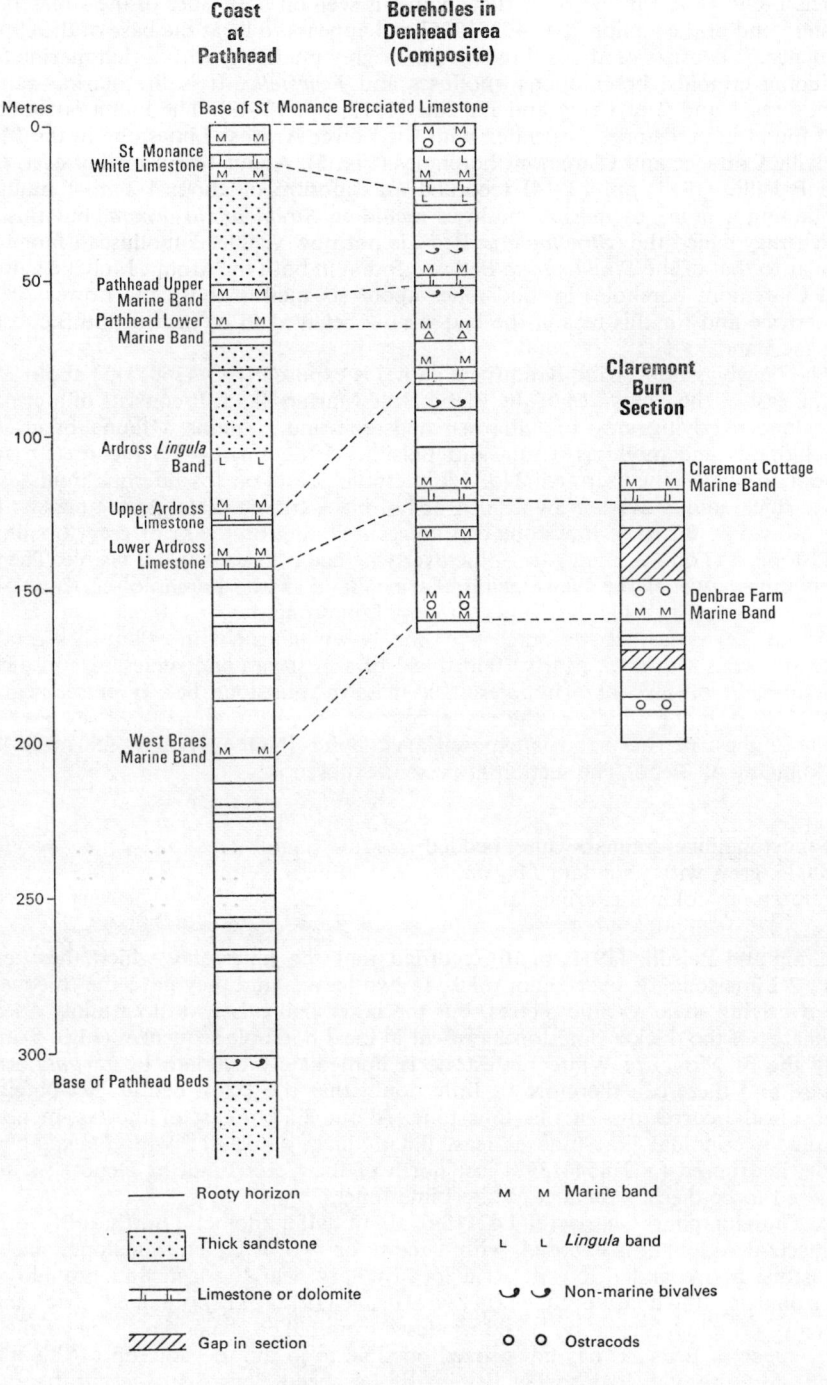

FIG. 6. *Vertical sections in the Pathhead Beds*

The Blebo Hole Marine Band (new name) is seen on both sides of the stream at the western end of the section [NO 42351315] and appears to lie at the base of the exposed sequence. It consists of about three metres of grey mudstone with a rich marine fauna including crinoids, brachiopods, molluscs and *Fenestella*. It is the marine band at Exposures 1 and 2 of Craig and Balsillie (1912, pp. 12–13). The fauna is similar to that found in mudstones associated with the Lower Ardross Limestone in the Mount Melville Cottages and Claremont boreholes (Fig. 5). About 100 m farther east, Craig and Balsillie (1912, pp. 13–14) recorded an exposure of about 1·2 m of mudstone containing a fauna of marine molluscs including *Streblopteria ornata*, but this bed, which they called the '*Streblopteria* Bed', is not now visible. A molluscan fauna very similar to that of the *Streblopteria* Bed was found in both the Mount Melville Cottages and Claremont boreholes in mudstones about six metres above the Lower Ardross Limestone and for this reason the bed is now believed to be part of the Blebo Hole Marine Band.

The Newbigging Marine Band (new name) is exposed [NO 42601315] about 300 m to the east of the exposures of the Blebo Hole Marine Band. It consists of a crinoidal limestone overlying grey fossiliferous mudstone and contains a fauna of crinoids, brachiopods and molluscs. Craig and Balsillie (1912, pp. 14–15) recorded that the limestone, which they termed 'No. 1 Limestone', rests on fossiliferous mudstone, is 0·6 m thick, and is overlain by a thick bed of black barren shale. At the present time, however, only 0·2 m of limestone can be seen, resting on 0·4 m of grey fossiliferous mudstone, and only a small part of the overlying bed of black shale is visible. The most likely correlative of the Newbigging Marine Band in the borehole sequence of the Denhead area (Fig. 5) is the Upper Ardross Limestone.

About 250 m east of the outcrop of the Newbigging Marine Band lies a largely unfossiliferous sequence, nearly 30 m thick, of sandstones and shales. A conspicuous development of carbonate nodules in a massive mudstone bed is present [at NO 42821318] near the middle of the sequence.

The highest marine band is exposed at a point [NO 42911325] 250 m SSW of Newbigging of Blebo. The section at present exposed is:

	m
Sandstone, fine-grained, white, bedded	0·9
Shale, grey, with abundant *Lingula*	5·5
Limestone, dolomitic, crinoidal (seen)	0·3

Craig and Balsillie (1912, p. 16) recorded that the limestone, which they termed 'No. 2 Limestone', is over 1·2 m thick, in two leaves, and they gave the thickness of the overlying shale as nine metres, but the latter figure is almost certainly an overestimate. Of the thicker limestones present in local borehole sequences (Figs. 5 and 9), only the St Monance White Limestone is immediately overlain by *Lingula*-bearing shales, and there can therefore be little doubt that the 1·2-m bed at Newbigging of Blebo is the correlative of this limestone. About 2·5 km east of the last-mentioned locality a crinoidal limestone at least 0·8 m thick, associated with *Lingula*-bearing shale, is exposed [NO 45441397] just north of the Ceres Fault at Denork; it too is believed to be the St Monance White Limestone.

At Craiglumphart Quarry [NO 424139], about half a kilometre north of Blebo Hole, a quartz-dolerite sill is exposed, with a metre or two of baked sandstone, shale and mudstone above and below it. At a locality very near the last, and probably at a horizon below the dolerite sill, Craig and Balsillie (1912, p. 18) obtained a rich marine fauna of crinoids, brachiopods and molluscs from baked shales which they called the '*Pterinopecten* Beds'. It has not proved possible to locate the outcrop of this marine band, and although it probably lies in the Pathhead Beds, its exact stratigraphical position cannot be determined. Strata probably lying above the Craiglumphart sill have been recorded at several points in the fields south of the quarry and consist of baked shale and sandstone; Craig and Balsillie (1912, p. 17) recorded marine molluscs from two localities in that area.

Claremont Burn. Between the north-eastern end of the Claremont Burn section [NO 47751560] and the vicinity of Denbrae Farm [NO 47501505] a discontinuous sequence of sandstones and grey mudstones is exposed, dipping steadily south-westwards at moderate angles. The Denbrae House Marine Band (new name), consisting of a sparse fauna mainly of brachiopods, is found in ironstone bands in mudstone [NO 47731549] near the base of the sequence. Its facies and position suggest that it cannot be correlated with any of the known marine bands in the Pathhead Beds; it may lie in the Sandy Craig Beds (Table 2).

Apparently lying at a higher stratigraphical level than the Denbrae House Marine Band is the Denbrae Farm Marine Band (new name), which is best exposed in a small tributary to the Claremont Burn at a point [NO 47201503] about 300 m west of Denbrae Farm. The section at this locality is:

	m
Sandstone, siltstone and shale, bioturbated in places .. .. (seen)	2·7
Mudstone, grey, with ostracods; discontinuous exposure .. about	3·0
Mudstone, grey, with *Lingula* and marine molluscs in basal 45 cm ..	3·0
Shale, black, tough	0·4
Mudstone, grey, with silty bands and ironstone lenses; fish debris (seen)	0·8

The band is also exposed a few metres farther downstream in a slipped section [NO 47531512] just below Denbrae Farm.

To the west of Denbrae Farm low dips prevail and the section (Fig. 6) includes the Denbrae Farm Marine Band and apparently higher strata with the Claremont Cottage Marine Band (new name). The latter horizon is best exposed at a point [NO 46561485] 250 m north of Claremont Cottage, where the section is:

	m
Mudstone, grey, fossiliferous .. .. (seen)	0·2
Limestone, crinoidal	0·4
Mudstone, grey, with rich marine fauna including fenestellid bryozoa	1·0
Mudstone, grey, sandy, bioturbated	0·2
Coal	0·02
Mudstone, rooty	0·08
Sandstone, buff, bedded, rooty at top .. .. (seen)	2·1

Mudstones, with a limestone 0·3 m thick, which are exposed [NO 46131480] near the western end of the Claremont Burn section also contain a rich marine fauna with fenestellids and are believed to lie at the same horizon.

In its facies and fauna the Claremont Cottage Marine Band closely resembles a marine band encountered in the Claremont and Mount Melville Cottages boreholes at 80·2 m and 23·8 m respectively, a band believed (Figs. 5 and 6) to correlate with the Lower Ardross Limestone of the Pathhead section. The Denbrae Farm Marine Band in turn resembles the supposed correlative of the West Braes Marine Band of the Pathhead section. On the basis of these correlations the upper part of the Claremont Burn section can be assigned to the Pathhead Beds but the lower part of the section, with the Denbrae House Marine Band, cannot be correlated directly with any part of the Anstruther–Pathhead coast section and its facies, though sandy at the base, cannot be properly assessed; it may belong partly to the Sandy Craig Beds and partly to the Pathhead Beds.

Area south and east of Mount Melville. A discontinuous section is exposed in the Lumbo Burn between Lumbo Bridge [NO 48801481] and a point [NO 48331407] 400 m south of Mount Melville Farm. In the poorly exposed northern part of this section the strata belong to the Lower Limestone Group (p. 69); they dip in a generally southerly direction and are believed to be separated by the Maiden Rock Fault from northerly-dipping Calciferous Sandstone Measures strata which form the southern part of the section. The sequence south of the fault consists of alternations of sandstone and mudstone with rooty horizons and a few thin coals, and is very similar to that encountered in the Mount Melville Borehole (1966) [NO 48361410] which is sited near the south-western end of the section. This borehole reached a depth of 36·5 m but encountered no fossiliferous horizons.

The Feddinch Borehole (1966) [NO 49151434] lies midway between the Lumbo Burn and the Cairnsmill Burn (see below) and reached a depth of 44·9 m. It penetrated a sequence of sandstones and mudstones with rooty beds but no fossiliferous horizons. Between Cairnsmill [NO 497149] and the western margin of the Wester Balrymonth neck [NO 49641423] the Cairnsmill Burn reveals a discontinuous sequence of sandstones and mudstones dipping at moderate angles to the north-west. At the northern end of the section, at Cairnsmill, a limestone taken to be the St Monance Brecciated Limestone is exposed (p. 69) but this part of the section is probably separated from the rest by the Maiden Rock Fault.

The facies and topographical location of these borehole and stream sections suggest that the strata belong to the upper part of the Calciferous Sandstone Measures, but the lack of marine fossil horizons suggests that they lie in the lower part of the Pathhead Beds rather than in its fossiliferous upper part.

Small exposures of strata similar in facies to those described above are seen in streams south of Feddinch [around NO 483130], between Priorletham and North Lambieletham [around NO 501131] and south of Wester Balrymonth [NO 501140] but nowhere are more than a few metres of section visible, and the stratigraphical position of the rocks in these areas remains in doubt.

Grange and Priormuir. Soft white sandstone is or was exposed in old quarries around Balmungo [NO 528146], Grange [NO 519145] and Priormuir [NO 528134], and old records of coal seams up to 80 cm thick survive from the same area, but none of these seams is exposed at the present time. J.I.C.

Craighall. There are numerous scattered exposures of high Calciferous Sandstone Measures in the Craighall Burn around and for about one and a half kilometres downstream from Newbigging of Craighall [NO 416102]. They include the St Monance White Limestone and strata both above and below it, which are apparently disposed in an irregular anticline plunging to the north. The lowest beds are seen in the axial zone for distances up to 500 m west of Harleswynd [NO 412105]. They consist mainly of dark-grey to black mudstones and shales with ironstone nodules, a thickness of nine metres being visible in one exposure [NO 407105]. In another [NO 41201045], *Lingula* occurs near the base of about six metres of shale, just above a 23-cm crinoidal limestone which may be at the level of the Pathhead Upper Marine Band. Sandy mudstones and silty sandstones are visible between these strata and the St Monance White Limestone. The latter is exposed in two places [NO 40501054; 41371041] as a crinoidal limestone 0·6 to 0·8 m thick, with *Lingula* in the mudstone below. The mudstone above is about six metres thick and is overlain by sandstone. Downstream from the more westerly of these two outcrops of the St Monance White Limestone there are scattered exposures, mainly of sandstone. Near the top of the south bank of Craighall Den the following section [NO 40291051] is, however, visible.

		m
Sandstone, with mudstone bands near base .. .. .. (seen)		2·0
Shale, black, with scattered shells		0·9
Shale, calcareous; full of shells, mainly brachiopods		1·2
Shale with coal laminae		0·3
Seatclay, grey .. .. (seen)		0·1

This marine band is probably a detached part of the shale immediately below the St Monance Brecciated Limestone.

The shale and sandstone above the St Monance White Limestone are again exposed in the Craighall Burn, 200 m north-west of Newbigging of Craighall. They are succeeded upstream by an 18-cm dolomitic limestone, which is overlain by 1·1 m of dolomitic mudstone with ostracods and fish debris. The limestone, which is also exposed in a streamlet 280 m N.55°E. of Harleswynd, is taken to be the correlative of the 0·9-m dolomitic non-marine limestone in the Higham Borehole (p. 52). The sandstones exposed in the stream at Newbigging of Craighall lie above this bed of limestone, and are in turn overlain by the strata in the following section, seen in the stream about 80 m south of the farm:

	m
Sandstone with shale bands .. .. (seen)	2·0
Shale, black, with ostracods and fish debris	0·08
Coal, with shale laminae	0·15
Shale, carbonaceous, with coal laminae	0·23
Seatearth, mainly sandy	0·45
Sandstone with rooty top .. .. (seen)	3·0

The sandstones overlying this section are cut off to the south by a fault.

Three bores put down near Bandirran [NO 407103] all proved between 18 and 30 m of strata high in the Calciferous Sandstone Measures. These rocks are almost entirely argillaceous, and probably include some of the thick mudstones associated with the Pathhead Marine Bands and the Ardross Limestones. Not far away, Hall Teasses No. 1 Borehole (unsited, but probably put down somewhere north of Hall Teasses [NO 413090]), penetrated nearly 18 m of mainly argillaceous strata including a limestone 0·8 m thick, probably the St Monance White Limestone, below a thick limestone taken to be the St Monance Brecciated Limestone.

Area between Cameron, Lochty, Carnbee, Elie and St Monance

In the northern part of the area, the strata dip to the west and north-west, and probably include parts of the Pathhead Beds, the Sandy Craig Beds and the Pittenweem Beds. Easterly-trending faults are known to affect the outcrops of the Back and Fore coals, and the Ardross Fault cuts across the southern edge of the area, but the structure is not known in detail, and the distribution of outcrops shown in Fig. 2 is very conjectural. In the south, strata belonging to the Pathhead Beds are well exposed on the coast in the Ardross Fault Zone.

Details

North Bank. In an unnamed streamlet some 450 m west of North Bank there is an exposure [NO 47801074] of black shale with an 8-cm limestone and a fauna of *Lingula* and marine bivalves; this probably lies at the horizon of one of the Pathhead Marine Bands. Sandstones and shales with ostracods are visible in the same streamlet further north. Limestone is believed to have been quarried at one time beside North Bank but no trace now remains. The sandstones formerly exposed in old quarries 300 m to the north are also now obscured. A 90-cm coal, dipping to the north-west at 12°, was encountered (Geikie 1902, p. 141) at 25·6 m in a shaft [NO 48151038] near which subsidence was noted in 1937. This seam is said to outcrop on the south-east side of the Largo–St Andrews road.

Higham Borehole. The Higham Borehole [NO 47460941] penetrated about 76 m into the Calciferous Sandstone Measures (Forsyth and Chisholm 1968, p. 73 and plate iv) and probably stopped just above the Pathhead Upper Marine Band. The sequence consists of sandstones and shales with several seatearths and a few thin coals. Qualitatively it resembles the corresponding sequence in the Drumcarro Borehole (see Fig. 5) but the strata between the St Monance Brecciated and White limestones are almost 46 m thick compared with 21 m at Drumcarro. The Higham figure is the greatest known amount for these measures, suggesting that this was an area of maximum accumulation before the end of the deposition of the Calciferous Sandstone Measures, as well as during Lower Limestone Group times (p. 60). Both the limestones, on the other hand, are thinner at Higham than they are either at Drumcarro to the north or on the coast at Pathhead to the south, the St Monance White Limestone being only 1·1 m thick. A dolomitised non-marine limestone, 90 cm thick, lies almost midway between the marine limestones. A band of kaolinised tuff was recorded by Francis (1968, p. 124) from just below the St Monance White Limestone: other bands possibly of volcanic origin occur elsewhere in the sequence. The mudstone below the St Monance Brecciated Limestone is rich in marine shells, but that above the St Monance White Limestone contains little other than *Lingula*, while the one below it has only a few marine bivalves, *Lingula* and *Orbiculoidea*.

Arncroach, Kellie and Lochty: Back and Fore coals. The Back and Fore coals were worked to some extent during the 19th century around Arncroach and Kellie (Geikie 1902, p. 117). They were first described by Landale (1837, pp. 295–299), who regarded them as being of 'comparatively little value'. He stated that the Back Coal is 2·7 m thick, with three partings of shale, and has a roof of hard black shale, also 2·7 m thick. The Fore Coal was placed 40·4 m below the Back Coal and given a thickness of 1·4 to 2·1 m, with two shale partings. The roof was said to be a shell-limestone 90 cm thick. Landale obtained enough information about the disused workings to enable him to trace the outcrops of the seams from a locality 365 m north of Balcaskie House [NO 525036] northwards for about 9·5 km across several faults, with an interruption at the volcanic neck of Kellie Law, near which he found that the coals had been devolatilised and locally upended. The outcrops are finally cut off by a fault south of Kingscairn Mill (now demolished) [NO 535096]. Landale also sought to recognise the Back and Fore coals on the north side of this fault in the Gilmerton area (see p. 34), largely on the basis of the occurrence of black shale and shell-limestone in the waste heaps there. Such materials, however, commonly occur in the roofs of coals in east Fife and their presence is not necessarily indicative of the Back and Fore coals.

Since Landale's time the only additional information about these coals has come from exploratory drilling around Lochty [NO 526081]. Between 1890 and 1900 several boreholes were put down, and more were drilled in 1936–38, but no mining resulted from either series. The names and sites of the more important boreholes are as follows: Lochty 'C' Borehole (old) [NO 52400849], Lochty 'D' Borehole (old) [NO 52550804], Lochty 'E' Borehole (old) [NO 52750803 (approximately)], Chesters No. 1 Borehole (1892) [NO 52980878], Lochty No. 1 Borehole (1936) [NO 52270799] (which did not reach the coals), Lochty No. 2 Borehole (1937) [NO 52330754] and Lochty No. 3 Borehole (1937–38) [NO 52590787]. Fig. 7 shows a generalised sequence based mainly on the later series of boreholes, which were examined by the late W. Manson and provide the best data on the strata above the coals.

In Lochty Nos. 2 and 3 boreholes the coal taken to be the Fore Coal is 1·2 m thick in two leaves but the roof was found to be composed of hard brown well-cemented sandstone, not shell-limestone. In the older boreholes the Back Coal was recorded as being 2·0 to 2·25 m thick, in several leaves, with a black shale roof. In Lochty No. 3 Borehole, however, it is apparently represented by a 5-cm seam, with a black shale roof, 42·7 m above the Fore Coal, and in Lochty No. 2 Borehole it was not found at all.

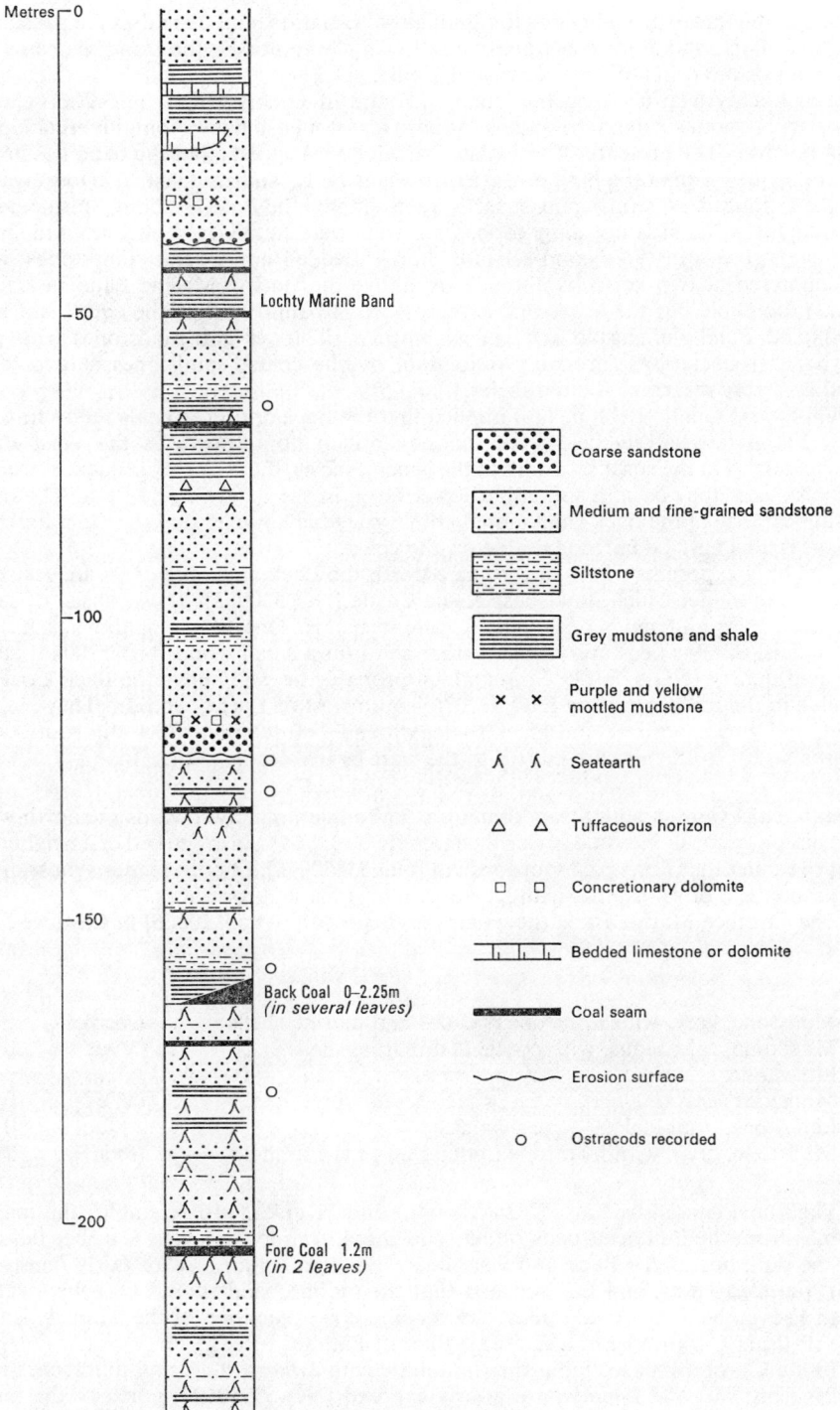

FIG. 7. *Generalised vertical section of strata proved in boreholes around Lochty*

The differences are probably due to rapid lateral variation, and this raises the possibility that the Back and Fore coals proved at Lochty may not be the same seams as the Back and Fore coals of Arncroach and Kellie.

The Lochty Marine Band (new name), found in Lochty Nos. 1 and 2 boreholes, consists of about six metres of mudstone with ironstone bands and contains brachiopods and bivalves. The presence of articulate brachiopods suggests that the band lies higher in the sequence than the base of the Pittenweem Beds, and its most likely equivalent in the Pathhead–Anstruther coast section seems to be the Boat Harbour Marine Band, although the fauna is not diagnostic. Ostracods were found in mudstones and shales at several levels in the second series of bores. Bedded non-marine 'limestones' were encountered at two horizons about 24 m above the Lochty Marine Band in Lochty No. 1 Borehole but the lower one is represented by sandstone at the equivalent level in No. 2 Borehole. Purple and yellow mottled siltstones and mudstones with pale carbonate concretions (probably dolomitic) overlie coarse sandstones at two levels and probably represent the red facies.

Kirkby (*in* Geikie 1902, p. 116) implied that the Back and Fore coals might lie 67 m and 104 m respectively below the marine horizon now known as the West Braes Marine Band in the coast section, but the general facies of the Lochty sequence strongly suggests that they lie at a somewhat lower level, in the Sandy Craig Beds. The rapid disappearance of the Back Coal, noted above, suggests however that the exact horizons of the seams may be unrecognisable on the coast.

The only exposures of strata associated with the Back and Fore coals are scattered along a streamlet which flows past Kellie Castle [NO 520052]. They consist of sandstones, shales and mudstones with ironstone bands. One of the shales has yielded ostracods. Similar beds are visible further south in a small wood [NO 520044] and a disused quarry [NO 523043]. Strata which probably lie well above the Back Coal are visible in the railway cutting [NO 517075] south-west of Lochty Station. They include pink and grey sandstone and black shale, with sills of dolerite, which dip south-westwards at 20° to 27° and are cut off to the west by a volcanic neck.

Carnbee and Gordonshall. Strata dipping at low angles, mainly towards the north-west, are discontinuously exposed in Carnbee Den [NO 532066] and in the bed of a neighbouring streamlet that flows past Gordonshall [NO 53450675]. In both sections most of the exposures are of sandstone, with siltstone, mudstone and root beds.

The Carnbee Marine Band (new name) is exposed [NO 53120666] in Carnbee Den. The section is:

	m
Mudstone, grey, with ironstone nodules and marine shells .. (seen)	0·6
Mudstone, calcareous, with corals and marine shells	0·4
Limestone, crinoidal	0·1
Mudstone, grey	0·1
Limestone, crinoidal	0·2
Mudstone, grey, with ironstone nodules and marine shells .. (seen)	2·6

The fauna consists mainly of brachiopods and bivalves, and resembles the marine faunas from the Pathhead Beds rather than those from lower levels. Carnbee lies east of the outcrops of the Back and Fore coals, however, in an area of fairly consistent north-westerly dips, and this suggests that the marine band should lie below, rather than above, the Sandy Craig Beds. The most likely explanation of the anomaly is that the strata at Carnbee lie in a faulted outlier of Pathhead Beds.

In the Gordonshall section a thin limestone with *Lithostrotion*, and mudstones with marine bivalves and *Lingula*, are poorly exposed [NO 53380694] north of the farm; they probably belong to the Carnbee Marine Band. Farther downstream a coal seam, which appears to have been worked, is exposed immediately west of the farm and is overlain by mudstones with poorly preserved marine bivalves—the Gordonshall Marine Band (new name). This band almost certainly lies below the Carnbee Marine Band, with perhaps 30 to 45 m of strata intervening.

Balmonth. About a kilometre east of Carnbee there are discontinuous exposures in the bed of a small stream for about 600 m on either side of Balmonth [NO 539070]. North of a fault zone [NO 54001717] the strata are mainly argillaceous but include some sandstones and siltstones as well as three thin coals. South of the fault they include sandstone, mudstone, seatclay and a 15-cm bed of weathered dolomite with ostracods. The Balmonth Water Borehole [NO 53980706] was put down to a depth of 122 m through a sequence of alternating sandstones, siltstones and mudstones, with three thin coals in the upper half. Three 'limestones' 0·3 to 0·6 m thick were also reported. Four shallow bores near the reservoir encountered similar sequences, with thin coals but no limestones. The Bonerbo Water Borehole [NO 54640733] penetrated a mainly argillaceous sequence about 21 m thick with a few thin sandstones and several coal seams up to 45 cm thick. Although no faunal evidence is available, the strata cut in these bores probably belong to either the Pittenweem Beds or the Anstruther Beds.

Ovenstone. There are scattered exposures, mainly of sandstone and siltstone, in a small stream near Ovenstone Convalescent Home [NO 53720480]. By the Water House [NO 53828496] there are also exposures of mudstone with a marine fauna, and numerous blocks of crinoidal limestone probably derived from excavations when the Water House was constructed. The presence of ribbed brachiopods in the fauna from this marine band, here named the Ovenstone Marine Band, suggests that it lies at or above the level of the Pittenweem Marine Band. About 100 m downstream, black shale with a 5-cm ironstone band containing bivalves and ostracods is seen to overlie a 40-cm coal which has been invaded and coked by a basalt intrusion, now bleached. On the south side of a small fault lies the following section [NO 53840489]:

	cm
Shale, black, with 2-cm ironstone at top	29
Shell-bed, full of *Carbonicola?*	13
Shale, black, with *Lingula* ⎱ Ovenstone *Lingula* Band (new name) ..	8
Dolomite, with *Lingula* ⎰	15
Shale, black	8

The proximity of the *Lingula* band to the Ovenstone Marine Band suggests that it, too, lies at or above the level of the Pittenweem Marine Band, while the presence of *Carbonicola?*, a genus which has not been recorded from the Pathhead Beds or the Sandy Craig Beds, suggests that the band lies below these subdivisions. All the strata at Ovenstone are therefore thought to lie in the Pittenweem Beds.

Grangemuir. About 1900 a series of over 20 bores was put down to depths of up to 113 m in unexposed ground around Grangemuir House [NO 539041] and Easter Grangemuir [NO 547041]. One went into an unexposed volcanic neck. The others encountered sandstone–siltstone–mudstone sequences, usually with some coals and seatbeds. Many of the mudstones were reported to contain nodules and bands, up to 25 cm thick, of ironstone. Only a few 'limestones' were reported, and only one shell-bed. Three of the 'limestones' were cut in Grangemuir No. 5 Borehole [NO 55070358]: the sequence in this bore resembles the lower part of the Lower Limestone Group, whose presence here cannot be entirely discounted. The coals vary in thickness from a few centimetres up to as much as 1·2 m. A 60-cm seam which was cut at a depth of 41 m in Grangemuir No. 11 Borehole [NO 54310401] was worked to a small extent from a nearby shaft. It has a south-westerly dip of 10°. The waste heap from the shaft includes fragments of a shell-limestone (which was not recorded in the bore). The bores

E

were not examined by a geologist, so that nothing is known about the faunas apart from the occurrence of shell-beds, noted above. The bores lie in an area devoid of surface exposures except in streams along its northern margin (see below). The geological structure is unknown and may well be complex; the Ardross Fault cuts across the southern part of the area and associated folds and faults may well be present so that quite different parts of the succession may be represented in adjacent bores. The general nature of the sequences suggests that they may lie in parts of the Pittenweem Beds, the Anstruther Beds or the Pathhead Beds, and even the Lower Limestone Group could be present (see above).

A little to the north of the area in which the bores were put down, coal was worked from a mine [NO 53960485] known as Annie's Mine, and from a nearby shaft in which the seam was at a depth of 8·2 m, but the identity of the seam and the extent of mining are unknown. The stream which flows from Ovenstone Convalescent Home to join the Dreel Burn south-east of Easter Grangemuir reveals a series of discontinuous exposures over a stretch about a kilometre in length [NO 54480469 to NO 55290457]. They include sandstone, siltstone, mudstone, black shale and seatearth, with a few thin coals, two bands of dolomite and a carbonate conglomerate. A tributary on the north side has intermittent exposures, for 400 m up from the stream junction [NO 54650475], of sandstone and seatclay, with a thin coal and a blackband ironstone; farther upstream there are exposures west and south of Carvenom [NO 548057], mostly of white sandstone but including a clayband ironstone with shell debris, and a seatclay. Like the borehole sections, the surface exposures give little indication of their position in the succession.

Dreel Burn. There are discontinuous exposures in the Dreel Burn between Balcormo Mill [NO 509041] and Over Kellie [NO 510069], a distance of about three kilometres. The section runs obliquely across the outcrop of the strata between the Back Coal and the base of the Lower Limestone Group, so that parts of the Pathhead Beds and the Sandy Craig Beds should be exposed. No marine bands have been detected, however, and the only faunas so far discovered have been of ostracods, *Spirorbis*, fish remains and *Curvirimula*. Most of the exposures are of sandstone, siltstone, shale and mudstone with ironstone bands, but thin beds of dolomite also occur, as do root beds and thin coals. Geikie (1902, pp. 118–119) reported that several coal seams were formerly exposed, and Landale (1837, pp. 298–299) mentioned a 90-cm splint coal west of Wester Kellie [NO 511059] and a 110-cm seam west of Arncroach. A coal noted by Geikie (1902, p. 119) in a quarry west of Kellie Mill [NO 505046] is now obscured. Sandstones and shales exposed in a small stream by Gibliston House [NO 498050] probably lie near the top of the Calciferous Sandstone Measures. Exposures near the mouth of the Dreel Burn show mainly sandstones and mudstones but include a coal at least 30 cm thick and several bands of dolomite, one of which is full of *Carbonicola?*. The equivalent beds on the coast lie in the Anstruther Wester section or hidden under the sands of West Haven.

Balcarres Den. Strata very high in the Calciferous Sandstone Measures are exposed at the southern end of Balcarres Den [NO 486037]. They include fossiliferous calcareous mudstones just below the St Monance Brecciated Limestone and a carbonate bed 30 cm thick which is possibly tuffaceous.

Ardross coast section. Between Elie Ness [NT 496993] and the western end of St Monance [NO 524015] the rocks are strongly affected by the Ardross Fault and the tight folding associated with it. The strata are all believed to belong to the Pathhead Beds, the Ardross Limestones being named from this section. The description which follows is based partly on notes and maps supplied by Dr. E. H. Francis and partly on published sources (Geikie 1902; Cumming 1936; Francis and Hopgood 1970). The volcanic and tectonic features of the section are described in Chapters 12 and 13 respectively. Excursion guides to this important section are included in publications by Mitchell, Walton and Grant (1960) and MacGregor (1968).

On the north-west side of the fault, along the upper part of the foreshore from near Ardross Castle [NO 508007] to Newark Castle [NO 518012] the strata are only gently folded, and include exposures of the Ardross Limestones, the Ardross *Lingula* Band and a bed of siltstone with *Curvirimula scotica* [NO 51470106]. This part of the section is famous for the occurrence of the 'shrimp band', a calcareous layer containing fossil crustacea, in mudstone just below the Lower Ardross Limestone (Brown 1861, p. 393; Cumming 1936, pp. 346–347).

On the south-east side of the fault, exposures extend almost continuously from Elie Ness to St Monance, where the Pathhead Beds pass up into the Lower Limestone Group. The sedimentary rocks are intensely folded, and the succession is much interrupted by faults and volcanic necks. North-east and south-west of the Wadeslea neck [NT 503997] unfossiliferous sandstones, siltstones and mudstones, believed to lie below the level of the Ardross Limestones, are disposed on the limbs of folds with northerly to north-north-easterly trend. Adjacent to the neck, in a brecciated zone on its south-western side, there is a coralline limestone fully four metres thick which is presumed to be the coralline phase of the St Monance White Limestone (Cumming 1936, p. 349). Blocks of similar material occur near the south-western margin of the Coalyard Hill neck (Geikie 1902, p. 114). Further to the north-east, between a point [NO 508004] about 300 m south of Ardross Castle and the margin [NO 52250121] of the St Monance neck, the strata are thrown into a series of tight folds about north-easterly-trending axes. The sequence exposed in this area is very similar to the equivalent part of the Anstruther–Pathhead coast section. It extends from below a *Lingula* band at the horizon of the West Braes Marine Band up to a thick sandstone above the Upper Ardross Limestone. This sandstone (on which Newark Castle stands) is in the core of the Newark Castle Syncline.

East of the St Monance neck the highest strata of the Calciferous Sandstone Measures are exposed, dipping east at about 45°. The sequence below the St Monance White Limestone is similar to that at Pathhead. The limestone itself [NO 52390145] is crinoidal, with scattered shells and solitary corals, and is 3·7 m thick. This development is in strong contrast to its white coralline equivalent at high water mark at Pathhead, but resembles that seen further down the tidal zone there. The shale above the limestone, which appears towards low water mark at Pathhead (Tait and Wright 1923), is about six metres thick and the calcareous lower part is full of ribbed brachiopods, mainly rhynchonelloids and productoids. The overlying strata consist of a sandstone with a rooty top and the fossiliferous mudstones which lie below the St Monance Brecciated Limestone.

I.H.F.

References

BALSILLIE, D. 1920. Description of some volcanic vents near St Andrews. *Trans. Edinb. geol. Soc.*, **11**, 69–80.

BENNISON, G. M. 1960. Lower Carboniferous non-marine lamellibranchs from east Fife, Scotland. *Palaeontology*, **3**, 137–152.

—— 1961. Small *Naiadites obesus* from the Calciferous Sandstone Series (Lower Carboniferous) of Fife. *Palaeontology*, **4**, 300–311.

BROWN, T. 1861. Notes on the Mountain Limestone and Lower Carboniferous rocks of the Fifeshire coast from Burntisland to St Andrews. *Trans. R. Soc. Edinb.*, **22**, 385–404.

BURGESS, I. C. 1960. Fossil soils of the Upper Old Red Sandstone of south Ayrshire. *Trans. geol. Soc. Glasg.*, **24**, 138–153.

CHISHOLM, J. I. 1968. Trace-fossils from the Geological Survey boreholes in east Fife (1963–4). *Bull. geol. Surv. Gt Br.*, No. 28, 103–119.

CHISHOLM, J. I. 1970a. Lower Carboniferous trace-fossils from the Geological Survey boreholes in west Fife (1965-6). *Bull. geol. Surv. Gt Br.*, No. 31, 19-35.
—— 1970b. *Teichichnus* and related trace-fossils in the Lower Carboniferous at St Monance, Scotland. *Bull. geol. Surv. Gt Br.*, No. 32, 21-51.
CRAIG, R. M. and BALSILLIE, D. 1912. The Carboniferous rocks and fossils in the neighbourhood of Pitscottie, Fifeshire. *Trans. Edinb. geol. Soc.*, **10**, 10-24.
CUMMING, G. A. 1936. The structural and volcanic geology of the Elie-St Monance district, Fife. *Trans. Edinb. geol. Soc.*, **13**, 340-365.
CURRIE, ETHEL D. 1954. Scottish Carboniferous goniatites. *Trans. R. Soc. Edinb.*, **62**, 527-602.
FORSYTH, I. H. 1970. Geological Survey boreholes in the Lower Carboniferous of west Fife (1965-6). *Bull. geol. Surv. Gt Br.*, No. 31, 1-18.
—— and CHISHOLM, J. I. 1968. The Geological Survey boreholes in the Carboniferous of east Fife, 1963-4. *Bull. geol. Surv. Gt Br.*, No. 28, 61-101.
—— and WILSON, R. B. 1965. Recent sections in the Lower Carboniferous of the Glasgow area. *Bull. geol. Surv. Gt Br.*, No. 22, 65-79.
FRANCIS, E. H. 1961. Thin beds of graded kaolinized tuff and tuffaceous siltstone in the Carboniferous of Fife. *Bull. geol. Surv. Gt Br.*, No. 17, 191-215.
—— 1968. Pyroclastic and related rocks of the Geological Survey boreholes in east Fife, 1963-4. *Bull. geol. Surv. Gt Br.*, No. 28, 121-135.
—— and HOPGOOD, A. M. 1970. Volcanism and the Ardross Fault, Fife. *Scott. Jnl Geol.*, **6**, 162-185.
GEIKIE, A. 1902. The geology of eastern Fife. *Mem. geol. Surv. Gt Br.*
GEVERS, T. W., FRAKES, L. A., EDWARDS, L. N. and MARZOLF, J. E. 1971. Trace fossils in the Lower Beacon sediments (Devonian), Darwin Mountains, southern Victoria Land, Antarctica. *Jnl Palaeont.*, **45**, 81-94.
GREENSMITH, J. T. 1961. The petrology of the Oil-Shale Group sandstones of West Lothian and southern Fifeshire. *Proc. Geol. Ass.*, **72**, 49-71.
—— 1965. Calciferous Sandstone Series sedimentation at the eastern end of the Midland Valley of Scotland. *Jnl Sed. Petrol.*, **35**, 223-242.
KIRK, S. R. 1925a. Geology of the coast between Kinkell Ness and Kingask, Fifeshire. *Trans. Edinb. geol. Soc.*, **11**, 366-382.
—— 1925b. A coast section in the Calciferous Sandstone Series of eastern Fife. *Unpublished St Andrews University Ph.D. Thesis.*
KIRKBY, J. W. 1880. On the zones of marine fossils in the Calciferous Sandstone Series of Fife. *Q. Jnl geol. Soc. Lond.*, **36**, 559-590.
—— 1901. On Lower Carboniferous strata and fossils at Randerstone, near Crail, Fife. *Trans. Edinb. geol. Soc.*, **8**, 61-75.
LANDALE, D. 1837. Report on the geology of the East of Fife Coalfield. *Trans. Highl. agric. Soc. Scotl.*, **11**, (vol. 5, new series), 265-348.
MACGREGOR, A. R. 1968. *Fife and Angus geology: an excursion guide.* Edinburgh and London: Blackwood.
MITCHELL, G. H. and MYKURA, W. 1962. The geology of the neighbourhood of Edinburgh. 3rd edit. *Mem. geol. Surv. Gt Br.*
—— WALTON, E. K. and GRANT, D. (Editors). 1960. *Edinburgh geology, an excursion guide.* Edinburgh: Oliver & Boyd.
MUIR, A., HARDIE, H. G. M., MITCHELL, R. L. and PHEMISTER, J. 1956. The limestones of Scotland. Chemical analyses and petrography. *Mem. geol. Surv. spec. Rep. Miner. Resour.*, 37.
MYKURA, W. 1960. The replacement of coal by limestone and the reddening of Coal Measures in the Ayrshire Coalfield. *Bull. geol. Surv. Gt Br.*, No. 16, 69-109.
NEVES, R., GUEINN, K. J., CLAYTON, G., IOANNIDES, N. S., NEVILLE, R. S. W. and KRUSZEWSKA, K. 1973. Palynological correlations within the Lower Carboniferous of Scotland and northern England. *Trans. R. Soc. Edinb.*, **69**, 23-70.
OOMKENS, E. 1967. Depositional sequences and sand distribution in a deltaic complex. *Geologie Mijnb.*, **46**, 265-278.

REFERENCES

TAIT, D. and WRIGHT, J. 1923. Notes on the structure, character and relationship of the Lower Carboniferous Limestone of St Monans, Fife. *Trans. Edinb. geol. Soc.*, **11,** 165–184.

TRAQUAIR, R. H. 1901. Notes on the Lower Carboniferous fishes of eastern Fifeshire. *Geol. Mag.*, decade 4, **8,** 110–114.

WALKER, R. 1872. On a new species of *Amblypterus* and other fossil fish remains from Pitcorthie, Fife. *Trans. Edinb. geol. Soc.*, **2,** 119–124.

WATTISON, A. 1962. Temporary exposures in the Calciferous Sandstone Measures of east Fife. *Trans. Edinb. geol. Soc.*, **19,** 133–138.

WILSON, H. H. 1952. The Cove Marine Bands in East Lothian and their relation to the Ironstone Shale and Limestone of Redesdale, Northumberland. *Geol. Mag.*, **89,** 305–319.

WILSON, R. B. 1974. A study of the Dinantian marine faunas of south-east Scotland. *Bull. geol. Surv. Gt Br.*, No. 46, 35–65.

Chapter 4
LOWER LIMESTONE GROUP

Introduction

THE limits of the Lower Limestone Group coincide closely with those of the Upper Bollandian Stage (P_2) of the Upper Viséan (Currie 1954, pp. 532–534). In the Glasgow district, which may be regarded as the type area for the group, the base is drawn at the bottom of the Hurlet seam (i.e. the Hurlet Limestone and Coal with the intervening Alum Shale—see Clough and others 1925, p. 32). The correlation of the Hurlet Limestone with the Charlestown Station Limestone of west Fife (Macgregor 1930) is now generally accepted (Forsyth 1970), but in the absence of a thick and persistent coal immediately beneath the Charlestown Station Limestone, the bottom of the group in west Fife is drawn at the base of the limestone. In east Fife the correlative of the Charlestown Station Limestone is now believed to be the St Monance Brecciated Limestone (Forsyth and Chisholm 1968; Forsyth 1970). The top of the Lower Limestone Group in Fife is drawn at the top of the Upper Kinniny Limestone. In east Fife this bed is of intermittent occurrence, but its position is commonly represented by a bed of mudstone with a marine fauna, lying just above the Marl Coal. Where the Upper Kinniny Limestone is absent, therefore, the upper limit of the Lower Limestone Group in east Fife has been drawn (Forsyth and Chisholm 1968, p. 73) at the base of this marine mudstone.

The thickness of the Lower Limestone Group in east Fife varies between 120 and 150 m around Drumcarro in the northern part of the outcrop, and in the Elie–St Monance district in the south. In the intervening Lathallan–Radernie area no single borehole section cuts the whole of the group, but the aggregate thickness of separate parts of the sequence amounts to about 240 m. It may be however that the group as a whole is nowhere quite as thick as this. Nevertheless, even when allowance is made for a possible miscorrelation of the Marl Coal in the Lathallan area (see p. 80), the thickness of the group must still exceed 225 m. This figure is close to the maximum known thickness for a non-volcanic sequence of the Lower Limestone Group in the Midland Valley of Scotland (Goodlet 1957, fig. 2). The new figures make some revision of the north-eastern part of Goodlet's generalised isopach map necessary, but the available data are insufficient to allow detailed isopachytes to be drawn for the whole or any significant part of the group in east Fife. It appears probable, however, that the Lathallan–Radernie zone of maximum thickness is flanked to north-west and south-east by areas of lesser thickness which lie near, and are perhaps elongated along, the Ardross and Ceres faults. The thickening in the Lathallan–Radernie area appears to affect the sandstones and the mudstones equally. The coal seams are also generally thicker in this area, and some coal horizons appear which are not universally present in east Fife. The limestones on the other hand are if anything thinner in the area where the group as a whole is thickest.

The outcrop of the Lower Limestone Group crosses the western margin of the present area near Teasses and extends north-eastwards almost as far as St Andrews. It also runs eastwards into the Largoward area, whence it continues southwards to the coast at Elie. The group is also present in the St Monance Syncline at St Monance and Pittenweem. The outcrop is complicated throughout by folds, faults, igneous intrusions and volcanic necks.

Lithology

The Lower Limestone Group in east Fife consists of sedimentary cycles of a type familiar in other parts of the Scottish Midland Valley (see p. 4 and Goodlet 1959, pp. 218–225, where a full discussion may be found). These cycles are similar to the Yoredale cycles in the same part of the Carboniferous succession in the north of England. The sequence within a 'typical' cycle may be summarised as follows:

Coal
Seatearth
Sandstone, generally fine-grained and cross-bedded or ripple-laminated
Alternations of sandstone, siltstone and mudstone
Mudstone, commonly with marine fossils
Limestone, crinoidal, often dolomitised
Mudstone, generally with marine fossils
Coal (of underlying cycle).

One or more of the above members may be absent in any particular instance. Cycles as thus defined range in thickness from 6 m to over 60 m. The cycle containing the Charlestown Main Limestone is usually the thickest, although that containing the Mid Kinniny Limestone is of similar dimensions in places. Minor rhythms of coal and seatearth, with some beds of generally unfossiliferous mudstone and sandstone, are also found, chiefly between the St Monance Little and Charlestown Main limestones, between the Mill Hill and Seafield marine bands and between the Lower and Mid Kinniny limestone positions.

Four limestones (Fig. 8) are persistent throughout east Fife, namely the St Monance Brecciated (3 to 3·7 m thick), St Monance Little (0·6 to 0·8 m), Charlestown Main (about 1·5 m) and Mid Kinniny (about 0·6 m) limestones. The Upper Kinniny Limestone (up to 0·45 m) is only intermittently developed. Locally there is a thin sandy limestone or calcareous sandstone at the level of the Mill Hill Marine Band. Partial or complete dolomitisation of these limestones is common. The Lower Kinniny Limestone is not developed at all in east Fife but its horizon has been recognised (Forsyth and Chisholm 1968, p. 70) in a persistent bed of marine mudstone.

Some of the coal seams maintain their thicknesses throughout east Fife, but others are very variable. The Largoward Black and Largoward Splint coals (Fig. 9) are generally thick (more than 120 cm in places) and the latter is the most extensively worked seam in east Fife. On the other hand the coals below the Mid Kinniny Limestone and the Mill Hill Marine Band rarely exceed 30 cm in thickness and are locally absent. The Marl Coal and the Radernie Brassie Coal are rather variable seams, which have locally been worked; the former reaches 105 cm in thickness and the latter 75 cm but both are thin or absent in places. There is usually more than one coal between the St Monance Little and Charlestown Main limestones. These coals vary greatly in thickness, but in the Largoward–Radernie area, where they are usually three in number (Radernie Duffie, Radernie Main and Radernie Marl coals), they attain thicknesses in excess of 1·2 m and have been mined. The lowest of these seams, the Radernie Duffie Coal, persists into the Denhead area where it contains a layer of ironstone, known and worked in the 19th century as the Winthank or Denhead Blackband Ironstone.

The mudstones in the group vary greatly in the degree of siltiness and in the richness of the fauna. The mudstone above the Charlestown Main Limestone is consistently thick, is often richly fossiliferous, and contains the Neilson Shell Bed (Wilson 1966). The Seafield Marine Band and the mudstone above the St Monance Brecciated Limestone locally yield rich faunas, but in some places the mudstones are silty and only scantily fossiliferous. Sparse marine faunas are found in mudstone at the horizon of the Lower Kinniny Limestone and in the Mill Hill Marine Band. Mudstones in the 'minor rhythms' (see above) are generally barren or contain restricted faunas including, rarely, *Lingula*.

Trace-fossils are present in some of the alternations of sandstone, siltstone and mudstone (Chisholm 1968, 1970a, 1970b); the coast sections at St Monance provide the best natural exposures of these structures.

Stratigraphy

In early accounts of the beds which include the Lower Limestone Group (Landale 1837; Maclaren 1839, pp. 102–104), little attempt was made to correlate individual horizons over great distances, but the distinctive character of this part of the sequence was recognised. The earliest known detailed geological map of east Fife was produced by Landale (1837), who delineated the outcrops of various limestones and coals which are now known to lie within the Lower Limestone Group. His notes on the inland limestone quarries are particularly valuable because many of these have long since become obscured. He indicated the existence of several exposures of limestone on the coast between Elie and St Monance, and provided an account of the strata in the St Monance Syncline. He also provided sections through the strata in the Largoward, Teasses and Radernie coalfields, noting the close association of coals and limestones in the two last-mentioned. Landale's work has been the basis of all subsequent studies in these areas. In the St Monance shore section Brown (1859) distinguished a group of strata with marine limestones, as 'Mountain Limestone', from the rocks above and below. In both his papers (Brown 1859, 1861) he included in this group beds which are now classified as part of the Calciferous Sandstone Measures in addition to what is now known as the Lower Limestone Group.

Kirkby (*in* Geikie 1902, pp. 149–151) prepared a detailed and generally accurate section of the strata on the shore at St Monance; his account served for many years as a basis for the work of others, and has been modified only to a small extent by the recent re-examination of the coast section. Geikie himself (1902) added some information based on various field surveys carried out between 1860 and 1900, but drew heavily for coalfield details on Landale's (1837) paper. The term Lower Limestone Group was already in use by the time Geikie's account was published, the base of the group being drawn at the

'Hurlet' Limestone. Geikie believed that the 'White Limestone' of the St Monance coast section—bed 58 of Kirkby's measured section (*in* Geikie 1902, p. 151); now known as the St Monance White Limestone—was the correlative of the Hurlet Limestone, a view later challenged by Macnair (see below) and now thought to be erroneous. Inland in east Fife however, Geikie adopted the widely-quarried limestone now called the St Monance Brecciated Limestone as the equivalent of the Hurlet, a correlation supported by more recent evidence (Forsyth and Chisholm 1968). The top of the Lower Limestone Group in east Fife was not closely defined by Geikie, who included some of the coal seams in the group, such as the Largoward Black and Splint, in the Limestone Coal Group. The term 'Kinniny Limestone' had not then been introduced and all limestones above the Hurlet were referred to as Hosie Limestones. This term is now restricted to the group of limestones at the top of the Lower Limestone Group in the Central Coalfield.

On the basis of faunal and lithological resemblances Macnair (1917) suggested that the 'White Limestone' of St Monance should be correlated with the Blackbyre Limestone of the Glasgow district, and that the next higher limestone in the St Monance sequence (Bed 53, the 'pseudo-brecciated bed' of Kirkby's measured section, now named the St Monance Brecciated Limestone) should be correlated with the Hurlet Limestone. The status of the limestones exposed inland was not discussed. Other workers on the coast sections either accepted Geikie's correlation of the White Limestone with the Hurlet Limestone (Wright 1914; Cumming 1928) or expressed no opinion on this correlation (Tait and Wright 1923). None of these authors attempted to define the top of the group in east Fife, or discussed the correlation of the limestones quarried inland.

Forsyth and Chisholm (1968) recorded the results of the Geological Survey boreholes put down in east Fife in 1963–64, and described the stratigraphy of the Lower Limestone Group in much greater detail than had previously been possible. The limestone formerly worked in many of the inland quarries was shown to be the equivalent of the St Monance Brecciated Limestone of the coast section, and it was suggested that this limestone should be correlated with the Charlestown Station Limestone of west Fife and hence with the Hurlet Limestone of the Glasgow area. The current correlation of the Hurlet Limestone into east Fife therefore agrees with Macnair's (1917) opinion rather than with Geikie's (1902), at least as far as the coast section is concerned; but for the inland quarries Geikie's original correlation is regarded as correct. The Upper Kinniny Limestone was identified, for the first time in east Fife, in a bed of marine mudstone at a depth of 256 m in the Dunotter Borehole (Forsyth and Chisholm 1968, p. 69), and its horizon is now recognised in a marine mudstone, locally containing a limestone, above the Marl Coal of the Largoward area. The top of the Lower Limestone Group can now, therefore, be recognised in east Fife.

The best available exposures of the Lower Limestone Group are on the coast at St Monance, where all but the topmost beds are seen on the limbs of the St Monance Syncline. Much of the group is also visible at Elie, but the exposures are less continuous and are more affected by faults and volcanic necks. Balcarres Den provides the only inland section where an appreciable part of the group is exposed, but certain parts of the succession are visible in other stream sections such as those of the Craighall Burn, Lumbo Den and Cairns Den. The St Monance Brecciated Limestone and associated strata can still be

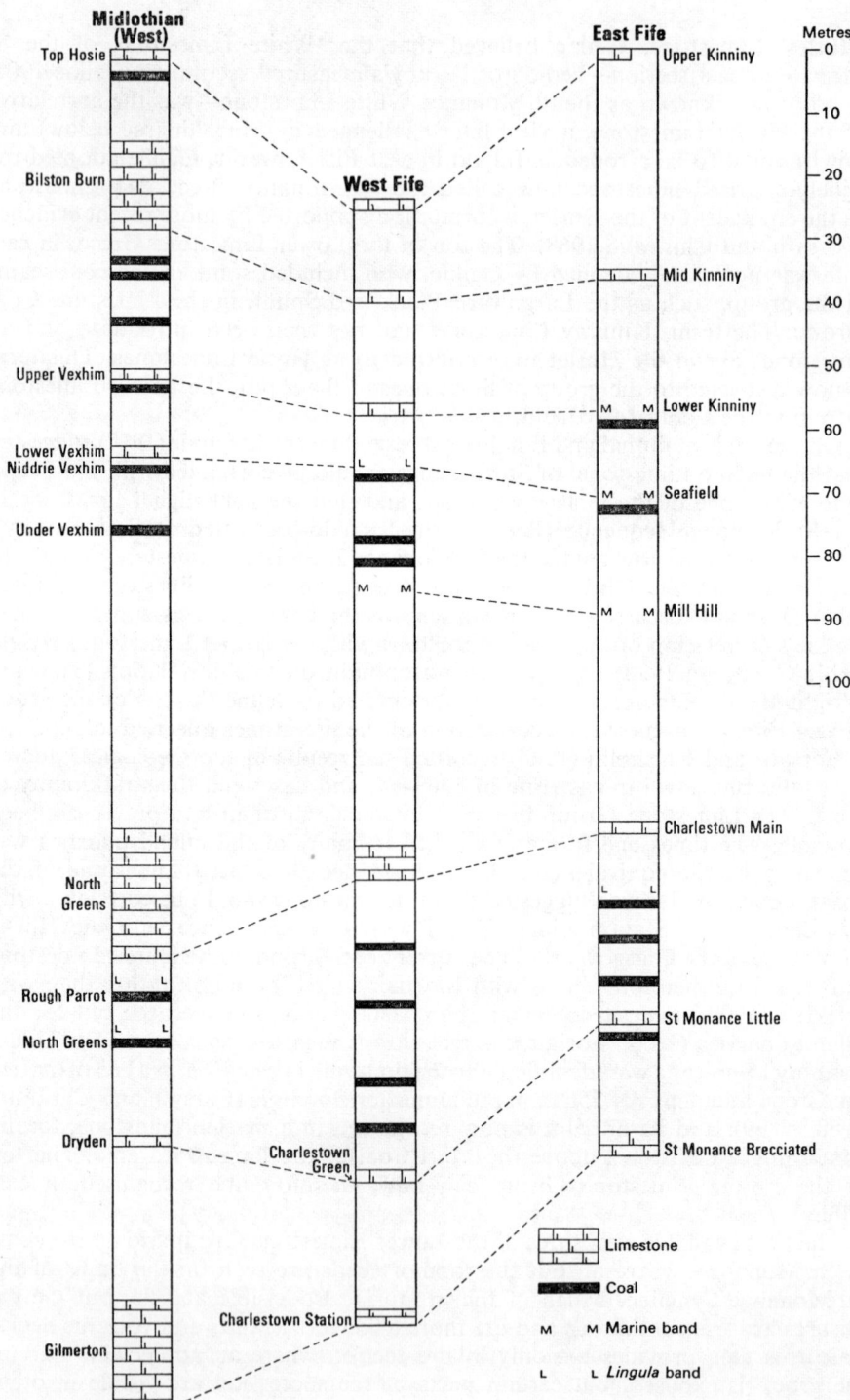

FIG. 8. *Generalised vertical sections of the Lower Limestone Group in Fife and Midlothian*

seen in some of the quarries from which the limestone was formerly extracted, for example Teasses (several quarries) and Craighall. Chemical analyses of several of the limestones exposed in east Fife are available (Muir and others 1956) and are referred to in the appropriate parts of the text.

The rest of the information presently available about the Lower Limestone Group in east Fife comes from boreholes, mine plans and reports by mining engineers (cf. Table 3). The mine plans are concerned mainly with workings in the Largoward Splint Coal, but plans are also available for the Marl Coal, the Largoward Black Coal, and the various Radernie seams. Boreholes which penetrate as deep as the Largoward Black Coal are quite numerous, but, except in the Radernie and Denhead areas, comparatively few bores cut the lower part of the group. Five of the recent series of Geological Survey boreholes cut various parts of the group (Forsyth and Chisholm 1968). The Drumcarro Borehole [NO 45911290] penetrated almost the whole of the group, the Higham Borehole [NO 47460941] cut the lower part, the borehole at Muircambus [NO 47220242] cut the upper part, and the Callange [NO 41711191] and Dunotter [NO 46500250] boreholes went down to the Mid Kinniny Limestone.

The Lower Limestone Group is now better known than any other part of the Carboniferous sequence in east Fife.

Comparison with west and central Fife

The correlation of the Lower Limestone Group (see Fig. 8) from east Fife to central and west Fife may now be regarded as established (cf. Forsyth and Chisholm 1968; Forsyth 1970) except for the lower part of the group in central Fife where the sequence is affected by volcanism. There are, however, certain contrasts between the successions in the different parts of the county. There is not much variation in the total thickness of the group but whereas the lower part is thinner in east Fife than it is in west Fife, the upper part is thicker. This applies in particular to the interval between the Mid and Upper Kinniny limestones, which certainly reaches 40 m in east Fife and may reach or even locally exceed 50 m there. The limestones are in general thinner in east Fife than elsewhere, the only exceptions being the St Monance Little Limestone, which more or less maintains the thickness of its west Fife equivalent the Charlestown Green Limestone, and the St Monance Brecciated Limestone, which is rather thicker on average than its correlative the Charlestown Station Limestone. All the marine bands known to occur in Fife are present in east Fife but in west Fife the Seafield Marine Band is reduced to a *Lingula* band (Forsyth 1970) and in central Fife the Mill Hill Marine Band is usually missing (Francis and others 1961, pp. 26–27). The faunas in the upper part of the group are rather impoverished in east Fife, but those in the lower part are abundant and the distinctive fauna of the Neilson Shell Bed (Wilson 1966) is present above the Charlestown Main Limestone. The coal seams are, on the whole, much thicker in east Fife than in the rest of the county; the Largoward Splint Coal has no equivalent of comparable thickness farther west and only locally in the Kirkcaldy area does the coal below the Seafield Marine Band attain a thickness similar to that of its east Fife equivalent, the Largoward Black Coal. The Marl Coal also is usually much thicker than its equivalents in west and central Fife. The tendency in east Fife for a number of coal seams to be developed between the St Monance Little and Charlestown Main limestones is evident locally in west Fife, but the seams there are usually thinner and have not been worked.

Comparison with Midlothian

The Lower Limestone Group successions in Midlothian and east Fife are of similar thickness. The correlation across the Firth of Forth (see Fig. 8) seems to be on the whole straightforward except that no equivalent of the Mill Hill Marine Band appears to have been recognised in Midlothian (cf. Tulloch and Walton 1958, p. 24). Its horizon may be present in the roof of the Under Vexhim Coal, but on the other hand it may lie near the top of the North Greens Sandstone, a situation similar to that which obtains in the Kirkcaldy area (Francis and others 1961, p. 26).

The most obvious difference between the Midlothian and east Fife successions lies in the fact that some of the limestones are much thicker in Midlothian, where the aggregate thickness of limestone in the Lower Limestone Group reaches its maximum (cf. Goodlet 1957, fig. 4, p. 60). In this respect Midlothian is markedly different not only from east Fife but also from the Central Coalfield, the other area of thick sedimentation during Lower Limestone Group times. The coals on the other hand are noticeably thinner in Midlothian, especially in the upper part of the group. In both respects the Midlothian sequence resembles that in west Fife rather than the succession in east Fife.

The comparisons with Midlothian and the rest of Fife confirm the impression given by Goodlet's (1957, fig. 5) generalised lithofacies map, and it is clear that during Lower Limestone Group times east Fife lay in an area of considerable, if variable, subsidence which allowed the accumulation of a thick, dominantly clastic sequence similar to sequences found in the middle of the Central Coalfield east of Glasgow (cf. Forsyth and Wilson 1965, plate iv) and in Midlothian (Tulloch and Walton 1958, fig. 8 and plate ii). I.H.F., J.I.C.

DETAILS

St Monance Brecciated Limestone. This limestone is exposed, on the eastern limb of the St Monance Syncline, on the shore near Pathhead [NO 538021], where it was first described as a 'pseudo-brecciated bed' by Kirkby (*in* Geikie 1902, p. 150). This locality is regarded as the type section of the bed. Its brecciated appearance there is now thought to be due mainly to local faulting (Tait and Wright 1923, pp. 170–171), and the name St Monance Brecciated Limestone is adopted in this account in preference to the older name 'pseudo-brecciated bed'. Brecciation is particularly marked in the exposures at the foot of the cliffs, but lower down on the shore the bed is not faulted and displays its normal character (Tait and Wright 1923). It is a crinoidal limestone, about 3·5 m thick, with a rather nodular texture which may have facilitated the development of the conspicuous brecciation near the faults. It appears to be largely non-dolomitic and is very susceptible to marine erosion; unlike the other limestones in the Lower Limestone Group it makes little or no feature in the tidal zone. A chemical analysis of the bed at this locality is available (Muir and others 1956, pp. 48, 113: sample no. SL 236).

The St Monance Brecciated Limestone is poorly exposed on the western limb of the St Monance Syncline, where it was apparently not noticed by Kirkby (*in* Geikie 1902), but was discovered by Tait and Wright (1923). It has a similar nodular texture and is about 2·5 m thick. It is variably dolomitised; the dolomitic parts of the bed appear to be more resistant to marine erosion than the unaltered parts.

The limestone formerly quarried [NO 500013, 500016 and 513020] south and east of Balbuthie was probably the St Monance Brecciated Limestone, and the latter may also have contributed the blocks of crinoidal limestone seen at or near the margins of the Coalyard Hill and Wadeslea necks (Geikie 1902, p. 114; Cumming 1936). The

STRATIGRAPHY 67

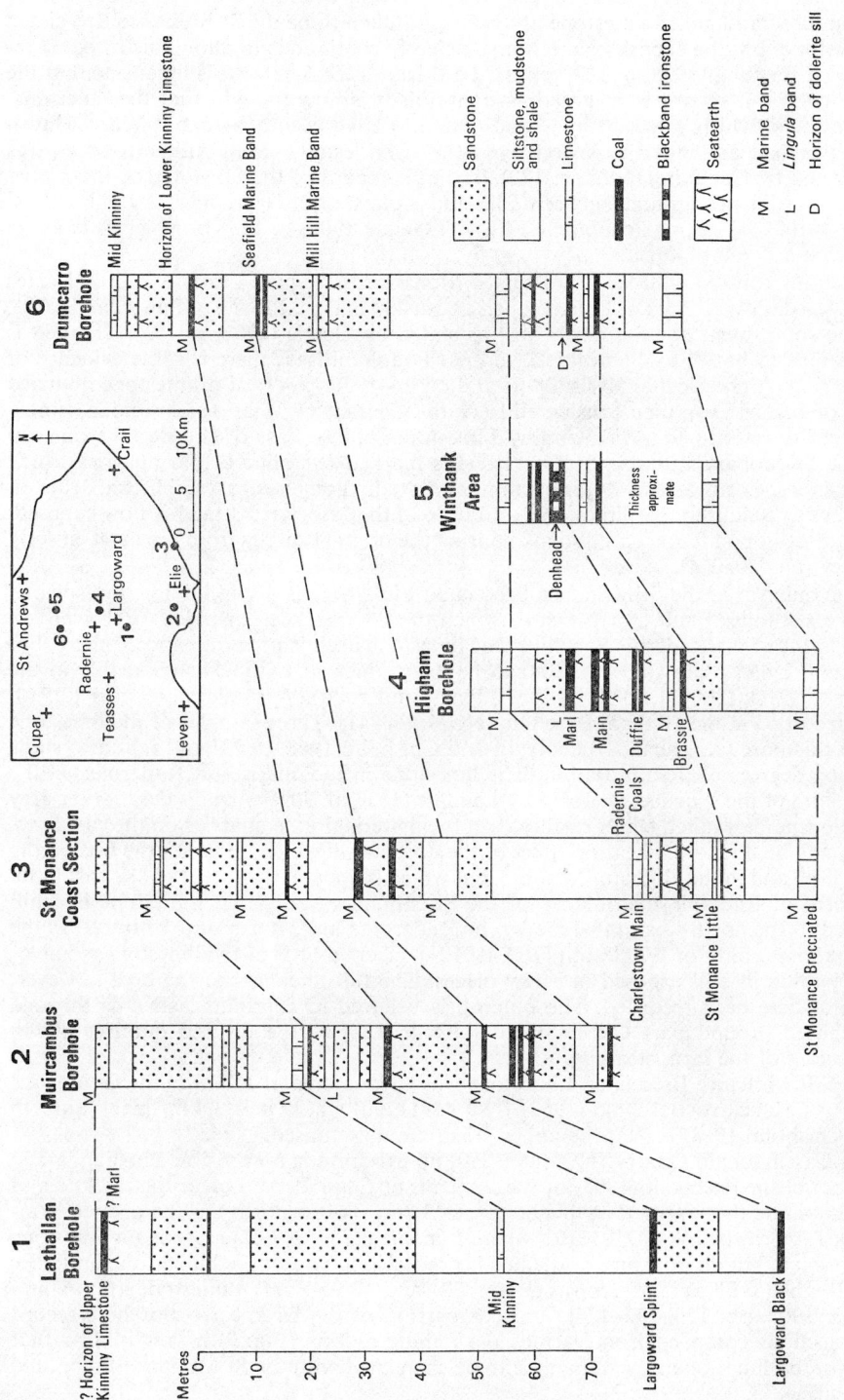

FIG. 9. Comparative vertical sections in the Lower Limestone Group

limestone exposed [NT 49619967] at Wood Haven, where at least 3·3 m of variably dolomitised and baked limestone are visible, is taken to be the St Monance Brecciated Limestone, but the exposures are not sufficiently continuous to allow this to be established. Cumming (1928, p. 127) reported a thickness of 6 m for this limestone, but the validity of this record is regarded as doubtful. It is now thought that the limestone, 3·3 m thick, which was said by Wood (1887) to have been quarried at Wood Haven was this bed and not the Woodhaven (i.e. Charlestown Main) Limestone as was suggested by Cumming (1928, p. 128). The only record of the St Monance Brecciated Limestone in the ground between Elie and Colinsburgh is provided by a borehole [NO 48100261] near Kilconquhar Mill (cf. Geikie 1902, p. 155) in which it is 3·3 m thick, at a depth of 40·5 m.

The outcrop [NO 48540379] of the St Monance Brecciated Limestone in Balcarres Den was discovered by Dr. E. H. Francis about 400 m from the site of Pitcorthie House, to the south-west, not the north-west as stated by Geikie (1902, p. 155). The bed is there seen to be 3·6 to 4·6 m thick, and partly dolomitised; there is some evidence of quarrying. A chemical analysis (Muir and others 1956, p. 47) of a limestone quarried west of Balniel Den, then considered to be the Charlestown Station Limestone, is now thought to refer to the Mid Kinniny Limestone (p. 79). East of the den the outcrop of the St Monance Brecciated Limestone is marked by a line of old quarries which extends east-north-eastwards for over one and a half kilometres past Gibliston, Belliston and South Baldutho. No limestone, and little of the associated strata, is now exposed, but a brachiopod fauna was obtained during the original survey from the most easterly quarry [NO 508064].

The outcrop of the St Monance Brecciated Limestone is displaced north-westwards by the Lathones Fault, north-east of which the bed has been quarried extensively on both limbs of the Radernie Syncline. The limestone itself is not now exposed there, but was said by Landale (1837, p. 308) to be 2·7 m thick at Lathockar (probably in the quarries [NO 480093] north of the Cupar–Crail road). A similar thickness (2·5 m) was recorded in the Higham Borehole [NO 47460941]. Three samples of the limestone from this bore were analysed (Forsyth and Chisholm 1968, p. 73) and found to show varying degrees of dolomitisation. The limestone in the quarries at Radernie, on the west limb of the syncline, was said by Landale (1837, p. 309) to be harder, darker grey in colour and of much better quality than the material in the quarries on the east limb. This may be due to the heating effect of the dolerite sill which underlies the bed on the west limb and in the Higham Borehole, but which is apparently absent on the east limb.

North of Radernie the outcrop of the St Monance Brecciated Limestone is again moved to the north-west, this time by the Radernie Fault. It can next be traced in the old quarries south of Wilkieston [NO 450121], where a little of the limestone, some of it dolomitic, is still exposed in a few places. The full thickness of the bed, however, has nowhere been recorded. The outcrop is believed to continue eastwards through unexposed ground past Cassindonald [NO 465122] to the disused quarries 400 m north-east of the farm steading. I.H.F.

The St Monance Brecciated Limestone was encountered at 118 m in the Geological Survey Drumcarro Borehole (1963) [NO 45911290] where it is 3·4 m thick (Forsyth and Chisholm 1968, p. 71); the upper 1·7 m are dolomitised.

At Cassindonald Quarry [NO 46691250] no exposure is now visible but Landale in an unpublished account (1895) of the minerals at Cassindonald gives the thickness of the limestone then worked at this quarry as 3·7 m. To the north lies the disused Winthank Limeworks [NO 47101310], where Landale (1837, p. 322) recorded a limestone 3·4 m thick, but no exposure of the bed is now visible. The worked limestone underlies the Denhead Blackband Ironstone (p. 72) and 'Five-Foot' Limestone (p. 76 and Geikie 1902, pp. 156, 173–174), and comparison of the Drumcarro Borehole record with sections compiled from various old mining reports (Fig. 9) makes it clear that the worked limestone at Winthank is to be correlated with the St Monance Brecciated Limestone.

Small folds associated with the Ceres–Maiden Rock Fault Zone bring the St Monance Brecciated Limestone to outcrop in Lumbo Den and at two points in Cairns Den near Cairnsmill. In Lumbo Den only a band of crinoidal limestone 0·15 m thick is now visible. At Cairnsmill an exposure [NO 49581477] at the side of a small reservoir shows 2·3 m of patchily dolomitised limestone dipping north-westwards; the upper part, 0·8 m thick, is composed of rubbly, shelly, crinoidal material and overlies a massive crinoidal bed, 1·5 m thick, whose base is not seen. In the Drumcarro Borehole only the St Monance Brecciated Limestone exceeds 1·5 m in thickness and on this evidence the limestone at Cairnsmill is correlated with that bed. In the den about 320 m to the north-west of Cairnsmill farm two limestones were formerly seen, the lower of which was regarded by Geikie (1902, p. 156) as the equivalent of the Hurlet Limestone. The section is now largely obscured but about 0·3 m of dolomitised crinoidal limestone, apparently in place, are still visible in the bed of the stream [NO 49401504]; it may be part of the St Monance Brecciated Limestone. A limestone 2·75 m thick was encountered at rockhead in Lumbo No. 4 Borehole, one of a series drilled about 1855 near Lumbo Farm in search of the Denhead Blackband Ironstone. This bore was sited 'in wood close on east side of farm' [probably about NO 48951475]; because of its thickness the limestone is taken to be the St Monance Brecciated.

At Ladeddie Limeworks [NO 44001360] near Backfield of Ladeddie, Landale (1837, p. 321) gave the thickness of worked limestone as 4 m, the top 0·6 m being inferior material composed of shells and crinoid fragments. No section of the worked bed is now visible but the details of the overlying section (p. 70) make it clear that the bed formerly worked is the equivalent of the St Monance Brecciated Limestone. Material from the tips has yielded a rich fauna, including corals, brachiopods, bivalves, trilobites and echinoderms, probably derived from the top part of the limestone and the mudstones above. Some parts of the bed at this locality are dolomitic (Robertson and others 1949, p. 110, where the limestone is wrongly regarded as the equivalent of the Charlestown Main Limestone), but a chemical analysis of limestone from this quarry, quoted by Muir and others (1956, pp. 47, 113: sample no. SL 47), shows it to be only slightly dolomitic.

About 800 m north of Ladeddie Limeworks, in a disused quarry at St Andrews Wells [NO 43831458], a partly obscured section shows 1·5 m of baked mudstone resting on 3 m of dolerite underlain by 1·5 m of crinoidal limestone. The first six-inch Geological Survey field map of the area records 3·3 m of limestone below the dolerite in this quarry while Robertson and others (1949, p. 110) record up to 6 m of ferrodolomitised limestone there. Such a thickness suggests a correlation with the St Monance Brecciated Limestone. About 800 m to the west of St Andrews Wells, in an old quarry [NO 43011440], a large mass of crinoidal limestone and baked calcareous mudstone, perhaps derived from the St Monance Brecciated Limestone, has been caught up in vent agglomerate. J.I.C

South of the Ceres Fault the outcrop of the St Monance Brecciated Limestone extends from Teasses Mill [NO 399109] along the north side of Craighall Den, where it can be traced quite accurately by means of old quarries, to beyond Craighall [NO 407106]. East of the latter, disused quarries provide a section of the overlying strata and of the top 1·2 m of the limestone itself. The dip swings from northwards to eastwards and the outcrop of the bed passes east of Newbigging of Craighall [NO 416102] through other quarries, where it was formerly seen to be 3 to 3·6 m thick and was mined to some extent (Geikie 1902, p. 157). The St Monance Brecciated Limestone is not exposed in the Craighall Burn south of Newbigging of Craighall, probably because of faulting, but obscure traces of quarries suggest that its outcrop runs southwestwards from the west side of the stream towards Simonden [NO 407096].

There is no doubt about the location of the outcrop of the St Monance Brecciated Limestone between Bankhead [NO 407092] and Backbraes [NO 401085] because the old quarries in it are nearly continuous. Of the full thickness of 3·2 m of limestone recorded here by Landale (1837, p. 315), up to 2·3 m are still visible in places. South-west of Backbraes a fault throws the outcrop south to Woodtop [NO 401080] where the disused quarries are now completely obscured. It appears that this is the locality of Landale's (1837, p. 315) 'Teasses Middle Limestone' for which he gave a thickness of 3 m.

Change of horizon by the quartz-dolerite sill has caused the outcrop of the St Monance Brecciated Limestone to be repeated south of Teasses [NO 407081], where it can be traced by means of quarries on both sides of the Ceres–Largo road. Teasses Quarry (the large one on the east side of the road) provides the best inland section of the limestone now available. The quarry was developed into mines, some of which can still be entered, and 3·3 m of limestone are visible. This must be almost the full thickness, but the base is no longer exposed. The bed is dolomitic in places; Muir and others (1956, pp. 47, 112: sample no. SL 37) quote an analysis of a crinoidal dolomite from this locality. The limestone is not now visible at all in the Teuchats Quarries on the west side of the road. The St Monance Brecciated Limestone was encountered at a depth of 183 m in the Teasses Moss Borehole (1897) [NO 42120864] where it is 3·2 m thick.

There is no entirely satisfactory record of the limestone in the Largoward and Lathallan districts. The Lathallan Borehole (1896–1902) [NO 46440579] passed through 2·4 m of limestone in three leaves at a depth of 404·8 m; this is about the level at which the St Monance Brecciated Limestone might be expected and the bed is therefore tentatively equated with that limestone.   I.H.F.

Strata between St Monance Brecciated and St Monance Little limestones. The known range of variation in the thickness of the interval between the St Monance Brecciated and St Monance Little limestones in east Fife is from 6·7 m at Ladeddie to 25·9 m at the Higham Borehole. The mudstone above the former limestone generally accounts for about two-thirds of the strata in this interval. It is exposed in nearly all the quarries and natural sections in which the limestone itself has been laid bare. This mudstone is generally silty and is locally rather barren, but quite a large suite of fossils has been obtained (see Chapter 9) from the various exposures and from the Higham and Drumcarro boreholes, consisting mainly of brachiopods, molluscs and bryozoans. The best exposures are on the coast at St Monance, in Balcarres Den and in the following quarries: Lathones, Radernie, Wilkieston, Winthank, Ladeddie, Craighall, Bankhead–Backbraes, Teasses and Teuchats.

The mudstone passes up into sandstone which is visible at Ladeddie Limeworks and on the coast at St Monance. The sandstone is overlain in turn by seatearth and then by the coal seam which lies just below the St Monance Little Limestone. This coal was said by Kirkby (*in* Geikie 1902, p. 150) to be 60 cm thick on the coast at St Monance, where it appeared to have been worked, but it is no longer visible. In the Radernie area it thickens to about 75 cm and was worked as the Radernie Brassie Coal. The same seam was formerly worked to a small extent as the Low Little or Low Lime Coal at Cassindonald Colliery [NO 46451255] and in the area between Winthank Limeworks and Denhead. According to old reports by mining engineers its thickness varies between 46 cm and 66 cm. There is no record of this coal at Lumbo, but at Ladeddie Limeworks it is seen in a small roadside exposure [NO 44001361] near the top of the quarry face, where it is 46 cm thick. The full section at this point is as follows:

	cm
St Monance Little Limestone	60
Coal	46
Seatearth	60
Sandstone, ripple-laminated	150
Mudstone (seen)	90

Landale (1837, p. 321) recorded a similar section at Ladeddie Limeworks giving a thickness of 1·2 m for the sandstone and 4·3 m for the mudstone, the latter resting on the worked limestone which is now correlated with the St Monance Brecciated Limestone. The total thickness of the interval between the St Monance Brecciated and St Monance Little limestones is therefore about 6·7 m at this locality, the thinnest known development in east Fife.

St Monance Little Limestone. The St Monance Little Limestone is exposed on the shore on both limbs of the St Monance Syncline: it is 0·8 m thick on the east limb, where it is bed 49 in Kirkby's section (*in* Geikie 1902, p. 150); the limestone is now named (Forsyth and Chisholm 1968, p. 72) from this locality [NO 53760210]. The bed is patchily dolomitised here, and lobate non-dolomitic areas can be distinguished by their lesser resistance to weathering. A chemical analysis of a non-dolomitic sample is available from this locality (Muir and others 1956, pp. 50, 116: sample no. SL 235). On the west limb the bed is 0·7 m thick. The St Monance Little Limestone was known to Macnair (1917) and Cumming (1928) as the *Productus giganteus* Limestone, and the Abercrombie Limestone of Landale (1837, p. 288) was probably the same bed. The limestone is not seen at Elie or in Balcarres Den, but it has been encountered farther north in boreholes and shafts in the Radernie area, where its thickness varies from 0·6 to 0·9 m; in the Higham Borehole it is 0·6 m thick, crinoidal, and largely non-dolomitic. In the Drumcarro Borehole the limestone is 0·7 m thick. At Ladeddie Limeworks (the only inland exposure), where it is seen at the top of the quarry face by the roadside [NO 44001361], it is an ochreous-weathering, crinoidal, dolomitised limestone 0·6 m thick. Details of the section at this point have been given above. The limestone was also recorded about 1910 in a small coal mine [NO 47061331] near Winthank Limeworks where it is 0·8 m thick. In the Teasses area the only available section through the St Monance Little Limestone is provided by the Teasses Moss Borehole, where it is 0·7 m thick. It has not been proved at all in the Ceres area.

I.H.F., J.I.C.

Strata between St Monance Little and Charlestown Main limestones. Between 13·7 and 16·8 m of strata lie between the St Monance Little and Charlestown Main limestones on the coast at St Monance. They include two coal horizons, occupied on the east limb by two coals each 13 cm thick, with well-developed seatearths. Very little of this part of the sequence is seen at Elie or in Balcarres Den, but the strata are much better known in the Radernie Syncline where they reach a thickness of 29 m in the Higham Borehole (Fig. 9; Forsyth and Chisholm 1968, plate iv). They include three coals which were formerly worked at Radernie and adjacent collieries. The lowest, the Radernie Duffie Coal, was cut by the Higham Borehole where it is 130 cm thick, in two leaves; it averages 115 cm in the Radernie area and was mined from a number of pits. The Radernie Duffie Coal lies about 7·5 m above the St Monance Little Limestone and some 6 m below the succeeding Radernie Main Coal. In this area the latter is also usually split into two leaves, totalling between 122 and 137 cm of coal; it was at one time quite extensively worked, especially from Radernie Colliery. Some 4·5 to 6 m of strata separate the Radernie Main and Radernie Marl coals. Both seams are underlain by thick seatearths. The Radernie Marl Coal is about 90 cm thick and is usually split into two leaves. A *Lingula* band, which was discovered in its roof in the Higham Borehole (Forsyth and Chisholm 1968, p. 72), is probably the correlative of one of the two bands (see Fig. 8) found in Midlothian in this part of the sequence. Between 9 and 12 m of strata intervene between this seam and the Charlestown Main Limestone in the Radernie area: they include a rooty bed with a mottled green, possibly tuffaceous, mudstone above it (Francis 1968, pp. 124–125).

Cassingray No. 1 Borehole (1859) [NO 48110798] cut two seams, separated by 5·2 m of fireclay, below the Charlestown Main Limestone. The upper one is 109 cm thick in two leaves, both described as 'foul'; the lower seam gave the following section (in descending order): parrot coal 30 cm, blackband ironstone 10 cm, coal 48 cm, very foul coal 61 cm, fireclay 33 cm, hard coal 84 cm. The bore stopped immediately below the lower seam and it is not certain which of the Radernie coals are represented here. The Lathallan Borehole probably passed through this part of the sequence but no coal was recorded.

The Radernie coals are said to be present east of Nether Radernie [NO 459106], and some of them at least have apparently been worked, but very little is known about their development in that vicinity. I.H.F.

In the Drumcarro Borehole the strata between the St Monance Little and Charlestown Main limestones are about 14·5 m thick but they contain a sill of basalt, so that the stratigraphical record may not be complete. The mudstones above the St Monance Little Limestone contain a rich marine fauna. A 25-cm coal, burnt by the overlying basalt sill, lies 3·4 m above the limestone; it is probably the Radernie Duffie Coal but it contains no trace of the Denhead Blackband Ironstone (see below). Another coal, 30 cm thick, is present 4·7 m below the base of the Charlestown Main Limestone; its equivalent in the Radernie sequence is in doubt, but a bed 0·9 m thick of green and cream mottled mudstone just above the coal probably correlates with a bed of similar lithology in the Higham Borehole (p. 71). About 3·4 m of mudstone with marine fossils lie between the mottled mudstone and the base of the Charlestown Main Limestone.

The coals and an ironstone in this part of the sequence were formerly worked about 800 m east of the Drumcarro Borehole, in an area between Cassindonald, Winthank and Denhead. A general section for this area (Table 3) has been compiled from unpublished mining reports and plans.

According to the old mining reports the equivalent of the Radernie Duffie Coal was worked for lime-burning on Craigton Common, between Denhead and Winthank, for many years before the value of its associated blackband ironstone was discovered. After about 1847 the ironstone and coal were worked together until the ironstone was exhausted, about 1866. The area in which the ironstone was workable is known from old plans and does not exceed about 20 hectares. The ironstone, known variously as the Denhead, Winthank, or Craigton Blackband Ironstone, and here named the Denhead Blackband Ironstone, is generally about 51 cm thick. It dies out southwards and was not found at Cassindonald Colliery [NO 46451255]. It is still present at the westerly limit of the Denhead workings, but no trace of it was found either in the Drumcarro Borehole (see above), or at Backfield of Ladeddie (see below). It was, however, found to be present in typical development at Lumbo (p. 74) about $2\frac{1}{2}$ km to the north-east of Denhead.

In an unpublished report dated 1885, Landale describes the ironstone as '. . . a light dun-coloured blackband with so much coaly matter in it that it lost half its weight in calcining . . .'. Carbonaceous shale with ironstone lenses collected from a tip [NO 47061331] near Winthank in 1962 contains ostracods and fish remains and is probably derived from some part of the ironstone or the overlying 'rums'. A list of fish remains collected from the ironstone at Denhead was given by Traquair (1901, p. 112).

The strata between the Radernie Duffie Coal and the Charlestown Main Limestone were reported (see Table 3) to contain two thin coals of poor quality in the Denhead–Cassindonald area; a bed of 'limestone' 0·6 m thick was often recorded between them. As suggested in Table 3 the lower coal may represent one or both of the two uppermost Radernie seams, while the 'limestone' may represent the distinctive cream and green mottled mudstone recorded at about this horizon in the nearby Drumcarro and Higham boreholes (see above). The upper coal seam, lying above this 'limestone', may be a seam not represented at Radernie; it was not always recorded in the mining reports and it is absent at the Drumcarro Borehole.

The Denhead Blackband Ironstone is present in the Radernie Duffie Coal in a small faulted syncline near Lumbo Farm [NO 489148]. Details of the ironstone and of the associated strata are known from records of the ironstone workings and from boreholes put down about 1855. The strata below the Radernie Duffie Coal were not penetrated

STRATIGRAPHY

TABLE 3

General section of the strata from the Charlestown Main Limestone to the St Monance Little Limestone in the area around Winthank and Cassindonald, based on unpublished mining reports dated between 1822 and 1910

		Local Name	Present Correlation
Limestone	about 3·3 m	150 cm	Charlestown Main Limestone
Strata (probably mainly argillaceous)			
Coal, poor (not always recorded)		38 cm	
Various strata including a 'limestone' 60 cm thick	1·3 m		Five Foot Limestone
			'Limestone' may be correlative of mottled cream and green mudstone in Drumcarro Bore (p. 72)
Coal, poor		51 cm	? Radernie Marl and Radernie Main Coals
Strata	2·5 to 3·7 m		Rums coal or Rums
Carbonaceous shale, canneloid especially in upper part	0·3 to 1·2 m	0 to 76 cm, usually 51 cm where worked	'Dunstane' = Winthank, Craigton or Denhead Blackband Ironstone
Blackband ironstone			Radernie Duffie Coal—with the Denhead Blackband Ironstone
Shale		5 cm	
Coal		48 to 66 cm	Ironstone Coal
Strata (probably mainly argillaceous)	about 3 m		
Limestone		60 to 80 cm	St Monance Little Limestone

in the boreholes, and the sediments in this part of the sequence are largely obscured at their outcrop in Lumbo Den nearby. The details of the Radernie Duffie Coal closely resemble those of the seam at Denhead (Table 3). The following general section is based on the borehole and pit records:

	cm
Cannel coal and carbonaceous shale	43 to 56
Blackband ironstone	13 to 58
Fireclay	0 to 10
Coal	46 to 51

In Lumbo No. 3 Borehole [NO 48871482] the seam is interleaved with thin basalt sills, and in a bore put down from the bottom of No. 2 Pit on Mount Melville estate (exact site unknown) 12 m of dolerite were encountered. These occurrences suggest that the Lumbo dolerite sill (p. 142), seen at outcrop [NO 48821490] in Lumbo Den, lies just below the Radernie Duffie Coal. The sill was also encountered, at least 6·4 m thick, in Mount Melville No. 1 Borehole [NO 48711488]. Its top lies 0·9 m below a coal 41 cm thick which is taken to be an altered equivalent of the Radernie Duffie Coal.

As in the Denhead area (Table 3) two thin coals, separated by a 'limestone', are present between the Radernie Duffie Coal and the Charlestown Main Limestone. The 'limestone' varies from 0·1 to 0·9 m in thickness and in places is divided into two leaves. It may represent the cream and green mottled mudstone at the same horizon in the Drumcarro and Higham boreholes. The lower of the two coals varies considerably in thickness. It is thicker (48 cm to 82 cm) where it overlies argillaceous rocks and thinner (less than 20 cm) where it overlies sandstone (see below). The higher coal is thin and is not always present.

The measures between the Radernie Duffie Coal and the lower of the two coals just described are subject to a marked lateral variation within the Lumbo area. An abrupt change of facies occurs across a line running north-eastwards through the farm steading at Lumbo. To the south-east of this line lies an area in which the sequence consists mainly of argillaceous rocks, 4 to 6·7m thick, a development similar to that at Denhead (Table 3). To the north-west of this line the sequence is 12·8 to 14·3 m thick and consists mainly of sandstone. The sandstone is probably that exposed in the east side of Lumbo Den about 64 m north of Lumbo Bridge, where approximately 3 m of white sandstone with convoluted bedding are visible. The sandstone lies directly on top of the Radernie Duffie Coal in Lumbo No. 7 Borehole [NO 48891482] but in Lumbo Nos. 2 and 3 boreholes nearby about a metre of shale is present above the coal. The sandstone is probably a 'washout' or channel-fill, the line of facies change described above representing the south-eastern margin of the channel. The thinning of the overlying coal in the area of the sandstone has already been mentioned.

To the east and south-east of Ladeddie Limeworks [NO 44001360] a small outcrop of beds above the St Monance Little Limestone has been proved by mining operations and by shallow boreholes. A limestone 0·5 m thick which was encountered at the bottom of Denork No. 2 Borehole (1882) [NO 44381381] is taken to be the St Monance Little Limestone. A coal 51 cm thick, lying 6·7 m above the limestone, is regarded as the probable equivalent of the Radernie Duffie Coal, and the same seam was encountered, 53 cm thick, in Denork No. 1 Borehole [NO 44561389], and in Denork No. 3 Borehole [NO 44441374] where its thickness is 74 cm. In no case was any trace of the Denhead Blackband Ironstone recorded. About 3 m above the supposed Radernie Duffie seam in all three bores is a thin coal 15 to 33 cm thick, and about 4·5 m above this Nos. 1 and 3 boreholes encountered 23 cm of 'limy fakes' and 90 cm of 'green fireclay'. The last-mentioned beds are presumed to correlate with the cream and green mottled mudstone found in the Drumcarro Borehole (p. 72). The Charlestown Main Limestone, although met with during mining operations, was not recorded with certainty in the boreholes; the exact thickness of the strata between it and the St

Monance Little Limestone cannot therefore be given, although it must be between 16 and 22 m. J.I.C.

Two seams, known locally as the Marl and Main coals, were formerly worked from the several pits which comprised Craighall Colliery. They are both 69 cm thick, and the upper seam, the Marl, is in two leaves. The thickness of strata separating the two seams is not known. These coals certainly lie above the St Monance Brecciated Limestone and must be presumed to lie below the Charlestown Main Limestone, but their positions in the sequence are not exactly known and their identification as the Radernie Marl and Main coals respectively must be regarded as tentative and unproved.

The only complete section of the strata between the St Monance Little and Charlestown Main limestones in the Teasses area is provided by the Teasses Moss Borehole [NO 42120864], where the sequence has been invaded by a dolerite sill 60 m thick which to some extent obscures its nature. The beds are at least 27 m in thickness and include three burnt coals, each less than 30 cm thick. Other bores which cut parts of this sequence show, however, that where unaffected by the sill some of these coals are, at least locally, somewhat thicker than that. The exact horizon of the Teasses Main Coal is uncertain but it must lie in this part of the succession. The seam was extensively worked south of Teasses House [NO 407081]. Landale's original account (1837, p. 314) gave a thickness of 112 cm of coal in two leaves for this seam, but in an unpublished report dated 1879 he amended this to the following: splint coal 30 cm, parting 33 cm, cherry coal 48 cm, and stated that the quality of the seam decreases eastwards. Other boreholes in the area corroborate these latter figures closely. Landale (1837, p. 314) also recorded a Teasses Under Coal, an unspecified amount lower in the succession, for which he gave a thickness of 61 cm, but this was altered to 71 cm in his 1879 report. The Teasses seams must be regarded as the equivalents of some of the Radernie coals, but the precise correlation has not been established.

Charlestown Main Limestone. The Charlestown Main Limestone is prominently exposed on the shore on both limbs of the St Monance Syncline, where it is between 1·5 and 1·6 m thick. On the east limb of the syncline it is bed 34 of Kirkby's measured section (*in* Geikie 1902, p. 150). It appears to be mainly dolomitised at this point [NO 53700205] but contains some non-dolomitic patches and layers. A chemical analysis, probably from the Charlestown Main Limestone at this locality, is available (Muir and others 1956, pp. 54, 121: sample no. SL 234); the sample analysed is a ferriferous dolomite. At Elie the bed was known as the Woodhaven Limestone (Cumming 1928). Wilson (1966, p. 113), in his description of the fauna of the Neilson Shell Bed, indicated that the Woodhaven Limestone of Elie is the correlative of the Charlestown Main Limestone of west Fife; the latter name is therefore used throughout the present account, in preference to the local east Fife names Woodhaven Limestone and Five Foot Limestone. In Wood Haven itself the bed is dolomitised and is about 1·2 m thick; the upper part is also seen in a small dome [NT 49309996] at the eastern end of Elie Harbour. Cumming (1928, p. 128) quoted Wood's (1887) reference to a limestone 3·3 m thick which was at one time quarried in Wood Haven and regarded this bed as being the Woodhaven Limestone. It is now thought, however, that it is more likely to have been the St Monance Brecciated Limestone (see p. 68).

In Balcarres Den the Charlestown Main Limestone is disturbed by an adjacent volcanic neck, but at least one metre of dolomitised limestone is present. It is 1·7 m thick in the Lathallan Borehole [NO 46440579], 1·6 m in three leaves in Cassingray No. 1 Borehole (1859) [NO 48110798] and 1·14 m in Cassingray No. 1 Borehole (1910) [NO 48930726]. There are several records of the limestone in the Teasses area. The Teasses Moss Borehole gave a thickness of 1·2 m and other unsited bores recorded amounts between 0·9 and 1·1 m. An unsited bore near Newbigging of Craighall cut 1·1 m of limestone at this horizon, but otherwise the Charlestown Main Limestone has not been proved in the Craighall and Ceres areas. It was cut in three bores near Radernie, including the Higham Borehole, where it is largely dolomitised and carries

15 cm of sandstone in the middle: its thickness in all these bores is between 1·2 and 1·5 m.   I.H.F.

In the Drumcarro Borehole (1963) the Charlestown Main Limestone is a crinoidal dolomitised bed 1·4 m thick (Forsyth and Chisholm 1968, p. 71). In the nearby Denhead area the limestone was known as the Five Foot or Five Feet Limestone (Geikie 1902, p. 173) during the period of coal and ironstone working, when it was encountered in boreholes and mining operations. According to old reports it is usually between 1·4 and 1·6 m thick but in an old borehole near Denhead village a thickness of 2·2 m was recorded. The bed is not now exposed in the Denhead area, although it is said to have been quarried at two places near Mount Melville Lodge (Geikie 1902, pp. 156, 174).

Boreholes to the Denhead Blackband Ironstone in the Lumbo area (p. 72) show the Charlestown Main Limestone varying between 1·5 m and 2·5 m in thickness. The bed is seen dipping south at about 30° in the stream at Lumbo Bridge [NO 48811484] where it is a dolomitic crinoidal limestone at least 1·1 m thick, overlain by at least 3 m of grey mudstone.

The 'Five Foot Limestone' is reported to have been encountered during coal-mining operations at Backfield of Ladeddie (p. 74) but it cannot be identified with certainty in the records of the nearby Denork boreholes (1882).   J.I.C.

Strata between Charlestown Main Limestone and Largoward Black Coal. The mudstone above the Charlestown Main Limestone contains the Neilson Shell Bed (Wilson 1966), which has yielded a rich and varied fauna in the Higham and Drumcarro bores and in surface exposures both in Balcarres Den and on the coast at St Monance and Elie (see Chapter 9). This mudstone is between 18 and 37 m thick: it gradually becomes silty upwards, some of the silty bands being graded, and passes by alternation into the overlying sandstone. The latter varies in thickness from 6 to 15 m and is succeeded by a group of mixed strata, usually including one or more thin coals or rooty horizons in the lower part and the Mill Hill Marine Band in the upper part. This marine band is often found in a rusty-weathering nodular calcareous sandstone, or sandy limestone, e.g. on the shore at Elie [NT 48909998] and on the eastern limb of the St Monance Syncline; in the latter area it dies out completely westwards and is not represented on the western limb at all. The marine nature of this band at St Monance was first recognised by Wood (1887, p. 501), who provided a succinct and accurate description, and confirmed by Wright (1914). The Mill Hill Marine Band was recorded in the Drumcarro and Muircambus boreholes and also in most of the other geologist-examined boreholes in east Fife which passed through this part of the sequence. The beds containing this marine band are separated from the lowest leaf of the Largoward Black Coal by $7\frac{1}{2}$ to $13\frac{1}{2}$ m of mainly sandy strata which at St Monance are strongly bioturbated (Chisholm 1970a, table 1).   I.H.F., J.I.C.

Largoward Black Coal. The tendency of the Largoward Black Coal to occur in widely separated leaves is exhibited on the coast at St Monance, where the upper leaf, worked as the St Monance Harbour Coal, was said to be 180 cm thick (Geikie 1902, p. 153): 60 cm of coal are still visible. It lies 4·5 to 6 m above the lower leaf which is 30 to 40 cm thick in this section. The seam was also worked at Elie, where one of the leaves may be the lowest coal in Landale's (1837, map) section: scarcely any coal can now be seen and the two leaves are at least 6 m apart. The Largoward Black Coal in the Muircambus Borehole comprises 165 cm of coal in four leaves distributed over 7·6 m of mixed strata. The position of the top leaf has been located in Balcarres Den, but only 20 cm of coal are now visible.

In the Largoward area itself, the Largoward Black Coal has been proved by many boreholes and was mined from several collieries including Largoward, Lathallan, Largobeath and Falfield, but the workings were nowhere extensive because there was no great demand for this seam as long as supplies of the Largoward Splint Coal were available. In this area also it is usually split into as many as four leaves, and the name tends to be applied only to the highest and thickest leaf, which is usually between 100 and 120 cm thick. One of the other leaves was worked to a very small extent as the Scarrat Loft Coal from Largoward No. 1 Pit, where it is 90 cm thick.

North-west of Largoward the outcrop of the Largoward Black Coal is twice repeated, by the Cassingray and Cadger's Bridge faults. South of the former it extends westwards past Falfield House [NO 445088], and one leaf, 30 cm thick, is seen in a streamlet [NO 43810893]. Farther west the location of the outcrop is uncertain, but it is thought to pass north of the 'rums' quarry [NO 429085] near Clockmadron (cf. Geikie 1902, p. 177). In the Appleton Basin the Largoward Black Coal was proved by the Baldastard Borehole (1896) [NO 42240747] to be split, as at Largoward, into widely separated leaves, but the total amount of coal (109 cm) is less in this bore than is usual at Largoward; the top leaf (48 cm) is the thickest. No workings are known in this area. Both south and north of the Cadger's Bridge Fault the outcrops of the Largoward Black Coal can be traced for over one and a half kilometres westwards from near North Bowhill [NO 461088] and Lawhead [NO 462097] respectively, but not much is known about the seam. It was worked to a small extent from Drumhead Pit [NO 45430993], where it is between 84 and 124 cm thick in three leaves. Falfield No. 6 Borehole (1954) [NO 45550898] encountered a waste taken to be at the level of the top leaf of the seam; the other three leaves combine to give a total of 112 cm of coal: together with the waste they are spread over a vertical distance of 7·6 m.   I.H.F.

The Largoward Black Coal outcrops on the hillsides south of Drumcarrow Craig, dipping north-westwards at a low angle. It was at one time worked from a group of pits about 460 m east of Drumcarro Farm, but no details are available. In the Drumcarro Borehole (1963), put down to prove the sequence in this area, the seam section is coal 28 cm, on seatearth 74 cm, on coal 53 cm. The coal was also worked in the Denhead Syncline, a north-easterly-trending structure lying to the east of Drumcarrow Craig. According to old reports the coal is 107 cm thick in this area, but in Denhead No. 3 Borehole (date unknown) [NO 46731362] the section recorded is: coal 130 cm, on fireclay and 'light calmy bands' 3·4 m, on coal 30 cm.

At Lumbo the upper leaf of the coal was proved in Mount Melville No. 3 Borehole (1855) [approximately NO 486146] where it is 79 cm thick at a depth of 41½ m.   J.I.C.

Strata between Largoward Black and Largoward Splint coals. The mudstone roof of the Largoward Black Coal is generally well-developed and includes the Seafield Marine Band (Francis and others 1961, p. 27; Forsyth and Chisholm 1968, p. 71), which contains a rich fauna including brachiopods, molluscs and crinoid ossicles (see Chapter 9). It is exposed in Balcarres Den and on the coast at Elie and St Monance, being especially well-developed on both limbs of the St Monance Syncline. The mudstone is overlain by sandstone, locally silty, which is seen on the coast at St Monance to have been extensively reworked by burrowing organisms (Chisholm 1970a, table 1). The total thickness of these strata is known mainly from boreholes and varies from about 12 m in the Lumbo–Denhead area in the north to almost 23 m in the Largoward area. On the shore at St Monance it is between 12 and 19 m.   I.H.F., J.I.C.

Largoward Splint Coal. The Largoward Splint Coal is not well displayed on the coast at either Elie or St Monance, probably as a result of crop workings. It is provisionally correlated with the second-lowest coal (69 cm thick) in Landale's (1837) Elie section (see Table 5). In the Muircambus Borehole this seam is 89 cm thick, and 90 cm of coal were formerly exposed at its horizon in Balcarres Den.

The Largoward Splint Coal was by far the most extensively worked seam in the Largoward area, where it was raised at several collieries, including Lathallan, Falfield, Largoward, Cassingray and Largobeath. Its thickness of 114 cm is on the whole very constant and its high quality made it a much valued subject in the 19th and early 20th centuries, the last sizeable working ending in 1914. West of Falfield the outcrop of this seam can be traced with reasonable accuracy as far as the 'rums' quarry [NO 429085] near Clockmadron. The available information about this area is very ambiguous and the identification of the 120-cm seam overlying the 'rums' (Geikie 1902, p. 177) as the Largoward Splint Coal is very tentative. The 'rums' are carbonaceous, slightly silty mudstones with fusainous debris, which were formerly used for lime burning. A thickness of 6 m of this material is now visible in the quarry and it seems unlikely that the full thickness, of what must be a local development, exceeds 9 m. The estimates of the thickness given by Landale (1837, p. 317) and Geikie (1902, p. 177), of fully 24 m and 45 m respectively, appear to be considerably exaggerated. The seam worked to a small extent from Boghall Mine [NO 428075] has also been tentatively identified as the Largoward Splint Coal: it was said to be 99 cm thick, the bottom 20 cm being of inferior quality.

The Largoward Splint Coal has been proved in several bores to be between 60 cm and 90 cm thick in the Appleton Basin. At least two pits were sunk to it near Baldastard [NO 421069], but it does not appear to have been worked there. The seam was, however, mined at Teasses Colliery, where the recorded thickness varies from 71 cm (in two leaves) to 122 cm.

Both south and north of the Cadger's Bridge Fault outcrops of the Largoward Splint Coal have been followed for more than 1·6 km, partly by means of traces of old opencast workings. An attempt to work the seam in 1937 from Falfield Mine [NO 45720902] was quickly abandoned although it is 147 cm thick there. The only other known workings in this vicinity were from Drumhead Pit [NO 45430993], abandoned in 1900, where the seam is 114 cm thick. It is likely, however, that it was also extracted from another old pit [NO 45041015], 400 m to the north-west. The nature of the Largoward Splint Coal in the Ceres area is unknown: it is burnt by dolerite in the Callange Borehole [NO 41711191], the only bore in this area to reach its horizon. I.H.F.

Like the underlying Largoward Black Coal, the Largoward Splint Coal was formerly worked south of Drumcarrow Craig and in the Denhead Syncline. Landale, in an unpublished report dated 1847, recorded that according to old colliers the seam (then known as the Parrot Coal) at Denhead consists of 101 cm of 'parrot' coal overlying 51 cm of splint coal, but he expressed some doubt about the accuracy of these figures. The Drumcarro Borehole (1963) encountered old workings in this seam and the nearby Ladeddie Nos. 8 and 10 boreholes (1953) proved 107 cm and 63 cm of coal respectively at this horizon, suggesting that the figures reported to Landale were exaggerated.

In the Lumbo area Mount Melville No. 3 Borehole (p. 77) encountered a full section of the seam, at a depth of 28·78 m. The details are:

	cm
Cherry coal	48
Rums	15
Parrot coal	46
Cherry coal	20

These details compare well with those recorded in a small working (1866) nearby. J.I.C.

Strata between Largoward Splint Coal and Mid Kinniny Limestone. The mudstone above the Largoward Splint Coal has been penetrated in numerous boreholes: it is generally silty and contains a rather sparse marine fauna, consisting mainly of *Lingula* and bivalves (see Chapter 9). The horizon of the Lower Kinniny Limestone lies within this mudstone but the limestone itself is not known in east Fife. The mudstone is exposed on the coast on both limbs of the St Monance Syncline and at Elie; inland it is certainly exposed in Balcarres Den, and a mudstone with marine fossils noted in the Craighall Burn [NO 42010958] is probably at this horizon.

The mudstone is succeeded by a generally sandy sequence which, on the coast at St Monance, includes a band of mudstone with marine bivalves and some beds in which the trace-fossil *Teichichnus* is very well developed (Chisholm 1970b). Burrowed strata were found at the same level in the Muircambus Borehole (Forsyth and Chisholm 1968, plate iv). This bore also yielded a *Lingula* band 10 m above the Largoward Splint Coal. In the Largoward area the sandy strata are overlain by up to 9 m of mainly argillaceous beds including a few coal seams. These are usually less than 30 cm thick, but Falfield No. 3 Borehole (1954) [NO 45270864] cut a 46-cm seam, and one of 61 cm was reported in Cassingray Shaft [NO 47880814]. The thickness of the whole interval varies between about 12 m in the Denhead area and a maximum of 30 m in parts of the Largoward area. I.H.F., J.I.C.

Mid Kinniny Limestone. This bed, which is generally present throughout east Fife, is a crinoidal, often dolomitic, limestone. In the many boreholes which have passed through it, it is usually between 0·5 and 0·7 m thick. On the coast at Elie, however, where it was called the *Chonetes* Limestone by Cumming (1928), it reaches a thickness of 1·2 m; it is dolomitic in this area. In the Balcarres Den section a limestone 3 m thick, believed by Geikie (1902, p. 155) to be the Hurlet Limestone, is now considered to be an unusually thick representative of the Mid Kinniny Limestone. An exposure [NO 48590445] showing 0·9 m of dolomitised limestone is still visible in the Den itself and 1·7 m of limestone, also mainly dolomitic, are exposed in an old quarry [NO 48640434] at the top of the south-west bank of the Den. A chemical analysis of a crinoidal dolomite from an 'Old Quarry, W. side of Balniel Den, ¾ mile [1·2 km] N.E. of Colinsburgh' is quoted by Muir and others (1956, pp. 47, 112: sample no. SL 34). The sample probably came from the quarry in the Mid Kinniny Limestone mentioned above. The bed can be traced around the axis of the St Monance Syncline on the coast [NO 52790158] (Geikie 1902, fig. 17, p. 154), where it is a hard, dolomitised crinoidal limestone with well-developed 'cauda-galli' markings (*Zoophycos*) in the upper part. A chemical analysis of a ferriferous dolomite from a 'Hosie' Limestone 0·7 m thick on the shore east of St Monance Harbour, quoted by Muir and others (1956, pp. 57, 125: sample no. SL 233) probably relates to the Mid Kinniny Limestone. A crinoidal dolomitic limestone exposed in a small stream about 140 m S.S.E. of Baldastard [NO 421069] is also taken to be the Mid Kinniny Limestone; in the coal mining area between here and Largoward the bed was formerly known by the local name of Baldastard Index Limestone. I.H.F., J.I.C.

Strata between Mid Kinniny Limestone and Marl Coal. The thinnest known sequences between the Mid Kinniny Limestone and the Marl Coal in east Fife are in the Callange and Falfield areas, where the thickness is about 26 m. The sequence certainly reaches 38 m in the Dunotter Borehole and 40 m in Cordies Mealling No. 8 Borehole [NO 46340723]. There is some doubt (see below) about the identification of the Marl Coal in the Lathallan and Appleton [NO 423076] areas, but if the seam worked under this name at Lathallan (and its presumed equivalent at Appleton) has been correctly identified the thickness of strata between the Mid Kinniny Limestone and the Marl Coal must be not less than 70 m there (44 m at Appleton). Even if the identification is incorrect, the thickness of this interval must still be considerable at Lathallan—about 50 m (29 m at Appleton)—unless the relevant borehole evidence, which is quite old, is unreliable.

Except at Lathallan and Appleton, the strata in this part of the sequence all belong to a single cycle, the mudstone member of which varies from 3 to 18 m in thickness and contains marine fossils. It becomes silty in the upper part and in places passes upwards by alternation into sandstone; the sandstone bands, especially in the Dunotter Borehole [NO 46500250], are conspicuously burrowed (cf. Forsyth and Chisholm 1968, plate iv; Chisholm 1968). Elsewhere, for example in the Muircambus Borehole, there is a disconformity at the base of the sandstone, which varies from 7½ to 30 m in thickness. The Dunotter and Muircambus boreholes both have a canneloid shale associated with fish debris in a group of argillaceous strata 6 to 7½ m below the horizon of the Marl Coal, but no seatearth was found.

Marl Coal. In the Largoward district, which is taken as the type area, the Marl Coal is generally about 90 cm thick, but in a few bores its thickness is reduced to about 10 cm. The outcrop is marked in places by traces of old workings, and the seam was worked to a small extent at Largoward and Cassingray collieries. The coal is rather thinner in the Callange and Drumcarro districts, where it varies from 30 to 60 cm, and it is very thin or absent in the Elie district. The seam worked as the 'Marl Coal' in a small area at Lathallan is usually between 70 and 110 cm thick; this seam is provisionally equated with the Appleton Splint Coal, which was said to be 120 cm thick in its small outcrop in the centre of the Appleton Basin.

In the Lathallan and Appleton areas, none of the borehole sections through the supposed Marl Coal or Appleton Splint has been examined by a geologist; consequently it is not known whether or not the marine band which in other areas occurs in the roof of the Marl Coal (and locally includes the Upper Kinniny Limestone) is present. Positive identification of these seams as the Marl Coal is therefore not possible. In these areas there is a coal, 38 to 69 cm thick, which lies some 15 to 18 m below the supposed Marl Coal. It was known at Lathallan as the Little Coal (Landale 1837, p. 304) and worked at Appleton as the Ell Coal. Elsewhere in east Fife no coal seam is known at this horizon, and it is possible that the Lathallan Little and Appleton Ell coals are in fact the equivalent of the Marl Coal of the rest of east Fife and that the 'Marl Coal' of Lathallan and the Appleton Splint Coal lie at a somewhat higher horizon, in the lower part of the Limestone Coal Group.

Upper Kinniny Limestone. Most of the sections through the horizon of the Upper Kinniny Limestone in east Fife recorded only a bed of mudstone with a rather sparse marine fauna, but in some of them, in the Largoward and Elie areas, the limestone itself is present. It is up to 0·3 m thick except on the shore at Elie where it exceeds 0·4 m and was called the Red Limestone by Cumming (1928). I.H.F.

References

BROWN, T. 1859. On a section of a part of the Fifeshire coast. *Q. Jnl geol. Soc. Lond.*, **15**, 59–62.

—— 1861. Notes on the Mountain Limestone and Lower Carboniferous rocks of the Fifeshire coast, from Burntisland to St Andrews. *Trans. R. Soc. Edinb.*, **22**, 385–404.

CHISHOLM, J. I. 1968. Trace-fossils from the Geological Survey boreholes in east Fife, 1963–4. *Bull. geol. Surv. Gt Br.*, No. 28, 103–119.

—— 1970a. Lower Carboniferous trace-fossils from the Geological Survey boreholes in west Fife (1965–6). *Bull. geol. Surv. Gt Br.*, No. 31, 19–35.

—— 1970b. *Teichichnus* and related trace-fossils in the Lower Carboniferous of St Monance, Scotland. *Bull. geol. Surv. Gt Br.*, No. 32, 21–51.

CLOUGH, C. T., HINXMAN, L. W., WILSON, J. S. G., CRAMPTON, C. B., WRIGHT, W. B., BAILEY, E. B., ANDERSON, E. M. and CARRUTHERS, R. G. 1925. The geology of the Glasgow district. 2nd edit., revised by MACGREGOR, M., DINHAM, C. H., BAILEY, E. B. and ANDERSON, E. M. *Mem. geol. Surv. Gt Br.*

CUMMING, G. 1928. The Lower Limestones and associated volcanic rocks of a section of the Fifeshire coast. *Trans. Edinb. geol. Soc.*, **12**, 124–140.

—— 1936. The structural and volcanic geology of the Elie–St Monance district, Fife. *Trans. Edinb. geol. Soc.*, **13**, 340–365.

CURRIE, ETHEL D. 1954. Scottish Carboniferous goniatites. *Trans. R. Soc. Edinb.*, **62**, 527–602.

FORSYTH, I. H. 1970. Geological Survey boreholes in the Lower Carboniferous of west Fife, 1965–6. *Bull. geol. Surv. Gt Br.*, No. 31, 1–18.

—— and CHISHOLM, J. I. 1968. Geological Survey boreholes in the Carboniferous of east Fife, 1963–4. *Bull. geol. Surv. Gt Br.*, No. 28, 61–101.

REFERENCES

FORSYTH, I. H. and WILSON, R. B. 1965. Recent sections in the Lower Carboniferous of the Glasgow area. *Bull. geol. Surv. Gt Br.*, No. 22, 65–79.

FRANCIS, E. H. 1968. Pyroclastic and related rocks of the Geological Survey boreholes in east Fife, 1963–4. *Bull. geol. Surv. Gt Br.*, No. 28, 121–135.

—— ALLAN, J. K. and KNOX, J. 1961. The economic geology of the Fife Coalfields, Area II. 2nd edit. *Mem. geol. Surv. Gt Br.*

GEIKIE, A. 1902. The geology of eastern Fife. *Mem. geol. Surv. Gt Br.*

GOODLET, G. A. 1957. Lithological variation in the Lower Limestone Group in the Midland Valley of Scotland. *Bull. geol. Surv. Gt Br.*, No. 12, 52–65.

—— 1959. Mid-Carboniferous sedimentation in the Midland Valley of Scotland. *Trans. Edinb. geol. Soc.*, 17, 217–240.

LANDALE, D. 1837. Report on the geology of the East of Fife Coalfield. *Trans. Highl. agric. Soc. Scotl.*, 11, (vol. 5, new series), 265–348.

MACGREGOR, M. 1930. Scottish Carboniferous stratigraphy: an introduction to the Carboniferous rocks of Scotland. *Trans. geol. Soc. Glasg.*, 18, 442–558.

MACLAREN, C. 1839. *A sketch of the geology of Fife and the Lothians.* Edinburgh: Black.

MACNAIR, P. 1917. The Hurlet Sequence in the east of Scotland and the Abden Fauna as an index to the position of the Hurlet Limestone. *Proc. R. Soc. Edinb.*, 37, 173–209.

MUIR, A., HARDIE, H. G. M., MITCHELL, R. L. and PHEMISTER, J. 1956. The limestones of Scotland. Chemical analyses and petrography. *Mem. geol. Surv. spec. Rep. Miner. Resour. Gt Br.*, 37.

ROBERTSON, T., SIMPSON, J. B. and ANDERSON, J. G. C. 1949. The limestones of Scotland. *Mem. geol. Surv. spec. Rep. Miner. Resour. Gt Br.*, 35.

TAIT, D. and WRIGHT, J. 1923. Notes on the structure, character and relationship of the Lower Carboniferous Limestones of St Monans, Fife. *Trans. Edinb. geol. Soc.*, 11, 165–184.

TRAQUAIR, R. H. 1901. Notes on the Lower Carboniferous fishes of eastern Fifeshire. *Geol. Mag.*, Dec. 4, 8, 110–114.

TULLOCH, W. and WALTON, H. S. 1958. The geology of the Midlothian Coalfield. *Mem. geol. Surv. Gt Br.*

WILSON, R. B. 1966. A study of the Neilson Shell Bed, a Scottish Lower Carboniferous marine shale. *Bull. geol. Surv. Gt Br.*, No. 24, 105–130.

WOOD, W. 1887. *The East Neuk of Fife: its history and antiquities.* 2nd edit. Edinburgh: Douglas.

WRIGHT, J. 1914. Additions to the fauna of the Lower Carboniferous Limestones of Leslie and St Monans, Fife. *Trans. Edinb. geol. Soc.*, 10, 132–147.

Chapter 5

LIMESTONE COAL GROUP

INTRODUCTION

THE Limestone Coal Group is probably the approximate equivalent in Scotland of the Pendleian (E_1) Stage of the Namurian (Currie 1954, pp. 532–535). The base of the group is drawn in Fife at the top of the Upper Kinniny Limestone. This bed is only locally present in the eastern part of the county but its horizon can be identified in most sections. The top of the group is placed at the base of the Index Limestone, but this bed has not yet been found in east Fife nor can its horizon be recognised with certainty. The thickness of the Limestone Coal Group in this district cannot therefore be accurately determined. It is probably, however, about 275 m in the Elie–Largoward area. This is rather more than is generally found in east-central Fife (Knox 1954, pp. 19, 23). It is comparable with the average thickness (see Fig. 10) in west-central Fife (Francis and others 1961, fig. 5, p. 32) but exceeded by the figure of 430 m reported in the Cowdenbeath–Lochore district (Francis and others 1961, p. 31). In the Ceres–Denhead area of east Fife, where the succession includes the Ceres Coals, the thickness appears to be rather less than it is in the Elie–Largoward area and is possibly between 240 and 260 m.
I.H.F., J.I.C.

The Limestone Coal Group is present at depth in the Largo area but has not yet been reached, except perhaps by the Geological Survey Borehole [NO 44810322] at Dumbarnie (Forsyth and Chisholm 1968, p. 68), which may have passed through a fault into the upper part of the group. The main outcrop begins on the coast at Earlsferry and Elie, where a good deal of rock is exposed but the nature of the succession is much obscured by volcanism. (The Elie Coalfield is discussed separately from the rest of the Elie–Largoward district, see below.) The outcrop extends northwards through drift-covered ground to Muircambus [NO 468024], where several boreholes have proved the presence of the group. From there it is believed to continue north-eastwards past Balcarres House [NO 474044] to include the Rires Coalfield (p. 89), and the scattered exposures, mainly of sandstone, in the Den Burn. The lower beds are preserved in the Baldutho Syncline to form a north-easterly extension of the outcrop from the Den Burn to North Baldutho [NO 496072]: a certain amount of information concerning this area is available from boreholes, from plans of the workings of Cassingray and Largobeath collieries and from surface exposures.

The main outcrop of the group continues northwards to Lathallan and Largoward, where many boreholes cut the lower beds and a few plans of workings are available, but there are scarcely any exposures. The middle and upper parts of the group are little known and the upper part is probably absent north of Balcarres House. West of Largoward the lower part of the Limestone Coal Group is very poorly exposed but the outcrop probably extends, much interrupted by basic sills, to the vicinity of Gilston Mains [NO 432072]. The basal beds survive in small isolated outcrops south of Woodside [NO 423079] and north of Bonnyton [NO 412069]. They probably also occur in an inferred outcrop in a syncline round the quartz-dolerite sill at Bruntshiels [NO 435102] but

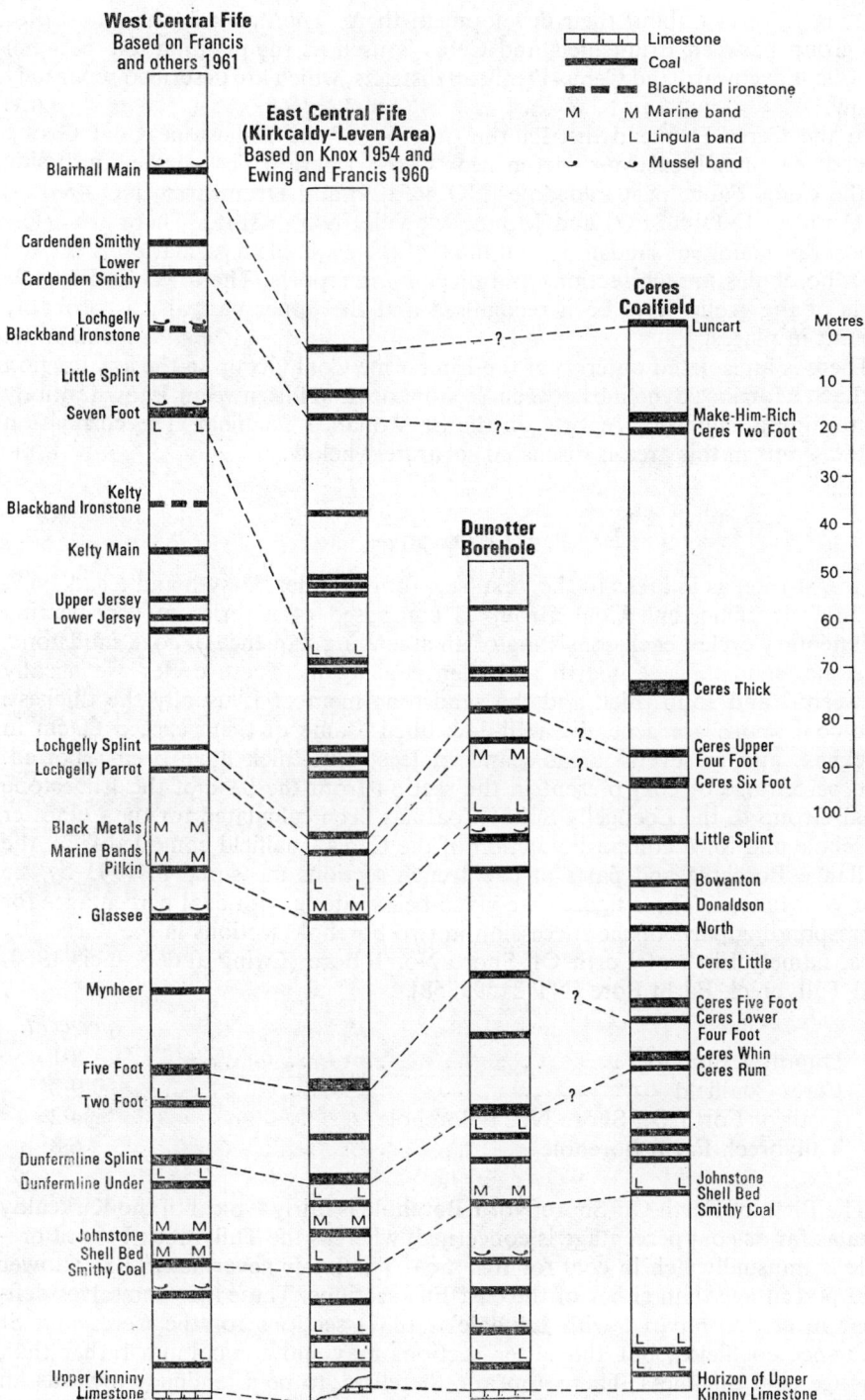

Fig. 10. *Comparative vertical sections of the Limestone Coal Group in Fife*

nothing is known about their development there. The break in the outcrop of the group between Bruntshiels and Ceres is used as the dividing line between the Elie–Largoward and Ceres–Denhead districts, which are described separately below. I.H.F.

In the Ceres–Denhead district the outcrop of the Limestone Coal Group extends east-north-eastwards from near Ceres, where it lies on the south side of the Ceres Fault, past Ladeddie [NO 443129] and Drumcarro [NO 453129] to Denhead [NO 468137] and Mount Melville [NO 483145]. There are a few exposures, mainly of sandstone, but most of the available information is derived from boreholes, trench sections and old mining reports. The lower and middle parts of the group have been recognised and the upper part also is probably present in places. J.I.C., I.H.F.

There is an isolated outcrop of the Limestone Coal Group in the axial region of the St Monance Syncline between St Monance and Pittenweem, known entirely from mining engineers' reports on the St Monance Coalfield. The correlation of the seams in this area is discussed separately below. I.H.F.

LITHOLOGY

In east Fife, as in areas to the west (see, for example, Forsyth and Read 1962, p. 34), the Limestone Coal Group is composed of a series of coal-bearing sedimentary cycles, each consisting of an ascending sequence of coal, mudstone, siltstone, sandstone, seatearth and then coal again. These cycles are usually between 3 and 15 m thick and the sandstone member is usually the thickest. The coal seams are generally well-developed; some of them exceed 60 cm in thickness and in several areas seams at least 3 m thick have been reported. The percentage of coal present in the section from the base of the Limestone Coal Group to the Lochgelly Splint Coal has been calculated for the Dunotter Borehole and for a composite section of the Ceres Coalfield compiled from the Callange Borehole and parts of two trench sections measured in 1943 by the late W. Manson. These figures are given below along with coal percentages for corresponding parts of the succession in two borehole sections in the Kirkcaldy area, namely Firth of Forth Off-Shore No. 1 Bore (Ewing and Francis 1960) and Tullybreck Farm Bore [NT 31599858].

per cent.

Dunotter Borehole	8·1
Ceres Coalfield	9·2
Firth of Forth Off-Shore No. 1 Borehole	4·2
Tullybreck Farm Borehole	6·8

The Firth of Forth Off-Shore No. 1 Borehole is fairly typical of the Kirkcaldy area as far as coal percentage is concerned, whereas the Tullybreck Farm Borehole is unusually rich in coal for this area. The latter nevertheless has a lower coal percentage than either of the east Fife sections. These are themselves deficient in coal compared with Landale's (1837) sections for the Ceres and St Monance coalfields, but the latter sections may show maximum rather than average coal thickness. There appears, therefore, to be a tendency towards an increase in the amount of coal in the Limestone Coal Group in east Fife as compared with east-central Fife.

There is little evidence of major volcanism in east Fife during the deposition of the Limestone Coal Group. The tuff and volcanic detritus in the lower part of the Geological Survey Little Pilmuir Borehole may have begun to accumulate in late Limestone Coal Group times (cf. Forsyth and Chisholm 1968; Francis 1968), but most of the activity took place during the deposition of the Upper Limestone Group. The upper part of the Lathallan Borehole (1896–1902) [NO 46440579] consists largely of 'ash' in several bands totalling some 55 m. If this material is interbedded tuff or volcanic detritus, a considerable outburst of volcanic activity is indicated at about the time of deposition of the strata between the Dunfermline Splint Coal and the Black Metals. The lack of 'ash' in adjacent bores, however, suggests that the Lathallan Borehole began in a neck with steeply inclined sides, and passed downwards out of it into a sedimentary sequence with several 'ash' bands, possibly intrusive tuffs. Bands of this type recorded in the basal beds of the group in several other bores near Largoward may have a similar origin.

Graded tuffaceous siltstones were recorded from two horizons in the Limestone Coal Group in the Callange Borehole, one 1·8 m below, and the other 27·5 m above, the Dunfermline Splint Coal. The higher of these may be the equivalent of a 3·7 m bed of tuff with accretionary lapilli in the cycle below the Glassee Coal in the Dunotter Borehole (Francis 1968, p. 125). Dr. Francis also noted, probably from horizons in the upper part of the group, several bands of tuff and tuffaceous sediments in the coast section immediately west of the Kincraig neck. A 60-cm band of coarse basaltic tuff seen in a ditch 550 m north-north-west of Gibliston [NO 494055] probably lies near the base of the group.

I.H.F., J.I.C.

Stratigraphy

Landale (1837) recorded sections for most of the coalfields in east Fife, many of which are now known to lie in the Limestone Coal Group. Geikie's brief account (1902, pp. 160–178) is based largely on Landale's report because little additional information had become available during the intervening period. Geikie was forced to treat each local succession separately because no overall stratigraphy had been established, and in fact he included descriptions of strata now known to lie in the Lower Limestone Group. The Geological Survey borehole [NO 46500250] put down in 1964 at Dunotter near Elie (Forsyth and Chisholm 1968, pp. 68–70) provided a section through the middle and lower parts of the group and showed them (Fig. 11) to be generally similar to the corresponding strata in central Fife (cf. Knox 1954; Francis and others 1961). Other Geological Survey boreholes, at Callange near Ceres (Forsyth and Chisholm 1968, p. 70) and at Craigtoun Park near St Andrews (p. 100), provided short sections, mainly in the lower parts of the group; and boreholes put down by the Opencast Executive of the National Coal Board proved the existence of strata belonging to the group at Callange, Ladeddie, Drumcarro and Largoward. The upper part of the group remains largely unknown, but there is no reason to doubt that the similarity noted in the lower part extends to this part also, except perhaps locally where volcanic rocks may be present as, for example, around Little Pilmuir near Largo (Forsyth and Chisholm 1968, p. 66).

Three marine bands are known in the Limestone Coal Group in east Fife.

The lowest is at the base of the group, in the mudstone above the Upper Kinniny Limestone, and the other two are the lower leaf of the Johnstone Shell Bed and the upper of the two marine bands in the Black Metals (cf. Read 1965, pp. 76–77). The last-mentioned band has been found only in the Dunotter Borehole. None has yielded a rich fauna and the Johnstone Shell Bed is usually represented by a *Lingula* band. *Lingula* bands have been recognised at four other horizons in the group (Forsyth and Chisholm 1968, p. 69), two close together near the base, one above the coal taken to be the Dunfermline Under Coal and the fourth at the horizon of the lower marine band in the Black Metals. Reference to Fig. 10 shows that there is no significant decrease in the number of marine and *Lingula* bands in east Fife as compared with central Fife. Records of non-marine bivalves are few in number. *Naiadites* was found in the Dunotter Borehole above the Glassee and Largoward Thick coals, and on the coast in a detached exposure east of Kincraig. Septate trace-fossils (including forms resembling *Diplocraterion*) were found in the Dunotter and Callange boreholes at various horizons in the Limestone Coal Group, and in many cases were associated with *Lingula* bands (Forsyth and Chisholm 1968, plate iv; Chisholm 1968). I.H.F., J.I.C.

St Monance Coalfield

Geikie (1902, p. 161) quoted a skeleton section of the strata in the St Monance Coalfield from a report by Landale dated 1854 (i.e. over 50 years after mining ceased). As no further information has since become available and there are no surface exposures, the accuracy of the section cannot be verified. It is reproduced in Table 4, together with a tentative correlation with central Fife. This suggests that most of the Limestone Coal Group is represented at St Monance in a coal-rich sequence about 120 m thick. The possibility that some of the seams in fact belong to the Lower Limestone Group cannot be entirely excluded, but the close spacing, especially of the lower seams, suggests that they all belong to the Limestone Coal Group. The workings are at least 300 years old and were abandoned about 1800. Considerable reserves, extending to some 13 million tonnes of coal (cf. Geikie 1902, p. 161) have been estimated to remain, but not much is known about the extent of the old workings and this estimate could be very inaccurate. I.H.F.

Elie Coalfield

Even less is known about the Elie Coalfield than about the St Monance Coalfield. Geikie (1902, p. 163) gave a brief description of the area and quoted a partial section of the strata, obtained locally a long time after mining had ceased; this section is reproduced in Table 5. Earlier, Landale (1837, map) had listed 17 seams totalling some 18 m of coal, which he termed the Earl's Ferry seams (see Table 5). There are points of similarity between the two sections, but as neither provides a complete record of the nature and thickness of the strata between the seams the sections cannot be correlated in detail either with each other or with the Dunotter Borehole, sited about three kilometres distant. Estimated positions of the outcrops of some of the seams on the shore were, however, indicated on the original Geological Survey six-inch field-map

TABLE 4

Section of strata in St Monance Coalfield

St Monance name	Thickness m	Thickness cm	Possible correlative
SPLINT COAL		46	BLAIRHALL MAIN COAL
Strata	7·0		
MY LORD'S COAL (3 leaves)		201	CARDENDEN SMITHY COAL
Strata	20·2		
FOULHOUSE COAL (2 leaves)		152	SEVEN FOOT COAL
Strata	17·2		
PARROT COAL (parrot, splint and cherry)		295	JERSEY and SWALLOWDRUM COALS
Strata	1·5		
BACK COAL (splint and cherry)		122	LOCHGELLY SPLINT COAL
Shale	1·8		
FORE COAL (splint)		152	LOCHGELLY PARROT COAL
Strata	17·1		
MID COAL (soft cherry)		99	GLASSEE COAL
Strata	11·2		
WANDERER COAL		86	MYNHEER COAL
Strata	2·9		
FIRST of FOUR COALS (splint and cherry)		188	FIVE FOOT COAL
Strata	0·6		
SECOND of FOUR COALS (splint)		178	TWO FOOT COAL
Strata	1·27		
THIRD of FOUR COALS (splint)		127	DUNFERMLINE SPLINT COAL
Strata	1·8		
FOURTH of FOUR COALS (splint)		91	DUNFERMLINE UNDER COAL
Strata (chiefly sandstone)	11·1		
THIRTEENTH COAL		46	SMITHY COAL
Strata	10·1		
FOUR FEET COAL (household)		107	LARGOWARD THICK COAL

of the Earlsferry and Elie area, and Landale showed outcrops for all 17 seams on his map. The positions of the outcrops of the Johnstone Shell Bed and the Upper Kinniny Limestone on the coast section are now known and this allows an attempt to be made to identify the seams in Landale's section (see Table 5) in terms of the names used elsewhere in Fife. Indeed this can be done reasonably satisfactorily. Geikie's section does not fit so well and the order of the seams does not tally exactly with that indicated on the original geological field-map. Comparison of the latter with the six-inch field-map and measured section compiled recently by Dr. E. H. Francis shows that the horizon of the Elie Salt

TABLE 5
Correlation of seams in Elie Coalfield

Landale's (1837) section	Thickness cm	Geikie's (1902) section	Thickness cm	Possible Correlatives
COAL	240			CARDENDEN SMITHY COAL
THICK COAL	340	TOP COAL	340	SEVEN FOOT COAL
		Strata	550	
COAL	150	UNDER COAL	120	KELTY MAIN COAL
HARD SPLINT COAL	137	THREE FEET COAL	90	JERSEY COALS
COAL	190	CHERRY COAL	90	LOCHGELLY SPLINT COAL
COAL	79			LOCHGELLY PARROT COAL
COAL	51			PILKIN COAL
COAL	59			GLASSEE COAL
COAL	71	CHERRY AND SPLINT COAL	150	FIVE FOOT COAL
COAL	109	TWO FEET COAL	60	TWO FOOT COAL
COAL	137	MAIN COAL	150	DUNFERMLINE SPLINT COAL
COAL	46			DUNFERMLINE UNDER COAL
SALT COAL	84	SALT COAL	90	Unnamed coal
COAL	41			SMITHY COAL
COAL	69			LARGOWARD THICK COAL
COAL	69			LARGOWARD SPLINT COAL
COAL	38			LARGOWARD BLACK COAL

Coal lies about 3·7 m above the Johnstone Shell Bed. This seam is therefore taken to be the same as the unnamed 97-cm seam at 207·4 m in the Dunotter Borehole (Forsyth and Chisholm 1968, p. 69). The suggested correlation of the rest of the section in Table 5 is based on this equivalence. The thickness which Geikie (1902, p. 163) quotes for the strata between this seam and his Elie Main Coal is now regarded as probably an error.   I.H.F.

Rires Coalfield

The small basin which forms the long-abandoned Rires Coalfield is situated west of Sprattyhall [NO 471048]. Landale (1837, pp. 301–303) gave the following section for this small coalfield, which he said takes the form of a trough, not more than 720 m wide, closing towards the south-west:

Coal	90 cm
Strata	26·5 m
Coal	200 cm
Strata	7·3 m
Coal	50 cm
Strata	1·8 m
Main Coal, with 35 cm of black shale	170 cm

The four coal seams probably lie in the middle or upper part of the Limestone Coal Group. The original six-inch Geological Survey field-map shows several pits, including one known as Keddie's Pit [NO 46780486] in which the Main Coal lies at a depth of 44 m. The only available plan of the workings from this pit shows only a limited area of extraction and is probably incomplete.   I.H.F.

Elie–Largoward District

Details

Strata below Largoward Thick Coal. The strata between the Upper Kinniny Limestone and the Largoward Thick Coal vary in thickness from 18 to 40 m, being thickest in the vicinity of Lathallan [NO 461062]. These beds generally contain four, or locally five, coal horizons, which are occupied by rather variable seams usually from 10 to 30 cm in thickness. Locally however, thicker coals occur: the top seam for example is about 60 cm thick around a shaft [NO 45850678] in the Lathallan district, from which it was worked to some extent as John Barnet's Coal. The roof mudstone of the Upper Kinniny Limestone rarely exceeds 1·8 m in thickness and contains a sparse marine fauna. Two persistent *Lingula* bands are present, in the roofs of the coals which are usually the first and third in upward succession: the upper one was found by Dr. E. H. Francis on the shore at Earlsferry. Ewing and Francis (1960, p. 17) recorded two such bands near the bottom of Firth of Forth No. 1 Off-shore Borehole and pointed out that it had been customary to place the lower one at the horizon of the Upper Kinniny Limestone, there being then no section showing both *Lingula* bands and the limestone in east-central Fife. Such sections are now available in east Fife, for example in the Dunotter Borehole (see Fig. 11), and they suggest that locally in east-central Fife the Upper Kinniny Limestone is not represented even by a *Lingula* band. Francis and others (1961, fig. 3, p. 32) showed only one *Lingula* band in this part

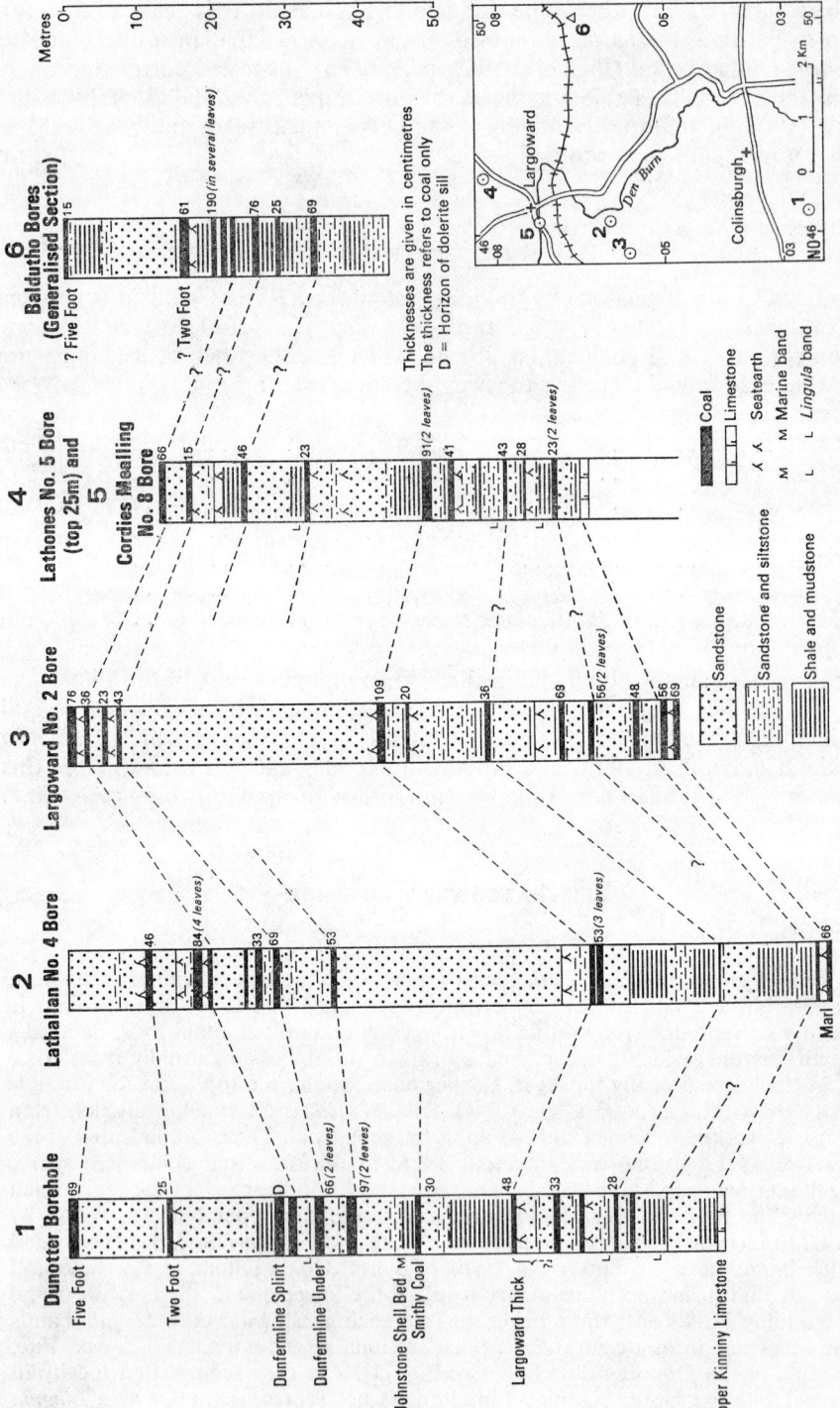

FIG. 11. *Comparative vertical sections in the lower part of the Limestone Coal Group in the Elie–Largoward area*

of the succession in west-central Fife: in the Fordell–Cowdenbeath district it is clearly the lower of the two bands known in east Fife. In the extreme west of the county, however, at Blair Mains No. 2 Borehole, $1\frac{1}{2}$ km west of Culross, Francis (1956, p. 5) recorded two *Lingula* bands near the base of the Limestone Coal Group, and also, between these bands and the Upper Kinniny Limestone, a band with *Lingula* and bivalve fragments. Trace-fossils of types usually associated with marine or quasi-marine conditions were found at several horizons in this part of the sequence in the Dunotter Borehole (Forsyth and Chisholm 1968, plate iv; Chisholm 1968).

Surface exposures are almost confined to the Elie–Earlsferry shore—where the section, measured by Dr. E. H. Francis, is similar to the equivalent part of the Dunotter Borehole—and to the Den Burn, where some at least of the scattered exposures between the Sawmill and Balmaken bridges [NO 47710488–48440459] belong to the same part of the group. They consist mainly of sandstone, including one bed at least 6 m thick, but there are two coal seams, 41 cm and 46 cm thick respectively. The exposures are too scattered to allow these seams to be identified, and they may lie higher in the succession.

Largoward Thick Coal. The Largoward Thick Coal is a very variable seam, usually in leaves, which was worked from several pits around Largoward and Lathallan. It rarely reaches the thickness of almost 4 m attributed to it by Landale (1837, p. 304) but records of 2 to 2·5 m of coal are known, e.g. the '14-fathom' Pit at Largoward, which may be taken as the type locality, showed 236 cm in two leaves. On the other hand, even in the Largoward and Lathallan districts, it is locally less than 60 cm thick. Several of the boreholes put down by the Opencast Coal Executive near Cordies Mealling [NO 462073] showed that the top leaf of the seam is overlain by a thick bed of siltstone and mudstone with ironstone bands, which is taken to be the equivalent of the bed described by Francis and others (1961, p. 41) as being 'persistent as a lithological unit' and 'a useful marker horizon throughout Fife'. No other bed of this type and thickness is known in this part of the sequence in east Fife. In the Dunotter Borehole this bed, which was found to contain numerous trace-fossils (Forsyth and Chisholm 1968, plate iv; Chisholm 1968), is at the same horizon as it is elsewhere in Fife, i.e. a short distance below the Johnstone Shell Bed. The horizon of the top leaf of the coal may therefore be regarded as established but it is possible that some of the leaves locally grouped as the Largoward Thick Coal belong to the cycle below.

Smithy Coal. The strata between the Largoward Thick Coal and the Smithy Coal (Sulphur Coal of Knox 1954—cf. Francis 1961, fig. 5, p. 32) are between 10·7 and 13·7 m thick on the shore at Earlsferry, and about 12 m thick in the Dunotter Borehole. Locally in the Largoward district the thickness increases to between 16·7 and 18·3 m, and a bed of siltstone with a thin seatearth below it appears in the middle of the sandstone which forms most of this sequence. The Smithy Coal was found to be 30 cm thick at Dunotter. It is usually thinner in the Largoward district. The roof mudstone contains the lower leaf of the Johnstone Shell Bed (the only part known in east Fife): it has usually yielded only *Lingula*, but in the Dunotter Borehole *Sanguinolites costellatus* and *Streblopteria ornata* (Etheridge jun.) were also obtained and the Earlsferry shore section yielded *S. ornata* and *Myalina*. Trace-fossils were found to be abundant at this horizon in the Dunotter Borehole (cf. Forsyth and Chisholm 1968, plate iv; Chisholm 1968).

In the Lathallan district the Smithy Coal and the associated strata appear to be highly variable, but none of the relevant boreholes was examined by a geologist and their records may not be reliable. The positions of the faunal bands are not known, and the correlation of the seams is doubtful. At the Cairn Pit [NO 45800635] a coal seam 59 to 79 cm thick lies only 3·7 m above the seam worked there as the Largoward Thick Coal, and in another pit [NO 45770678], 400 m to the north, a 71-cm seam was worked as the Upper or Two Foot Coal: it lies about 9 m above a waste taken to be that of the Largoward Thick Coal. Both seams are provisionally regarded as the Smithy Coal. Locally the Smithy Coal appears to be cut out by a sandstone which at its maximum exceeds 30 m in thickness.

The cycle overlying the Smithy Coal is usually about 7·6 m thick: its uppermost member is a rather variable coal seam, which may be the Elie Salt Coal (pp. 87–89). This seam is 97 cm thick in the Dunotter Borehole, where it has been invaded by tuffisite. It is provisionally correlated with the one worked as the Six Feet Coal at the Cairn Pit, Lathallan: other records in that district show a coal at this horizon varying in thickness from 53 to 101 cm, the latter figure being the total of five separate leaves of coal. The 89-cm coal in two leaves recorded at 11 m in Balcarres Shaft [NO 47990667] may be this seam. Its development in the Largoward district is obscured by the occurrence of bodies of tuffisite, but it is 46 cm thick in Lathones No. 5 Borehole (1954) [NO 47240832]. The Largoward Parrot Coal, the top seam in Landale's (1837, p. 304) section of the Largoward Coalfield, may be at this horizon.

Several bores were put down in 1902 around North Baldutho [NO 498072] and are here referred to as 'the Baldutho boreholes' (cf. Fig. 11). Some of them encountered a close-set group of thick coals. The structure of the area is inadequately known and certainly complex, but the most likely place for this group of coals appears to be near the position of the Dunfermline Splint Coal (see below). If this is correct, the seam under discussion here is represented in four of the bores by a coal between 59 and 79 cm thick.

Dunfermline Under Coal. In the Dunotter Borehole the Dunfermline Under Coal is 66 cm thick in two leaves: it lies about $4\frac{1}{2}$ m above the underlying coal and has *Lingula* in its roof. The lack of certainty about the correlation of the strata above the Largoward Thick Coal in the Lathallan district applies to this seam also: it is tentatively identified as a coal between 60 and 90 cm thick, generally in two leaves, which was worked at the Cairn Pit as the Parrot Coal. Several bores around Cordies Mealling, on the other hand, recorded less than 30 cm of coal at what appears to be this horizon. The seam tentatively identified as the Dunfermline Under Coal in the Baldutho boreholes is, however, 71 to 81 cm thick.

Dunfermline Splint Coal. In the Dunotter Borehole the Dunfermline Splint Coal is spoiled by a dolerite sill, but it clearly has been a thick seam (150 to 180 cm); it is in several leaves, the upper two of which were probably reached by Muircambus No. 7 Borehole [NO 46940223] where they are 25 cm and 76 cm thick respectively. It may be the equivalent of the Elie Main Coal (see p. 88), which is not now visible on the shore at Earlsferry. The only two bores in the Lathallan district which appear to have begun high enough in the succession to cut the Dunfermline Splint Coal are Largoward No. 2 Borehole (1897) [NO 45700564], in which it is 112 cm thick in two leaves, and Lathallan No. 4 Borehole (1895) [NO 46380593] where it is in six leaves totalling 107 cm of coal. There are a few records of a coal up to 66 cm thick at the level of this seam in the Largoward district, but this may not represent the full thickness of the seam there. A 104-cm seam of inferior coal encountered in a mine sunk immediately north of Balcarres Shaft may be the Dunfermline Splint Coal. The thickest seam in the Baldutho boreholes is provisionally equated with the latter coal: it is recorded as being between 120 and 210 cm in several leaves.

Two Foot Coal. A 25-cm seam at 166·4 m in the Dunotter Borehole has been taken to be the Two Foot Coal (Forsyth and Chisholm 1968, p. 69). It lies about 15 m above the Dunfermline Splint Coal, the intervening strata being almost all sandstone. This sandstone may, however, be a local development: at Muircambus No. 7 Borehole, 550 m away, 4·5 m of seatearth separate the two seams, the supposed Two Foot Coal being 20 cm thick. In the Lathallan district only Lathallan No. 4 Borehole began high enough in the sequence to cut the Two Foot Coal, which is believed to be represented by a 46-cm seam. In the Baldutho boreholes, the coal regarded as the Two Foot Coal is, however, usually between 56 and 74 cm thick and in one bore reaches 145 cm. No evidence of the *Lingula* band which occurs above this seam in west and central Fife has been found.

Five Foot Coal. The 63-cm seam at 153·54 m in the Dunotter Borehole, which is separated from the Two Foot Coal by nearly 13 m of strata consisting mainly of sandstone, has been equated (Forsyth and Chisholm 1968, p. 69) with the Five Foot Coal. The seam in this horizon at Muircambus No. 7 Borehole was recorded as being only 33 cm thick and no coal at all was recorded near the top of Lathallan No. 4 Borehole, where the Five Foot Coal should have occurred if the rather tentative correlation of this section is correct. The top 24 m of strata in this bore are mainly sandstone, which may belong to the same bed as the thick sandstone known from several boreholes to occur south of East Cassingray [NO 491070] in the axial region of the Baldutho Syncline. This sandstone is believed to cut out the Five Foot Coal in that vicinity. Farther east Baldutho No. 21 Borehole [NO 49960692] was probably the only one of the Baldutho boreholes to start high enough in the sequence to cut this seam, which proved to be only 15 cm thick. On the other hand, the Five Foot Coal is tentatively equated (see pp. 87, 88) with a 71-cm seam at Elie and one of 188 cm at St Monance. The available evidence, if correctly recorded and interpreted, suggests therefore that in east Fife the Five Foot Coal is a rather variable seam.

Five Foot Coal to Lochgelly Splint Coal. The only completely satisfactory record of the strata from the Five Foot Coal up to and including the Lochgelly Splint Coal in the Elie–Largoward district is provided by the Dunotter Borehole (see Fig. 10), where they are some 55 m thick. In this section the lower of the Black Metals marine bands is represented by a *Lingula* band, and the upper by 30 cm of bioturbated sandstone with marine shell debris. The Lochgelly Parrot Coal is 81 cm thick and the Lochgelly Splint Coal 97 cm: both occur in two leaves. The nearby Muircambus No. 7 Borehole provides a similar record but both the coals and the other strata are somewhat thinner. Several of the coals have been tentatively equated with seams in the Elie Coalfield (see Table 5), but these correlations must be regarded as provisional pending the availability of new sections through these measures.

Lochgelly Splint Coal to top of the Limestone Coal Group. The sections in the Dunotter and Muircambus No. 7 boreholes are similar and both extend about 30 m above the Lochgelly Splint Coal, but doubt exists about the identification of the seams (see Forsyth and Chisholm 1968, p. 69). The upper seams in the Elie Coalfield have been placed in this part of the succession, and very tentatively equated with individual seams therein. Some of the Rires coals may also be in the upper part of the Limestone Coal Group. Landale (1837, p. 344) recorded the finding of three coal seams at Balchrystie [NO 460031], lying irregularly and distorted: they probably lie near the top of the group. The lowest strata encountered in the Geological Survey Dumbarnie Borehole [NO 44810322] (see Forsyth and Chisholm 1968, p. 68) may also belong to the upper part of the group: after passing through a fault, the borehole cut about 15 m of sandstone, siltstone and seatearth with two coal seams about 3 m apart. The upper seam, in three leaves, is 84 cm thick and the lower 104 cm. Below the lower seam the bore proved fully 6 m of sandstone without reaching its base. The disturbed strata exposed on the shore on the east side of Shell Bay, west of the Kincraig neck, probably lie in the upper part of the Limestone Coal Group. The visible beds consist of sandstone, mudstone, siltstone and seatearth in thin beds. Some of the siltstones and seatearths were described by Dr. E. H. Francis as being tuffaceous. Two coals, 15 cm and 30 cm thick respectively, were also noted. Geikie (1902, p. 180) recorded a marine band at this locality but it is now thought that the strata concerned are probably not in place. I.H.F.

Ceres–Denhead District

A belt of complex structure associated with the Ceres and Maiden Rock faults runs in an east-north-easterly direction from near Ceres to the coast near St Andrews. In folds and faulted blocks within this belt lie several small areas of coal-bearing strata now assigned mainly to the Limestone Coal Group. The largest is the Ceres Coalfield; the others, grouped together in the following account as 'Small Coalfields', are named, from south-west to north-east, the Ladeddie Backfields, Ladeddie Frontfields, Drumcarro West, Drumcarro East, Denork–Denhead, and Mount Melville coalfields. None of the Small Coalfields exceeds 0·65 square kilometres in area. Surface exposures are few and the nature of the sequence is known mainly from boreholes, trench sections, old mining reports and old maps.

J.I.C., I.H.F.

CERES COALFIELD

Landale (1837, pp. 318–319) drew up a section of the strata in the Ceres Coalfield (Fig. 12) which was avowedly based on reports which even then were old, because a long time had already elapsed since mining had ceased. He found that 'there are few who know anything about' the Ceres Coals, but none the less he produced a section which compares on the whole very well with those obtained (Fig. 12) from trenches dug in 1943 by the Directorate of Opencast Coal Production, and from the Callange Borehole (1963) [NO 41711191]. These have provided the only other available information about the Ceres Coals.

The Ceres Coals outcrop in an elongate wedge-shaped area which begins at the eastern end of Ceres village and extends, widening eastwards, for about 2 km to Kinninmonth [NO 424124], where the outcrop is about 400 m across. The strata dip north-north-westwards at high angles, increasing to almost vertical near the Ceres Fault, which cuts off the field to the north. Several south-south-easterly-trending faults cut the area and the strata have been invaded by volcanic necks and intrusions of dolerite. The latter are so extensive east of Kinninmonth that they effectively limit the Ceres Coalfield in that direction. The few surface exposures have been distorted by slipping and subsidence caused by the collapse of old workings. Neves (*in* Forsyth and Chisholm 1968, p. 70) established, by means of the spore content of the seams in the Callange Borehole, that the Ceres Coals belong to the lower part of the Namurian and the general nature of the sequence clearly indicates that most if not all of them belong to the Limestone Coal Group. The horizons of the Upper Kinniny Limestone and the Johnstone Shell Bed were identified in the bore (see Fig. 12) and some of the lower coals in Landale's section were recognised.

Details

Strata below Smithy Coal. The only complete section through the beds below the Smithy Coal is provided by the Callange Borehole, in which their thickness is fully 40 m. This is slightly more than the figure for the corresponding strata in the Dunotter Borehole. The lowest 12 m at Callange are very similar to the equivalent beds at Dunotter: they contain two coals, each with a *Lingula* band and trace-fossils in its roof. The coals are 33 cm and 61 cm thick respectively: the upper one may be Landale's

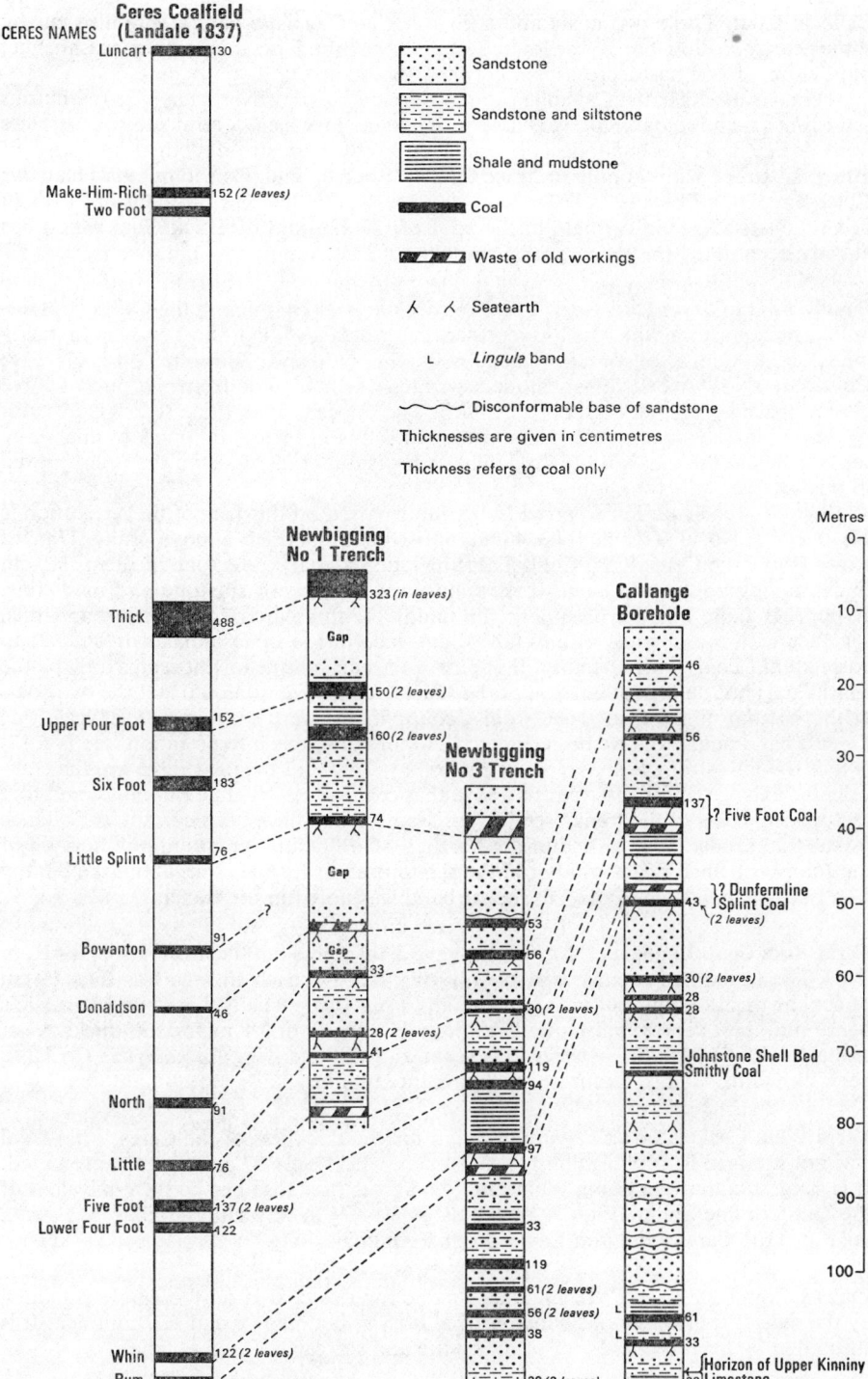

Fig. 12. *Comparative vertical sections in the Limestone Coal Group in the Ceres area*

Ballfield Coal. These two coals and their associated *Lingula* bands were also cut by some nearby shallow bores; the lower seam is about the same thickness as at Callange, but the upper one has decreased to between 30 and 40 cm.

These basal beds in the Callange Borehole are overlain by over 20 m of garnetiferous sandstone, mainly coarse or very coarse in grain, in which several erosion surfaces were detected. This sandstone is overlain by sandy seatearth and then by a bed of sandy siltstone with abundant trace-fossils (Forsyth and Chisholm 1968, plate iv; Chisholm 1968). This bed is believed to be a sandy, attenuated equivalent of the thick bed of siltstone and silty shale in the roof of the Largoward Thick Coal which has already been noted (p. 91) to persist throughout Fife.

Smithy Coal to Ceres Rum Coal. The Smithy Coal is 23 cm thick in the Callange Borehole. The Johnstone Shell Bed is represented by a *Lingula* band in the overlying silty mudstone, which also contains trace-fossils (Forsyth and Chisholm 1968, plate iv; Chisholm 1968). A bed of sandstone separates this mudstone from a group of three closely-spaced coals, all about 30 cm thick. The uppermost of these seams is overlain by 8 m of mainly coarse-grained sandstone and this in turn is followed by the 1·5-m seatbed below the Ceres Rum Coal. The latter is about 23 m above the Smithy Coal in this section.

The only trench section to provide any information on this part of the succession is Newbigging No. 3 Trench [NO 40951163–41021152], which exposed fully 27 m of strata below the Ceres Rum Coal. These include six coal seams (one of them, 119 cm thick, having apparently been worked) and several beds of siltstone and mudstone. A possible fault was detected near the middle of this part of the section and this, together with the absence of any faunal evidence in the trench, makes it difficult to correlate the coals with certainty. It is possible that the three lowest seams may be the Marl Coal (just below the horizon of the Upper Kinniny Limestone) and the two coals at the bottom of the Limestone Coal Group; if this were so, however, at least one *Lingula* band ought to have been detected. A more probable interpretation (see Fig. 12) is that the lowest of the six seams is the Smithy Coal and the next three are the equivalents of the group of three seams above that coal in the Callange Borehole; and that the two top coals of the trench section are seams which have been cut out at Callange by the 8-m sandstone recorded in the borehole. On this interpretation, the horizon of the Johnstone Shell Bed should be in the siltstone and silty mudstone above the Smithy Coal, but, as indicated above, no faunal band was noted in the trench.

Ceres Rum Coal. Landale (1837, p. 318) gave a thickness of 90 cm for the Ceres Rum Coal. The Callange Borehole and Newbigging No. 3 Trench showed less than 60 cm of coal at this horizon, but in both cases the seam was partially mined and Landale's figure may best represent its true thickness. His figure of 1·8 m for the thickness of strata (mainly seatearth) between this seam and the overlying Ceres Whin Coal has been shown to be fairly accurate by the recent sections.

Ceres Whin Coal. Landale's figure of 1·2 m for the thickness of the Ceres Whin Coal was not attained in Newbigging No. 3 Trench, where only 97 cm of coal were noted. It is suggested that the Ceres Whin and Rum coals may together be the equivalent of the Dunfermline Splint Coal. A thickness of 6 to $7\frac{1}{2}$ m of mainly argillaceous strata separates the Ceres Rum and Lower Four Foot coals.

Ceres Lower Four Foot Coal. Landale's name for this coal is on the whole validated by the several trench sections which proved it to be between 90 and 120 cm thick. It is succeeded by from 0·6 to 3·7 m of mudstone and seatearth.

Ceres Five Foot Coal. Several sections of the Ceres Five Foot Coal have shown it to be between 120 and 150 cm thick. Together with the Ceres Lower Four Foot Coal, this seam is perhaps the equivalent of the Five Foot Coal of the rest of Fife.

Ceres Five Foot Coal to Little Splint Coal. According to Landale (see Fig. 12), the Little Coal, 76 cm thick, lies 9 m above the Ceres Five Foot Coal. In the borehole and trench sections the latter seam is overlain by $4\frac{1}{2}$ to $7\frac{1}{2}$ m of mainly sandy strata, followed by a coal in two leaves separated by 1 to 3 m of seatearth. The lower leaf is 23 to 56 cm thick and the upper is up to 30 cm. This split seam is presumed to be the Little Coal despite the discrepancies between its development in the recent sections and Landale's description. The next higher seam of Landale's section is the North Coal, 90 cm thick and 7·3 m above the Little Coal. In the borehole and trench sections, the most likely correlative of the North Coal is a seam 30 to 56 cm thick which lies $5\frac{1}{2}$ to $7\frac{1}{2}$ m above the upper leaf of the presumed Little Coal. The intervening strata, which are mainly sandy, in places include a poor coal up to 30 cm thick, but Landale makes no reference to it.

In Landale's section the Donaldson Coal, 46 cm thick, is separated from the North Coal below by 12·8 m of hard sandstone. The sandstone above the supposed North Coal in the Callange Borehole is the highest bed encountered in that section; in the trench sections this sandstone forms most of the $4\frac{1}{2}$ to $7\frac{1}{2}$ m of strata between the supposed North Coal and a seam consisting of about 60 cm of coal, in two leaves close together, which must be taken as the Donaldson Coal. The latter is said by Landale to be separated from the next higher seam, the 90-cm Bowanton Coal, by 7·3 m of hard stone. Two trench sections (Nos. 2 and 3) cut this part of the sequence. No. 2 Trench cut 9 m of sandstone, siltstone and seatearth, with a coal 46 cm thick which is taken to be the Bowanton Coal. The coal is not present in No. 3 Trench, probably being cut out by the coarse sandstone that forms the bulk of the 14 m of strata between the supposed Donaldson Coal and the waste of the Little Splint.

Little Splint Coal. Only Newbigging No. 1 Trench [NO 40701165–40791158] cut the Little Splint Coal, whose thickness of 74 cm approximates closely to Landale's figure of 76 cm. A sequence of silty sandstone and sandy siltstone, 12 m thick, was encountered in the trench between this seam and the Ceres Six Foot Coal. No trace was discovered of either of the Black Metals marine bands, which probably lie in this part of the sequence.

Ceres Six Foot Coal. A coal 160 cm thick in two leaves was found in Newbigging No. 1 Trench at the level of the Ceres Six Foot Coal, which is overlain in turn by 3 m of mudstone with ironstone nodules, followed by 3 m of silty sandstone and by the Ceres Upper Four Foot Coal. The close spacing of these two seams suggests a correlation with the Lochgelly Splint and Parrot Coals of central Fife.

Ceres Upper Four Foot Coal. Despite its name, Landale gave the thickness of this coal as 152 cm (5 ft), and in Newbigging No. 1 Trench it is only 2·5 cm short of that figure.

Ceres Thick Coal. Landale regarded the Ceres Thick Coal as a variable seam averaging 500 cm in thickness. This variability is illustrated by Newbigging Nos. 1 and 2 trenches, which recorded respectively 323 cm of coal in five leaves and 241 cm of coal in four leaves. Landale's average figure therefore appears to be an optimistic estimate, but it may be attained locally by this seam, which is probably formed by the coming together of several central Fife seams including, possibly, the Swallowdrum, Jersey and Kelty Main coals. The Ceres Thick Coal is at the top of the section in Newbigging No. 1 Trench: in No. 2 it is seen to be overlain by over 15 m of strata, mainly sandstone.

Ceres Two Foot Coal to Luncart Coal. Landale's section (see Fig. 12) still provides the only information concerning the top part of the Ceres succession, which probably lies in the upper part of the Limestone Coal Group, but could possibly include strata belonging to the lower part of the Upper Limestone Group. The Ceres Two Foot and Make-Him-Rich coals may be the equivalents of the Seven Foot and Little Splint coals of central Fife, and the Luncart Coal that of the Cardenden Smithy Coal. I.H.F.

SMALL COALFIELDS

A small part of the Limestone Coal Group sequence is present in each of the Small Coalfields but correlation between these coalfields and the sequences at Ceres (Fig. 12) and Dunotter (Fig. 10) is hindered by lack of recent detailed information. It has proved possible, however, to assign the beds present in most of the Small Coalfields to a stratigraphical position either below or above the supposed horizon of the Black Metals marine bands (Fig. 10).

Details

Strata below Black Metals. An unpublished report (1900) by the mining engineer John Gemmel on the minerals at Ladeddie states that the following seams of coal were proved about 1860 at Ladeddie Frontfields in a group of pits [NO 44451270] situated about 275 m south-east of Ladeddie Farm:

	cm	
Clay ironstone	6	
Coal	53	not worked
Strata	370	
Coal	46	not worked
Strata	120	
Coal	76	not worked
Black stone	5	
Wild parrot or rums	20	not worked
Fireclay	5	
Strata, thickness unknown but not great		
Coal	51 ⎫	
Parrot coal	11 ⎪	worked as
Coal	33 ⎬	'Six Foot' Coal
Strata	15 ⎪	
Coal	36 ⎭	
Strata, thickness unknown (now estimated at about 60 m)		
Splint coal	13 ⎫	
Gas coal ('parrot')	13–15 ⎪	worked as
Black fireclay	5–13 ⎬	'Parrot' Coal
Coal	30 ⎭	

A series of boreholes (Drumcarro Nos. 1–5), drilled by the Opencast Executive of the National Coal Board in 1953, proved wastes corresponding to the two worked coals of this sequence but failed to establish the spacing of the seams. The upper three seams of Gemmel's sequence were not proved satisfactorily. No. 3 Borehole [NO 44461256] did however encounter a pair of coals beneath the waste of the 'Parrot' Coal, and comparison of the full record of this bore with the Callange succession suggests that a 7-cm bioturbated band of sandy siltstone in the roof of the lowest of these seams may represent the horizon of the Upper Kinniny Limestone. The 'Parrot' Coal probably lies low in the Limestone Coal Group therefore, its most likely correlative being a

61-cm coal at 96·80 m in the Callange Borehole. A *Lingula* band in Drumcarro No. 2 Borehole [NO 44461263] probably lies between the 'Parrot' and 'Six Foot' coals; it may represent either the Johnstone Shell Bed or the *Lingula* band associated with the Dunfermline Under Coal. In either case the 'Six Foot' Coal and the unworked seams above it must lie in the region of the Five Foot, Lower Four Foot, Whin and Rum coals of Ceres, or the Five Foot and Dunfermline Splint coals of central Fife.

Coal seams probably lying in the same part of the sequence as the 'Six Foot' Coal of Ladeddie Frontfields were stated in Gemmel's 1900 report to have been worked about 1850 from a group of shafts [NO 45801290] about 460 m east of Drumcarro Farm. These workings are named for convenience the Drumcarro East Colliery in this account. The seams were described as a 120-cm coal overlying a 180-cm coal; no other details were given. Boreholes drilled by the Opencast Executive of the National Coal Board in 1953 (Ladeddie Nos. 6 and 7 boreholes) provided a section through these strata. No. 6 Borehole [NO 45751293] proved 188 cm of coal in five leaves, closely overlying a 168-cm cavity; together these probably represent the worked coals mentioned in Gemmel's report. A *Lingula* band was proved at a depth of 21 m in No. 7 Borehole [NO 45821293], the strata in which probably lie stratigraphically below those in No. 6 Borehole. Although coals in the Lower Limestone Group were also formerly worked from pits in this vicinity (p. 77) the general aspect of the strata in Nos. 6 and 7 boreholes indicates that the 120-cm and 180-cm coals, with the underlying *Lingula* band, probably all lie in the lower part of the Limestone Coal Group; they are separated from the Lower Limestone Group seams by a small fault.

Beds low in the Limestone Coal Group were encountered in shallow boreholes put down by the Geological Survey (1966) in the Mount Melville area, where no old workings are known. The Craigtoun Hospital Borehole [NO 48281464] was 27·5 m deep and started close below the Mount Melville dolerite sill. It encountered six seams of coal between 30 and 50 cm thick in some 25 m of strata. There is a *Lingula* band at 10·36 m and a 6-m bioturbated sandstone at 24 m. A coarse sandstone 3·5 m thick was encountered at a depth of 15·2 m. Comparison with the Callange Borehole suggests that the *Lingula* band represents the Johnstone Shell Bed, so that all the coals in this bore lie near the base of the Limestone Coal Group. Craigtoun Park No. 1 Borehole [NO 47751426] was drilled close to a major fault and probably provides an incomplete stratigraphical record. A condensed log of the bore is as follows:

	Thickness m	Depth m
Drift	2·59	2·59
Strata	1·70	4·29
Coal	2·82	7·11
Strata with shattered zones, probably at least one fault present	32·16	39·27
Coal	0·79	40·06
Seatearth	0·74	40·80
Coal	0·13	40·93
Seatearth	1·14	42·07
Coal	0·99	43·06
Strata	7·59	50·65
Coal	0·60	51·25
Strata, with *Lingula* band at 55·78 m	5·62	56·87
Tuffisite	3·94	60·81

Structural considerations suggest that the beds below the supposed fault in this bore lie stratigraphically above those cut in the Craigtoun Hospital Borehole; comparison with the Dunotter Borehole suggests that the *Lingula* band can be correlated either with the *Lingula* band above the Dunfermline Under Coal or with the upper marine band of the Black Metals. Comparison with the Ceres successions (Fig. 12) seems to indicate that the former correlation is more likely, and that the upper three of the four coal seams below the supposed fault should together be regarded as equivalent to the Five Foot Coal of central Fife, while the lowest seam should be correlated with the Dunfermline Splint Coal. The 282-cm coal at the top of Craigtoun Park No. 1 Borehole may well lie at a horizon higher than the Black Metals marine bands but will be discussed here for convenience. The only seam of comparable thickness known in the Ceres–Denhead area is the Ceres Thick Coal, and this is tentatively correlated with the 282-cm seam. If this correlation is correct, faults with a total throw of 30 to 45 m are present in the bore, between the 282-cm coal and the 79-cm coal.

Craigtoun Park No. 2 Borehole [NO 47721442] proved two coal seams separated by about 5 m of rooty beds. The upper seam, at a depth of 3 m, is 84 cm thick; the lower is 102 cm thick and is underlain, at 12 m, by a *Lingula* band. At 14·8 m the bore entered the Mount Melville dolerite sill and was stopped at 18·4 m without reaching the base of the sill. The sedimentary sequence in this borehole closely resembles that in the lower part of Craigtoun Park No. 1 Borehole. The upper coal of No. 2 Borehole probably correlates with the 99-cm coal at 43 m in No. 1 Borehole, and may therefore be a lower leaf of the Five Foot Coal; the lower coal in No. 2 Borehole probably correlates with the lowest seam in No. 1 Borehole, the presumed equivalent of the Dunfermline Splint Coal.

Strata above the Black Metals. A 282-cm coal encountered in Craigtoun Park No. 1 Borehole has already been mentioned. The correlation of that seam with the Ceres Thick Coal places it above the Black Metals but for convenience it has been discussed in the previous section.

A thick coal seam was formerly worked, as the 'Main' or 'Thick' Coal, at Drumcarro West Colliery [NO 44831295], a group of pits midway between the farms of Drumcarro and Ladeddie. The colliery closed down in 1850, and according to Gemmel's unpublished report of 1900 the last engine pit was 119 m deep with coal seams as follows:

Splint Coal	99 cm thick at 42 m
Main Coal	300 to 370 cm thick at 97 m
Four Feet Coal	120 cm thick at 111·5 m
Eight Feet Coal	180 cm thick at 119 m

This record is derived from a number of earlier reports which date from between 1753 and 1850. The highest or 'Splint' seam was also known as the 'Five Feet' seam in some reports; estimates of its thickness range from 100 to 150 cm. The 'Main' or 'Thick' seam was subject to spontaneous combustion which caused trouble throughout the life of the colliery. According to Gemmel's report the thickness of strata between the Four Feet and Eight Feet seams is 5½ m but in an unpublished report (1841) by Landale it is stated that this interval varies from 1 to 3½ m; Telfer (1842, unpublished report) gave the thickness of the interval as 2·3 m including 0·5 m of coal. In Ladeddie No. 1 Borehole (see below) the interval is 1 m. Early reports mention other workable coals above the 'Splint' seam, but these records are not substantiated by the later more detailed reports, and were ignored by Gemmel. A series of exploratory boreholes was put down in this area by the Opencast Executive of the National Coal Board in 1953, proving a sequence very similar to that quoted by Gemmel (above). Ladeddie No. 1 Borehole [NO 45051298] cut a seam 311 cm thick, in three leaves, at 21 m, a waste at 34·6 m and a 358-cm seam in four leaves just below the last at 39·8 m. The uppermost seam probably represents the 'Main' coal while the waste and lower coal probably represent the 'Four Feet' and 'Eight Feet' coals respectively. Stratigraphically higher beds were encountered in four other bores of the series which were sited a short distance to the west of No. 1 Borehole near the old engine pit. These bores did not provide a complete record of the strata above the 'Main' coal but if known dips are taken into account the logs can be pieced together to reveal an apparently unfossiliferous sequence 60 to 90 m thick containing at least four thin coal seams, of which two appear to have been worked. One of the worked seams probably represents the 'Splint' coal of Gemmel's account. The sequence at Drumcarro West Colliery may be compared with the higher part of the Ceres succession (p. 97). The preferred correlation links the 'Main' coal of the Drumcarro area with the Ceres Thick Coal, the 'Four Feet' coal with the Ceres Upper Four Foot Coal and the 'Eight Feet' coal with the Ceres Six Foot Coal. The beds above the 'Main' coal cannot easily be related to the Ceres sequence.

The final record of a 'thick' coal comes from Ladeddie Backfields about 600 m north of Ladeddie Farm, where an early report by Landale (1837, p. 322) notes the presence of two steeply-dipping seams, 180 cm and 300 cm thick, on the hillside south of Ladeddie Limeworks. Gemmel's report (1900) is more specific and notes that a 44-m shaft was sunk about 1855 to work a group of three 'edge coals', 180 cm, 120 cm and 340 cm in thickness. These seams are very disturbed by faults and igneous intrusions and the working was soon abandoned. They are now regarded as the probable equivalents of the Six Foot, Upper Four Foot and Thick coals of Ceres respectively; a major fault, the continuation of the Ceres Fault, must run between them and the gently-dipping strata of the Lower Limestone Group at Ladeddie Limeworks just to the north.

Strata probably belonging to the Limestone Coal Group. Between Ladeddie Hill [NO 446135] and Denhead [NO 468137] lies the Denork–Denhead Coalfield, a tract of poorly exposed ground with evidence of old coal workings. On the present structural interpretation of the area, this belt of ground is bounded northwards by an east-north-easterly fault, one of the Ceres–Maiden Rock group of faults. Coarse sandstones outcrop below Denork [at NO 45461394] and on Ladeddie Hill [at NO 44651362], while a bluff of coarse sandstone [NO 45611362] immediately to the south of Denork Craig contains the opening to a small coal mine. The recent Geological Survey Drumcarro, Callange, Craigtoun Park Nos. 1 and 2 and Craigtoun Hospital boreholes show that an association of coal and coarse sandstone is a characteristic feature of the Limestone Coal Group in this area, and the whole tract of ground in question has therefore been assigned to this group. Near Elderburn Farm there are waste tips indicating the site [NO 46231400] of an old working, and smaller tips are present about 365 m further east [NO 46581409]. The various old reports, boreholes, and plans provide information about the seams which were worked in this vicinity but no general section of the strata can be pieced together from these sources. The stratigraphical level of the seams must therefore remain in doubt. The seam extracted near Elderburn Farm was called the Denork Gas Coal and a plan of the workings, dated 1889, is available. The seam consists of 79 cm of coal on 5 cm of stone on 69 cm of 'parrot' coal. This seam is the lowest encountered, and is overlain by about 15 m of strata containing two coals each about 90 cm in thickness. All these coals were encountered in Denork No. 5 Borehole (about 1883) nearby. In the region of the more easterly tips mentioned above, old records refer to earlier workings in seams named, from below upwards, the 'Cave' Coal, 90 to 120 cm thick, the 'Scrumpie' Coal, 51 cm thick, and a second 'Parrot' Coal. If these seams are in unbroken sequence with those already described, they must lie at a stratigraphical horizon 90 to 120 m above them. J.I.C.

References

CHISHOLM, J. I. 1968. Trace-fossils from the Geological Survey boreholes in east Fife, 1963–4. *Bull. geol. Surv. Gt Br.*, No. 28, 103–119.

CURRIE, ETHEL D. 1954. Scottish Carboniferous goniatites. *Trans. R. Soc. Edinb.*, **62**, 527–602.

EWING, C. J. C. and FRANCIS, E. H. 1960. Nos. 1 and 2 Off-Shore Borings in the Firth of Forth (1955–1956). *Bull. geol. Surv. Gt Br.*, No. 16, 1–47.

FORSYTH, I. H. and CHISHOLM, J. I. 1968. Geological Survey boreholes in the Carboniferous of east Fife, 1963–4. *Bull. geol. Surv. Gt Br.*, No. 28, 61–101.

—— and READ, W. A. 1962. The correlation of the Limestone Coal Group above the Kilsyth Coking Coal in the Glasgow–Stirling region. *Bull. geol. Surv. Gt Br.*, No. 19, 29–52.

FRANCIS, E. H. 1956. The Stirling and Clackmannan Coalfield, Scotland: Area north of the River Forth. *Coalfld Pap. geol. Surv. Gt Br.*, No. 1.

—— 1968. Pyroclastic and related rocks of the Geological Survey boreholes in east Fife, 1963–4. *Bull. geol. Surv. Gt Br.*, No. 28, 121–135.

—— ALLAN, J. K. and KNOX, J. 1961. The economic geology of the Fife Coalfields, Area II. 2nd edit. *Mem. geol. Surv. Gt Br.*

GEIKIE, A. 1902. The geology of eastern Fife. *Mem. geol. Surv. Gt Br.*

KNOX, J. 1954. The economic geology of the Fife Coalfields, Area III. *Mem. geol. Surv. Gt Br.*

LANDALE, D. 1837. Report on the geology of the East of Fife Coalfield. *Trans. Highl. agric. Soc. Scotl.*, **11** (vol. 5, new series), 265–348.

READ, W. A. 1965. Shoreward facies changes and their relation to cyclical sedimentation in part of the Namurian east of Stirling, Scotland. *Scott. Jnl Geol.*, **1**, 69–92.

Chapter 6

UPPER LIMESTONE GROUP

INTRODUCTION

THE Upper Limestone Group lies in the Arnsbergian (E_2) Stage of the Namurian (Currie 1954, table 1, p. 532). Its base is drawn at the Index Limestone, which has not yet been found in east Fife, and its top at the Castlecary Limestone, the presence of which has been confirmed. The thickness of the group is therefore not established but it is thought to be about 400 m. In east Fife the Upper Limestone Group is confined to the Largo area, where it forms a poorly-defined outcrop round Largo Law.

LITHOLOGY

The Upper Limestone Group consists of an alternating sequence of sandstones, siltstones and mudstones, with a number of coal seams and their underlying seatearths, and a few beds of limestone. The pattern of sedimentation is cyclic (see Fig. 13), but there is so much variation in the composition of the cycles, and in the thicknesses of both the cycles and of individual beds, that no 'typical cycle' can be constructed.

STRATIGRAPHY

Geikie (1902) was able to provide only a brief account, based on inadequate data, of the Upper Limestone Group in east Fife. The stratigraphy of the group is now however reasonably well established from the Orchard Beds upwards (see Fig. 13), although much less is known about the bottom part. The main sources of information about the group are the three boreholes put down by the Geological Survey in 1963–64 (Forsyth and Chisholm 1968), namely Little Pilmuir [NO 40720392], Balcormo Wood [NO 41320575] and Dumbarnie [NO 44810322]. Two sets of old bores, one around Lundin Links and the other around Little Pilmuir Farm [NO 407038], have provided some information, mainly about the upper half of the group. This information was used by Knox (1954) for his account of the Upper Limestone Group in the Lundin Links area. Surface exposures of beds near the top of the group are abundant in the Keil Burn, but they are not continuous enough to yield good vertical sections. They do, however, include four fossiliferous beds. Near Hatton [NO 405044] the Hatton Burn provides two partial sections of the Castlecary Limestone and also exposures of the immediately underlying strata; other tributaries of the Keil Burn yield isolated exposures. On the coast at Shell Bay various strata are seen including a marine band possibly at the horizon of the Index Limestone. Considerable areas of the inferred outcrop of the Upper Limestone Group are, however, devoid of borehole sections or surface exposures.

The lower part of the group (below the Orchard Beds) is known only in two strongly contrasted sections, the Little Pilmuir and Dumbarnie boreholes, and even there the available information is incomplete. In the Dumbarnie Borehole

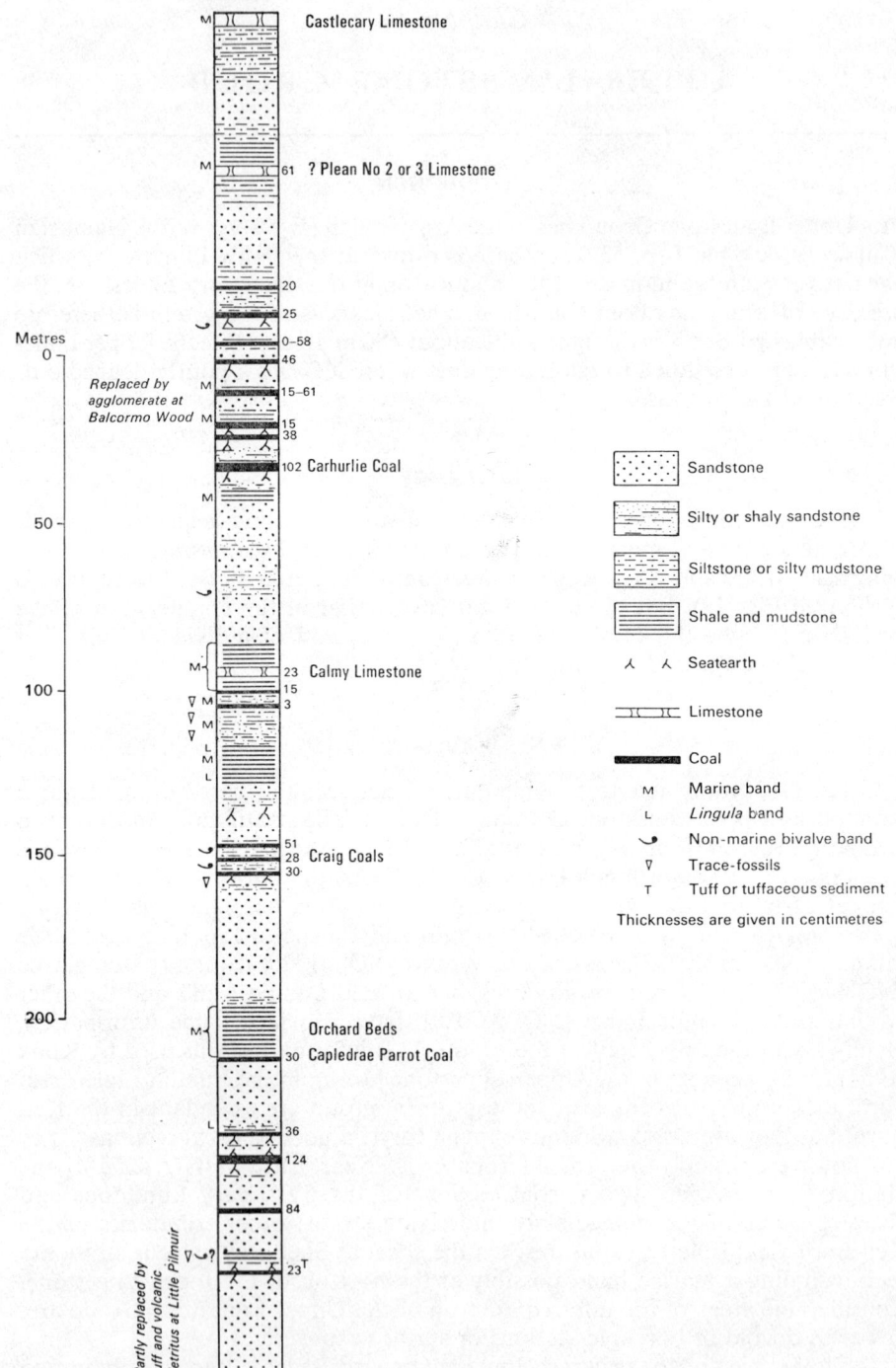

FIG. 13. *Generalised vertical section of the Upper Limestone Group*

(on which the lower part of the generalised section, Fig. 13, is based) the sequence is composed mainly of sandstone and includes several coal seams: in the Little Pilmuir Borehole it is largely made up of tuff and volcanic detritus. The Orchard Beds closely resemble in lithology and fauna their equivalents in Midlothian (Tulloch and Walton 1958) and elsewhere in Fife (Knox 1954; Francis and others 1961). They are overlain by a thick sandstone, followed by the equivalents of the Craig Coals (Francis and others 1961, p. 118) and a series of mainly argillaceous strata up to 55 m thick which are characterised by abundant trace-fossils (Chisholm 1968, p. 106). These beds, according to Francis and others (1961, pp. 118–119), are the equivalents of the strata between the Hirst Coals and the Calmy Limestone, which are locally only about 30 cm thick in the Central Coalfield. They reach their maximum thickness in the Balcormo Wood Borehole where the lower, sandier part is over 30 m thick. The higher, more argillaceous part, however, is thicker in the Little Pilmuir Borehole. Faulting prevents comparison with the Dumbarnie Borehole.

The Calmy Limestone is generally less than 30 cm thick: it lies between 50 and 80 m below the Carhurlie Coal, the intervening strata being mainly sandstone, especially in the Balcormo Wood Borehole, where the greatest thickness was recorded. The Carhurlie Coal is the thickest seam in the Upper Limestone Group that has been identified in more than one section, and is the only one definitely known to have been worked. It is succeeded in the Lundin Links area by a sequence including up to seven coal horizons, usually between 3 and 9 m apart. The coals at these horizons are rather variable, but some locally reach 60 cm in thickness. The intervening strata contain at least two bands with sparse marine faunas. These beds are the equivalents of the strata associated with the Plean Limestones in the Central Coalfield: at the top there is a marine limestone, possibly the correlative of Plean No. 2 or No. 3 Limestone. They are faulted out in the Dumbarnie Borehole and their nature south-east of Largo Law is consequently unknown. The Castlecary Limestone exceeds 4·5 m in thickness and is therefore the thickest known limestone in east Fife.

VOLCANIC ROCKS

In the Little Pilmuir Borehole [NO 40720392] much of the lower part of the Upper Limestone Group sequence is made up of tuff and volcanic detritus which form a continuous accumulation nearly 30 m thick, overlain by some 60 m of beds in which pyroclastic rocks alternate with sediments to within 40 m of the base of the Orchard Beds. The extent to which these sediments are affected by neck-margin phenomena (Francis 1968) strongly suggests that one of the necks is situated in the immediate vicinity of the bore: indeed it appears that the latter actually entered the neck in one place (Francis 1968, p. 131). The almost complete absence of such phenomena in the overlying strata suggests that this particular neck became extinct after the final accumulation of tuff and volcanic detritus, the top of which lies 38 m below the base of the Orchard Beds. These pyroclastic rocks presumably belong to the same volcanic episode as those at Balgrummo, 3 km to the west (Knox 1954, p. 50): in the Dumbarnie Borehole, $4\frac{1}{4}$ km to the east, they are represented only by a bed of siltstone-tuff with accretionary lapilli, overlying a possibly tuffaceous seatclay (Francis 1968, p. 126). Tuffaceous siltstones exposed on the shore near Ruddons Point probably lie in the lower part of the Upper Limestone Group and may also belong to this episode.

There is a single record of graded tuffaceous siltstone some 24 m below the base of the Orchard Beds in the Little Pilmuir Borehole (Francis 1968, p. 126). Tuffs or tuffaceous siltstones were encountered in all three Geological Survey boreholes, at about the same stratigraphical horizon, between 21 and 38 m below the Calmy Limestone. They appear to correlate with the highest band of such materials noted in the Upper Limestone Group in east-central Fife by Francis (1961, pp. 202–203).

Some of the strata associated with the Plean Limestones in the Balcormo Wood Borehole are replaced by at least 18 m of apparently unbedded coarse agglomerate (Forsyth and Chisholm 1968, p. 66), which has probably been derived from a nearby volcanic neck, perhaps the one exposed at Balmain [NO 417059]. The equivalent sediments at Little Pilmuir contain a band of tuff and tuffaceous sediments, including a bed of limestone (Forsyth and Chisholm 1968, p. 65). Several old bores in the Lundin Links area recorded limestones which may well be tuffaceous, and several tuffaceous bands high up in the Upper Limestone Group are seen in scattered exposures in the Keil Burn. A tuff band recently exposed behind a wall near Newburn Church [NO 448034] may lie near the top of the Upper Limestone Group or near the base of the Passage Group.

For detailed descriptions of the pyroclastic rocks encountered in the Upper Limestone Group in the Geological Survey boreholes at Little Pilmuir, Balcormo Wood and Dumbarnie, the reader is referred to Francis (1968).

Details

Index Limestone. The presence of a marine band at the base of the Upper Limestone Group has not as yet been established in east Fife. (The limestone formerly exposed in the Hatton Burn west of Carhurley [NO 398052] (Geikie 1902, p. 178), just outside the district under review, is now known to be the Calmy, not the Index.) In the Dumbarnie Borehole the trace-fossil *Diplocraterion*, which elsewhere is associated with marine or quasi-marine body-fossils (Chisholm 1968), was found at or near the base of the group, but a fault intervenes and the presence of a marine band could not be proved.

In Shell Bay, west of Elie, there are a few exposures [NO 461005] of a bed of calcareous sandstone, at least 76 cm thick, which has a lenticular 15-cm band of limestone with shell and crinoid debris at or near the top. As the exposures are isolated and occur in an area where the strata have been much disturbed by volcanic activity, the horizon of this marine band cannot be established, but general considerations and lack of likely alternatives suggest that it lies at or near the level of the Index Limestone. Francis and others (1961, pp. 112–116), however, have indicated that in west-central Fife there are two marine bands in the basal part of the Upper Limestone Group, and that it is the upper one, correlated by them with the Huntershill Cement Limestone of the Glasgow district (Clough and others 1925, p. 73), which locally contains a thin limestone. The Huntershill Cement Limestone was not recognised in east-central Fife by Knox (1954, pp. 48–49), who recorded only one marine band, with no limestone, near the base of the group and provisionally placed it at the level of the Index Limestone.

Horizon of Index Limestone to Orchard Beds. The burrowed bed in the Dumbarnie Borehole (see above) is overlain by some 100 m of strata composed mainly of sandstone (see Fig. 13), in bands up to 24 m thick, which is locally coarse-grained and contains quartz pebbles up to 2·5 mm across. The succession also includes several coal seams, including one 124 cm thick in five leaves and another 84 cm thick in two leaves. The only faunal band, about 40 m above the inferred position of the Index Limestone, is in a bed of mudstone which yielded indeterminate bivalve fragments possibly to be referred to *Naiadites*. This may well be a record of the persistent bed of mudstone, with *Naiadites* in places, which was noted by Francis and others (1961, p. 117) in west-central Fife and (under the obsolete name of Lochore Marine Band) by Knox (1954, p. 49) in the Kirkcaldy–Leven area.

The lowest strata encountered in the Little Pilmuir Borehole (see Fig. 13) are largely composed of tuff and volcanic detritus, mainly (but possibly not entirely) of Upper Limestone Group age (Forsyth and Chisholm 1968, p. 66). The total thickness of volcanic rocks penetrated was about 42 m and their base was not reached. These beds are grey or green in colour and range in grain-size from clay grade to coarse sand with small pebbles. They are for the most part water-laid and are in places cross-stratified. Accretionary lapilli occur in some bands. The interstratified sediments consist mainly of sandstone, mudstone and seatearth, but they have all been so much affected by neck-margin phenomena (Francis 1968) that the exact nature of the sequence has been obscured.

The strata between the highest bed of tuff and volcanic detritus and the base of the Orchard Beds, 38 m higher in the succession, are composed mainly of sandstone, with beds of mudstone in the lower part. Near the middle there is a shaly coal 36 cm thick with a hitherto unknown *Lingula* band in its roof. Lundin Mill F Borehole (1862) [NO 41370285] and the three Geological Survey boreholes all cut the Capledrae Parrot Coal (which lies immediately below the Orchard Beds); its thickness was found to be between 20 and 41 cm.

Orchard Beds. The term Orchard Beds was first used by Tulloch and Walton (1958, p. 62) to describe the bed of mudstone (the Orchard Blaes of Francis and others 1961, pp. 114, 118) which in the Central Coalfield of Scotland contains near its base the Orchard Limestone. The shelly sandstone band used to define the top of the Orchard Beds in Midlothian has not been found in east Fife, where the top is drawn at the highest stratum with marine fossils, and the base at the top of the Capledrae Parrot Coal. Thus defined, the Orchard Beds in east Fife are between 12 and 15 m thick and consist of mudstone with ironstone nodules. They are fossiliferous throughout, abundantly so in a few calcareous bands near the middle (the Capledrae Marine Band of Knox 1954, p. 50). The fauna is rich and varied (see p. 133) and establishes their identity. The Orchard Beds are nowhere seen at the surface within the district under review but they are well exposed in the Hatton Burn a short distance downstream from Houndshead Bridge [NO 396056], only 90 m west of the western limit of the district. These exposures were formerly allocated to the Lower Limestone Group, but both the lithology and the fauna identify them as the Orchard Beds.

Orchard Beds to Calmy Limestone. The Orchard Beds pass upwards into a sandstone, 21 to 38 m thick, which is interrupted in the Balcormo Wood Borehole by 3 m of mixed strata including two thin coals. Between 1·5 and 7·5 m of mixed silty and sandy strata, containing trace-fossils of marine type (Chisholm 1968), lie between this sandstone and the lowest of the Craig Coals. The latter are three in number and are spread over 6 m of strata at Little Pilmuir and nearly 15 m at Balcormo Wood. The lowest seam reaches 46 cm in thickness, the middle one 28 cm and the uppermost 51 cm, but they are all variable and tend to occur in leaves. There are records of *Naiadites* above the lowest seam, and of *Naiadites* and *Carbonicola* above the middle one (Wilson *in* Forsyth and Chisholm 1968, pp. 79–80).

The strata immediately above the Craig Coals are very variable. At Little Pilmuir they are only 9 m thick and consist mainly of siltstone and seatearth of silt grade, whereas at Balcormo Wood they are fully 30 m thick and include similar beds but are mainly sandy. At Dumbarnie the thickness again exceeds 30 m, and includes 6 m of sandstone at the base; this is succeeded by 15 m of strata which consist largely of siltstone, but include many poorly sorted beds of admixed sand, silt and clay, possibly derived from pre-existing laminated sediments. Burrows of marine type occur near the base of this sequence (Chisholm 1968). The upper part is 12 m thick and composed largely of sandstone, locally rooty, burrowed or convolute. The graded tuffaceous siltstones (Francis 1968), already noted (p. 106) as occurring in this part of the succession, are associated in the Balcormo Wood Borehole (Forsyth and Chisholm 1968, p. 67) with disturbed beds consisting of poorly sorted mixtures of sand, silt and clay; these contain contorted fragments of finely laminated sediments of similar bulk composition, which are taken to be remnants of the sediments from which the mixed rocks have been derived.

These variable strata are overlain by a bed of mudstone which is 7·5 to 10·5 m thick except at Lundin Mill F Borehole, where it increases to as much as 21 m. The constancy of this bed in central Fife was noted by Knox (1954, p. 52), and by Francis and others (1961, p. 119) who recorded a marine band in the upper part. Marine fossils have also been found in east Fife and there are records of *Lingula* both above and below the marine horizon. This mudstone is separated from the Calmy Limestone by 9 to 18 m of alternating beds of sandstone, siltstone and mudstone, many of which are abundantly and conspicuously burrowed. The trace-fossils are of types associated with marine conditions (Chisholm 1968), and *Lingula* and marine bivalves (Fig. 13) were found. Two thin coals occur near the top of this sequence. The upper one is separated from the Calmy Limestone by 1·5 to 5 m of marine mudstone.

Calmy Limestone. The Calmy Limestone is up to 46 cm thick in east Fife and is represented in the Dumbarnie Borehole by 23 cm of crinoidal sandstone. The overlying marine mudstone is from 1·8 to 12·2 m thick, its original thickness having clearly been reduced by erosion preceding the deposition of the overlying sandstone. Neither the limestone nor the mudstone is exposed within the district under review but both are visible on the east bank of a reservoir 180 m west-north-west of Carhurley [NO 398052] and about 30 m west of the western limit of the district.

Calmy Limestone to Carhurlie Coal. In the Little Pilmuir Borehole and adjacent old bores, between 30 m and 45 m of mixed clastic strata lie between the top of the mudstone above the Calmy Limestone and the Carhurlie Coal. They include a mudstone with *Naiadites?* and a burrowed band 4½ m thick with trace-fossils of marine type, 18 m and 23 m respectively above the Calmy Limestone, and a band with a sparse marine fauna (Fig. 13), 9 m below the Carhurlie Coal. The last-named band was also encountered in the Balcormo Wood Borehole, where this part of the sequence, almost 73 m thick, consists almost entirely of sandstones, among which several disconformities were detected.

Strata probably lying below the Carhurlie Coal are exposed in the Keil Burn between 180 and 640 m south of Balmain [NO 417059]; they consist largely of sandstone but include coaly seatclay and a bed of mudstone with indeterminate, possibly marine, bivalves. The outcrop of this faunal band lies within 90 m of that of the supposed Carhurlie Coal (see p. 109). It is probably one of the bands found in the boreholes. In the Keil Burn at and below its confluence [NO 41050296] with the Hatton Burn there are scattered exposures of sandstone and mudstone in this part of the sequence, which end downstream at a small cliff [NO 41390287] where 3·7 m of sandstone overlie 3·7 m of siltstone with tuffaceous bands.

Carhurlie Coal. The Carhurlie Coal (Fig. 13) is a rather variable seam which locally splits into two or more leaves, with a total thickness of coal varying from 81 to 163 cm. A small area was mined from a pit [NO 40960369] sited 400 m south-east of Little Pilmuir. An old shaft [NO 41730578] south of Balmain is said to have been sunk

7·3 m to a seam 120 cm thick, formerly exposed in the Keil Burn, which is tentatively identified (see p. 108) as the Carhurlie Coal. A 90-cm coal said to outcrop in a ditch by the roadside at Balcormo Mains Farm [NO 407057] may also be the Carhurlie seam, as may be one of similar thickness cut at 30 m in an old bore in the south-west part of Balcormo Wood. The Carhurlie Coal was also worked a short distance to the west of the district under review, in the Kinross (40) Sheet, where it reaches fully 2 m in thickness (Knox 1954, pp. 54–55).

Carhurlie Coal to Castlecary Limestone. There are up to seven rather variable coal seams in the 60 m of strata which overlie the Carhurlie Coal. Average thicknesses or ranges in thickness are shown in Fig. 13. Two of the coals have sparse marine faunas in their mudstone roofs and one of the others has a *Curvirimula* band. The base of the agglomerate which was encountered at the top of the Balcormo Wood Borehole lies just above the 15-cm coal (Fig. 13). A record of a waste of old workings at a depth of 8·8 m in an old bore at Lundin Links [NO 41520289] probably refers to the 46-cm seam or the one below it. No trace has been found of the shaft or adit used. The Little Pilmuir Borehole encountered tuffaceous beds in this part of the sequence, which are mainly sandy but include a 33-cm limestone with a coal breccia at its base. These tuffaceous strata are probably derived from the outburst of volcanic activity which produced the agglomerate at Balcormo Wood. Two other, possibly tuffaceous, limestones, 40 to 90 cm thick, were recorded slightly higher in the sequence in old bores near Lundin Links.

The 60-m sequence with seven coals, which has just been described, is overlain by a sandstone 12 to 18 m thick. This sandstone is probably the one exposed in the Keil Burn about 400 m south of Pitcruvie [NO 414046]. It is succeeded first by a series of mixed strata and then by a marine limestone, usually in more than one bed with a total thickness of about 60 cm. This limestone is known from the exposures in the Hatton Burn 180 m north-north-west of Thomsford Bridge [NO 40060438] and in two places in the Keil Burn, namely beside Pitcruvie and below Flour Mill Farm [NO 413029]; it has also been encountered in boreholes. It has already been indicated (p. 105) that this limestone may be the equivalent of either Plean No. 2 or Plean No. 3 Limestone, but it may be that it lies at a horizon, within the strata associated with the Plean Limestones, at which no limestone is developed in the Central Coalfield of Scotland. The limestone is overlain by 12 to 18 m of mudstone and siltstone, followed by about 9 m of sandstone and at least 12 m of mixed strata.

Castlecary Limestone. The Castlecary Limestone is still partly visible in the quarries at Thomsford Bridge and Hatton Den, on either limb of the Hatton Syncline. Landale (1837, p. 312) gave its thickness as $4\frac{1}{2}$ m. Muir and others (1956, pp. 61, 130) quoted a partial analysis of this limestone, which was described as a slightly ferriferous dolomite: it contains 16 per cent magnesia and 12 per cent acid-insoluble material. The only other limestone probably to be regarded as the Castlecary Limestone in east Fife occurs in the Dumbarnie Borehole, which encountered 4·85 m of dolomite at rockhead. A short distance below this bed a fault cut the borehole and the correlation cannot therefore be confirmed. This is unfortunate because the occurrence in this bed of *Semiplanus* cf. *latissimus* casts some doubt on its identification as the Castlecary Limestone, at which horizon this fossil has not previously been recorded. It is difficult, however, to put forward any alternative: even the St Monance Brecciated Limestone in the Lower Limestone Group is not known to reach as great a thickness. In any case, general structural considerations appear to rule out any possibility that Lower Limestone Group strata are present at outcrop at Dumbarnie and the fault in the bore would require to be a reversed one of considerable magnitude (about 600 m).

The Castlecary Limestone is nowhere seen in the Keil Burn, on the west side of Largo Law, although exposures are sufficiently abundant that such a hard bed should be prominent if it were present. It is possible that the limestone may have been cut out by a fault or that its outcrop may not extend westwards across the stream; but the most likely explanation for its absence—and one that fits best with the field evidence—is that it has been removed by penecontemporaneous erosion before the deposition of the thick sandstone which is prominently exposed in Pitcruvie Den between 365 and 640 m north of Pitcruvie [NO 414046]. The disappearance of the Castlecary Limestone in this manner is known elsewhere in the Midland Valley of Scotland, e.g. the Stirling district (Francis 1956, fig. 4), Midlothian (Tulloch and Walton 1958, p. 80) and the Glasgow area (Forsyth 1961).

I.H.F.

References

CHISHOLM, J. I. 1968. Trace-fossils from the Geological Survey boreholes in east Fife, 1963–4. *Bull. geol. Surv. Gt Br.*, No. 28, pp. 103–119.

CLOUGH, C. T., HINXMAN, L. W., WILSON, J. S. G., CRAMPTON, C. B., WRIGHT, W. B., BAILEY, E. B., ANDERSON, E. M. and CARRUTHERS, R. G. 1925. The geology of the Glasgow district. 2nd edit. revised by MACGREGOR, M., DINHAM, C. H., BAILEY, E. B. and ANDERSON, E. M. *Mem. geol. Surv. Gt Br.*

CURRIE, ETHEL D. 1954. Scottish Carboniferous goniatites. *Trans. R. Soc. Edinb.*, **62**, 527–602.

FORSYTH, I. H. 1961. The succession between Plean No. 1 Limestone and No. 2 Marine Band in the Carboniferous of the east Glasgow area. *Trans. geol. Soc. Glasg.*, **24**, 213–234.

—— and CHISHOLM, J. I. 1968. Geological Survey boreholes in the Carboniferous of east Fife, 1963–4. *Bull. geol. Surv. Gt Br.*, No. 28, 61–101.

FRANCIS, E. H. 1956. The geology of the Stirling and Clackmannan Coalfield, Scotland: Area north of the River Forth. *Coalfld Pap. geol. Surv. Gt Br.*, No. 1.

—— 1961. Thin beds of graded kaolinized tuff and tuffaceous siltstone in the Carboniferous of Fife. *Bull. geol. Surv. Gt Br.*, No. 17, 191–215.

—— 1968. Pyroclastic and related rocks of the Geological Survey boreholes in east Fife, 1963–4. *Bull. geol. Surv. Gt Br.*, No. 28, 121–135.

—— ALLAN, J. K. and KNOX, J. 1961. The economic geology of the Fife Coalfields, Area II. 2nd edit. *Mem. geol. Surv. Gt Br.*

GEIKIE, A. 1902. The geology of eastern Fife. *Mem. geol. Surv. Gt Br.*

KNOX, J. 1954. The economic geology of the Fife Coalfields, Area III. *Mem. geol. Surv. Gt Br.*

LANDALE, D. 1837. Report on the geology of the East of Fife Coalfield. *Trans. Highl. agric. Soc. Scotl.*, **11** (vol. 5, new series), 265–348.

MUIR, A., HARDIE, H. G. M., MITCHELL, R. L. and PHEMISTER, J. 1956. The limestones of Scotland. Chemical analyses and petrography. *Mem. geol. Surv. spec. Rep. Miner. Resour. Gt Br.*, **37**.

TULLOCH, W. and WALTON, H. S. 1958. The geology of the Midlothian Coalfield. *Mem. geol. Surv. Gt Br.*

Chapter 7

PASSAGE GROUP

Lithology and Stratigraphy

THE Passage Group, formerly known as the Scottish Millstone Grit (MacGregor 1960), comprises strata between the top of the Castlecary Limestone and the local base of the Coal Measures. In east Fife the group is confined to the Largo district where the main outcrop extends from the coast northwards to Largo Law. The Castlecary Limestone has been recognised in this area but the upper limit of the group is not known because an unknown thickness of strata spanning the Passage Group–Coal Measures boundary has been cut out by faulting or has been replaced by tuff. Passage Group strata presumably underlie the Coal Measures south of the Durie Fault but no borehole in this area has gone sufficiently deep to reach them. Much of the history of Passage Group times in east Fife remains obscure. The lower beds, where known, appear to be largely sedimentary but include a few bands of tuff and two intercalations of basalt lava. There is a passage by alternation (cf. Knox 1954, p. 61) both upwards and laterally into tuff and agglomerate. Pyroclastic rocks appear to make up most, if not all, of the upper part of the group and may extend into the Coal Measures. They occur around the volcanic centre of Largo Law, from which many of them were no doubt derived. The complexity of this centre and the scattered and inadequate nature of the exposures, however, have prevented the order of events there from being fully elucidated. The volcanic necks in east Fife are described in Chapter 12, in which the problem of their age is more fully discussed.

Sedimentary rocks

The sediments of Passage Group age fall conveniently into two parts. There is a poorly-known basal sequence, possibly between 30 and 60 m thick, which consists mainly of sandstone; it is not known to contain any faunal bands. It is succeeded by sediments which are dominantly argillaceous, with shale more abundant than seatclay. This assemblage is at least 60 m thick and in places probably reaches between 90 and 120 m. It includes at least two marine horizons, the Temple Upper and Lower marine bands (new names), the faunas of which (see p. 134) resemble those of the lower bands in the Passage Group of Midlothian (Wilson *in* Tulloch and Walton 1958). The lower one was discovered by Kirkby (1898) who tentatively placed it at the horizon of the Index Limestone (cf. Geikie 1902, p. 180). It was placed in the 'Millstone Grit' by Knox (1954, pp. 62–64). The upper band was first noted during the recent re-examination of the coast section at Largo by Dr. E. H. Francis.

The thick development of shale, some at least of which is seen in exposures to be black and carbonaceous, is most unusual in the Passage Group, which elsewhere in Scotland consists almost entirely of sandstone and seatearth. It is therefore unfortunate that so little is known about the stratigraphy of the group in the Largo area. Indeed it is not even absolutely certain that these shales are of Passage Group age.

Bedded tuffs and lavas

Geikie (1902, pp. 256–259) described the tuffs around Largo Law, noting many of the exposures mentioned below, and assigned them to the Permian period because he believed that they lie unconformably on strata belonging to various parts of the Carboniferous. Balsillie (1923) claimed that the tuffs are of 'Lower Carboniferous' age (i.e. Upper Limestone Group or older), and Knox (1954, p. 61) indicated that some of them are interbedded with sediments of Scottish 'Millstone Grit' age. Geikie himself admitted (1902, p. 258) that the evidence for unconformity was not 'as clear as could be wished' and detailed re-examination of the sections concerned has favoured the view that the tuffs either lie conformably on or are interbedded with the sediments (cf. Knox 1954, p. 61). The Strathairly coals, which certainly lie below the tuffs south of Largo Law, were likened by Landale (1837, p. 279) to strata now known to belong to the Middle Coal Measures at Lundin Links. Structurally, however, it would be difficult—though not impossible—to accommodate strata as young as this at Strathairly. It would greatly strengthen Geikie's argument in favour of a Permian age for the volcanism if these coals were of Middle Coal Measures age, but he himself (1902, p. 257) regarded them as belonging to, at latest, the Upper Limestone Group, because they lie on the northern, upthrow, side of the Durie and Temple faults. The stratigraphical position of the Strathairly coals, which unfortunately are nowhere exposed, remains uncertain, and they could be of Passage Group or Upper Limestone Group age. Both these groups are known to contain worked coals: there is, for example, a record of a waste in Passage Group strata west of Largo Law and the Carhurlie Coal near the top of the Upper Limestone Group has been worked.

The tuffs around Largo Law are now regarded as being most probably of Passage Group age, but they may locally have begun to accumulate during Upper Limestone Group times and probably extend up into the Coal Measures. They have been distinguished on the relevant six-inch Geological Sheets (mainly NO 40 N.W., NO 40 S.W.), and on the third edition of one-inch Geological Sheet 41, from the tuffs in the Largo Law and Rires necks (see pp. 203–205), but this distinction is, at least in places, largely academic, because they all belong to a complex volcanic centre. In the absence of other evidence, topographic features have locally been taken to demarcate the tuffs inside the necks from those outside, but in reality there is probably no clear distinction between bedded tuff which accumulated just outside the neck-margins and material which has begun to collapse into the volcanic pipes. Other subsidiary necks probably lie undetected within the area mapped as interbedded tuffs, and both main necks are doubtless more complicated than the map indicates (see Chapter 12 for further details).

The tuffs are basaltic throughout (except where some sedimentary material has been incorporated), are generally green in colour and vary in grain-size from fine to coarse tuffs and lapilli-tuffs with occasional bombs. Geikie (1902, p. 258) placed their thickness at fully 120 m. This would appear to be a minimum amount which may be considerably exceeded, but exposures are too poor to allow an accurate estimate to be made. The best exposures are on the coast south of Carrick Villa [NO 43760244], locally in the Strathairly Burn and on the side of a meltwater channel south-east of Monturpie [NO 433039].

Volcanic activity had already broken out before the end of Upper Limestone Group times (p. 106) on the north-west flank of Largo Law and possibly also on the south-east flank, although the age of the tuffs included in that group in the latter area is in fact uncertain. Tuffs are certainly present south-east of Largo Law just above the Castlecary Limestone. South-west of Largo Law, on the other hand, the only known volcanic rocks in the lower part of the Passage Group are two intercalations of basalt lava. The tuffs south-east of the hill interdigitate with sediments and similar interdigitation (cf. Knox 1954, p. 61) at rather higher horizons appears to take place to the south-west. The tuffs seem to have extended gradually so that the upper part of the Passage Group appears to consist almost entirely of tuff on the south side of Largo Law; the nature of the sequence on the north side is unknown.

One of the lavas mentioned above, an olivine-basalt of Dalmeny type, has been radiometrically dated by the Isotope Geology Unit of the Institute of Geological Sciences, using the 'whole-rock' potassium–argon method (Forsyth and Rundle, in press). Samples from the old quarry at Blindwells [NO 418040] and a quarry 250 m south-west of Largo House [NO 420034] gave minimum ages of 267 ± 6 m.y. and 292 ± 6 m.y. respectively (see Table 9, p. 179). Forsyth and Rundle consider that the apparently younger age of the Blindwells sample is certainly due to argon loss. The Largo House sample must also have lost argon if the lava is indeed of Passage Group age, but the field evidence for assigning it to that group is not conclusive. The basanitic intrusions in the Largo Law neck (see Chapter 12) are considered by Forsyth and Rundle to be of probable Stephanian age and it is not impossible that this lava (together with the overlying tuffs) is Stephanian also, i.e. that the minimum age obtained is close to the true age of emplacement. This interpretation, however, is difficult to reconcile with the fact that the lava is unlike any of the neck intrusions in petrographic character; moreover, it would be difficult structurally to accommodate post-Coal Measures rocks in the area where the lava crops out. On balance, therefore, it seems more probable that the lava is of Passage Group age and has been subject to argon loss.

Details

The lowest strata in the Passage Group in east Fife, which consist mainly of sandstone with some beds of siltstone, are exposed in the axial zone of the Hatton Syncline along the Hatton Burn below Hatton House [NO 405044], in the Keil Burn about 400 m north of Pitcruvie [NO 414046] and in scattered exposures south and south-west of Lahill House [NO 444038]. They were formerly seen, with tuff above them, in the disused quarry 230 m north-west of Dumbarnie [NO 449031]. The basalt lava of Dalmeny type in the old quarry at Blindwells (Knox 1954, p. 61), which can be traced southwards for some 800 m by means of exposures in other old quarries, probably lies near the base of the group.

The sediments exposed in the stream between West Coates [NO 447041] and Lahill Mains [NO 442044] consist mainly of sandstone and probably belong to the basal part of the Passage Group.

These basal sandstones are overlain by a sequence in which argillaceous beds predominate and are interbedded with tuffs. This sequence is believed to outcrop on the western slopes of Largo Law but there are no exposures and the only information is derived from a few old boreholes east of Pitcruvie, one of which cut an 89-cm coal 7 m from the surface, while another found a waste at 8·2 m. An intercalation of basalt lava occurs near the base of this part of the succession and is exposed by the roadside [NO 41920424]. Interdigitation of black shale and green basaltic tuff is visible at

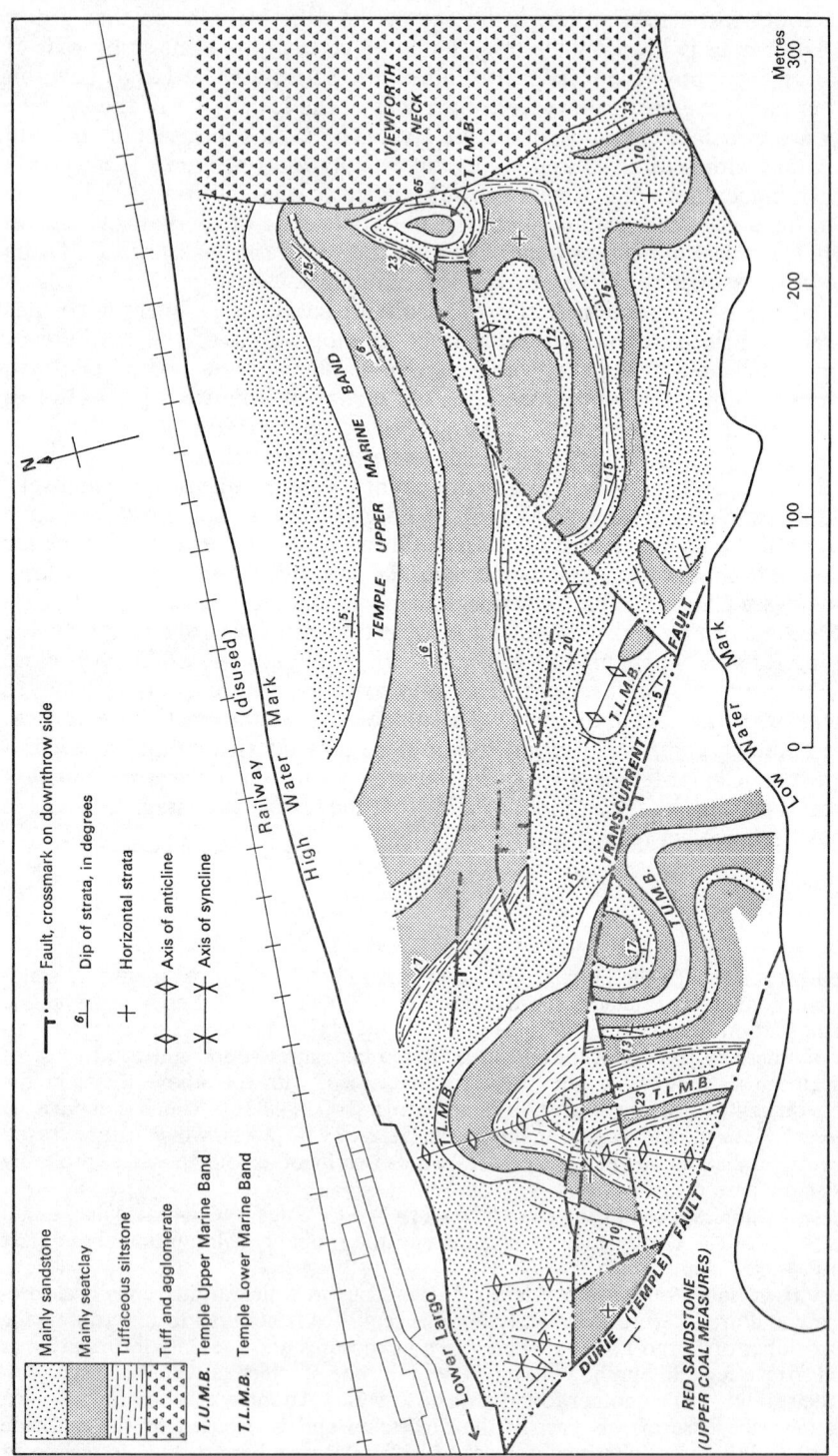

FIG. 14. Sketch-map of the shore section at Temple, Lower Largo

several places on the side of a meltwater channel 400 m west of Chesterstone [NO 428041] and locally in the Largo Burn for up to 320 m south of Largo Church [NO 424035].

Largo No. 1A Borehole (1907) [NO 42470293] yielded a section of 73 m of strata, nearly all argillaceous, in which shale is more abundant than seatclay and there are only a few thin inferior coal seams. No tuff bands were reported. This section probably includes the beds exposed in the lowest part of the Largo Burn and on the shore at Temple. There is one marine band in the stream and on the shore two are visible (the Temple Upper and Lower marine bands). The exposures on the shore are much affected by faulting and folding associated with the adjacent volcanic neck (see Fig. 14), but the following section has been worked out by Dr. E. H. Francis:

	m
Sandstone, red and yellow, with seatclay bands	5·5
Siltstone and shale with marine fossils (Temple Upper Marine Band)	1·5
Seatclay with two 8-cm coal seams	2·0
Sandstone, rooty	0·5
Seatclay, locally pale purple	1·5
Siltstone, tuffaceous, purple and grey	1·2
Sandstone with siltstone and seatclay partings	3·7
Seatclay with ironstone nodules	0·6
Siltstone with sandy ribs at top	0·9
Shale with marine fossils (Temple Lower Marine Band)	0·9
Coal	0·05
Seatclay	0·6
Sandstone, argillaceous, pale purple and grey mottled	0·9
Seatclay, pale purple and grey mottled	0·6
Siltstone, tuffaceous, purple and grey	1·2
Sandstone, argillaceous, pale purple and yellow	0·3
Seatearth, pale purple and yellow	0·6
Sandstone, argillaceous (seen)	3·0

Largo No. 10A Borehole (1907) [NO 42680304] cut nearly 30 m of strata consisting almost entirely of shale, with two intercalations of 'green whin' which are taken to be tuffs. These beds are thought to lie entirely above those cut by Largo No. 1A Borehole and those exposed on the shore at Temple. Black shales are also known to occur, associated with tuff, about 400 m north-east of Strathairly House [NO 433032].

Tuffs are well exposed in the Strathairly Burn between Strathairly House and Drumeldrie [NO 441033], where they are seen to consist of fine to coarse-grained green water-laid tuffs with lapilli up to 6 mm across. Generally similar material, with occasional bombs up to 46 cm long, is extensively laid bare in the same stream over a stretch some 550 m long between Lahill House [NO 444038] and West Coates [NO 447041]. Apparently unbedded green tuffs with rare bombs are visible on the north-east side of the meltwater channel which extends south-eastwards from Monturpie [NO 433039], but otherwise the bedded tuffs which are assumed to come to crop on the lower slopes of Largo Law are very poorly exposed. The largest area of exposed tuff is on the coast south-east of Carrick Villa [NO 43760244] where green, generally coarse-grained tuffs, and lapilli-tuffs, in places clearly water-laid, are prominently displayed. I.H.F.

References

BALSILLIE, D. 1923. Further observations on the volcanic geology of east Fife. *Geol. Mag.*, **70**, 530–542.

FORSYTH, I. H. and RUNDLE, C. C. *In press.* The age of the volcanic and hypabyssal rocks of east Fife. *Bull. geol. Surv. Gt Br.*, No. 60.

GEIKIE, A. 1902. The geology of eastern Fife. *Mem. geol. Surv. Gt Br.*
KIRKBY, J. W. 1898. On the occurrence of Carboniferous Limestone fossils at View-forth, near Largo, Fife. *Trans. Edinb. geol. Soc.*, **7,** 488–493.
KNOX, J. 1954. The economic geology of the Fife Coalfields, Area III. *Mem. geol. Surv. Gt Br.*
LANDALE, D. 1837. Report on the geology of the East of Fife Coalfield. *Trans. Highl. agric. Soc. Scotl.*, **11,** (vol. 5, new series), 265–348.
MACGREGOR, A. G. 1960. Division of the Carboniferous on Geological Survey maps of Scotland. *Bull. geol. Surv. Gt Br.*, No. 16, 127–130.
TULLOCH, W. and WALTON, H. S. 1958. The geology of the Midlothian Coalfield. *Mem. geol. Surv. Gt Br.*

Chapter 8

COAL MEASURES

Lithology and Stratigraphy

In Scotland the Coal Measures are now classified as comprising Lower, Middle and Upper subdivisions (MacGregor 1960). Strata belonging to the Middle Coal Measures are certainly present at the surface in east Fife and other strata at outcrop have been assigned to the Upper Coal Measures; the Lower Coal Measures however do not come to crop because of faulting, and only one borehole, sunk in 1862, is tentatively regarded as having reached their upper part. Neither the Queenslie Marine Band (which separates the Lower from the Middle Coal Measures) nor Skipsey's Marine Band (at the base of the Upper Coal Measures) has been found in east Fife, although the horizon of the former has been tentatively identified in the old borehole mentioned above, and the outcrop of the latter is believed to be located in a gap in the exposures on the coast near the mouth of the Keil Burn.

In east Fife strata of Coal Measures age are known to occur only in a very limited area at Lundin Links and Lower Largo, between the Durie Fault and the sea. In general the strata are of normal Coal Measures type, comprising cyclic sequences of sandstones, siltstones, mudstones, coals and seatearths. They are exposed on the shore and in the Keil Burn near its mouth, and are also known from a number of old boreholes in the long-since disused Lundin or Drummochy Coalfield. Little is known about the workings in this coalfield. The strata lie on the eastern limb of the Lundin Anticline, with eastward dips generally between 10 and 20 degrees. Surface exposures provide an almost complete sequence in the upper part of the Middle Coal Measures and the lower part of the Upper Coal Measures. Some of the boreholes extend further down the sequence and one probably penetrated into the Lower Coal Measures, where, after cutting about 36 m of sediments of normal Coal Measures type, it encountered variously-coloured strata tentatively interpreted as tuffs, tuffaceous sediments or volcanic detritus related to similar beds in Firth of Forth No. 3 Off-Shore Borehole (Ewing and Francis 1960). The coal-bearing sediments inside the Viewforth neck (see Chapter 12) contain spores indicating a middle Westphalian A (Lower Coal Measures) age (Dr. R. Neves, personal communication, 1974).

The Middle Coal Measures (see Fig. 15) are generally similar to the equivalent strata in the Kirkcaldy–Leven area (Knox 1954). They include a number of coal seams among which most of the seams worked in that area can be identified. The total thickness of this division is about 150 m, but only part of the Upper Coal Measures succession, also amounting to about 150 m, is preserved on land. Coal Measures strata must also be presumed to be present off-shore under much of Largo Bay, linking up with those proved in the undersea workings further south.

The possibility that some of the volcanic activity around Largo Law is of Coal Measures age is indicated in Chapters 7 and 12. Such activity certainly occurred in the Largo area after the deposition of the Middle Coal Measures, which are cut by a neck exposed on the shore at Lundin Links.

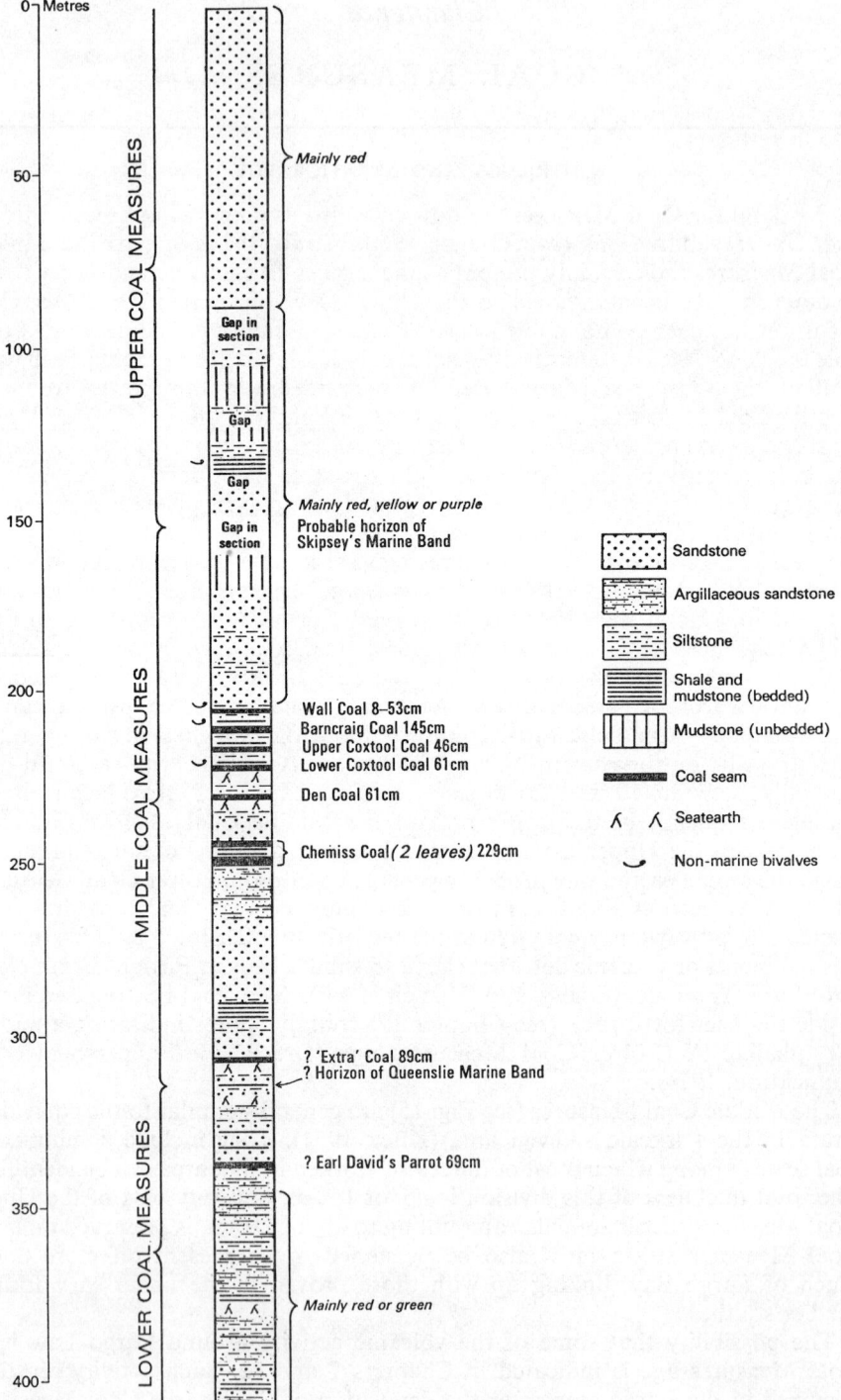

FIG. 15. *Generalised vertical section of the Coal Measures in the Largo district*

Landale (1837, pp. 277–281) described the Drummochy Coalfield and gave a section of the strata exposed on the coast at Lundin Links and Largo. In this, thicknesses are given only for the coal seams, the figures for the intervening strata referring to the width of outcrop. When thickness figures calculated on the basis of observed dips are substituted for outcrop widths this section bears a good resemblance to the one measured by Dr. E. H. Francis in 1961, upon which the detailed account below and Fig. 15 are largely based. Landale did not note any gaps in his section, which suggests that exposure of the beds may have been more complete in his day than it is now. This may account for the presence of an 'extra' coal seam in the Landale section. Knox (1954, pp. 115–116) described the Lundin Coalfield, drawing largely upon an old vertical section of the strata, the source of which is unknown. This section is referred to below as the Lundin generalised section.

Landale (1837, pp. 278–279) indicated his belief that strata of Coal Measures age, including some of the seams worked at Drummochy, are also present in the vicinity of Strathairly House [NO 433032] below the tuffs surrounding Largo Law. He stated that an attempt was made to work these coals but that they were found to be of very inferior quality. No other evidence of the existence of coal seams at Strathairly has since been obtained and general structural considerations suggest that the strata there should more probably be assigned to the Passage Group or to the top part of the Upper Limestone Group (see p. 112).

Details

LOWER AND MIDDLE COAL MEASURES

The lowest Coal Measures strata in the Largo district of which a record is available were cut by Drummochy D Borehole [NO 41300248] put down in 1862 near high water mark on the shore at Lundin Links. Under the 69-cm coal mentioned below, this bore (see Fig. 15) cut some 73 m of strata comprising an alternating sequence of sandstone, siltstone and mudstone, with no coals. Many of the beds were described as red or green in colour. The lack of coals is unusual, for this part of the succession includes several thick and persistent seams in the Kirkcaldy–Leven area (cf. Knox 1954). The reason for their absence, as is suggested by the coloration of the strata, may be that the sequence is largely made up of tuff or volcanic detritus produced by the volcanic activity which affected the same part of the succession in Firth of Forth No. 3 Off-Shore Borehole (Ewing and Francis 1960, pp. 52–54). This inference has been strengthened by the discovery (Dr. R. Neves, personal communication, 1974) that the spores in the sediments interbedded with the tuffs in the Viewforth neck (see Chapter 12) indicate a middle Westphalian A (Lower Coal Measures) age.

The 69-cm coal mentioned above, which is probably Earl David's Parrot (Knox 1954, pp. 96–99), was encountered at a depth of 103·3 m in Drummochy D Borehole. Thirty metres of mixed strata separate this seam from one 89 cm thick, which may be the equivalent of the so-called 'Extra' Coal of the Kirkcaldy–Leven area, referred to by Knox (1954, p. 102). One of the beds of mudstone below this coal probably contains the Queenslie Marine Band and therefore marks the top of the Lower Coal Measures, but no opportunity to search for this horizon in east Fife has as yet become available. About 55 m of strata separate this coal from the Chemiss Coal: they consist largely of red sandstone, some of which probably fills a washout at the level of the Bush Coal.

Chemiss Coal. The Lundin generalised section (cf. Knox 1954, p. 115) ends with a 137-cm seam which Knox provisionally equated with the Chemiss Coal. The latter, however, is generally much thicker than this in the Kirkcaldy–Leven area (Knox 1954, pp. 104–105) and often occurs in leaves. It is therefore suggested that the next higher seam in the section, which is 92 cm thick and is separated from the 137-cm seam by 2·7 m of argillaceous rock, is also part of the Chemiss Coal. The Chemiss seam is poorly developed in Drummochy D Borehole and is represented by a waste in Drummochy E Borehole [NO 41210244] (which proves that it was wrought in the Lundin Coalfield), but the two leaves are 94 cm and 129 cm thick respectively in Lundin No. 31 Borehole (1840) [NO 41260247]. The Chemiss Coal is overlain by some 12 m of mainly sandy strata.

Den Coal. The seam identified as the Den Coal in the Lundin generalised section is 61 cm thick, but there is no satisfactory borehole section through this seam and it is not exposed in the coast section. The overlying beds include sandstone, siltstone and seatclay.

Lower and Upper Coxtool Coals. Two seams 66 cm and 84 cm thick, lying 3 m apart in the Lundin generalised section, are provisionally regarded as the Lower and Upper Coxtool Coals respectively. Their positions have been located on the coast, but in neither case is the full thickness of the seam exposed. Landale's section (1837, p. 277) appears to have a coal too many in this part of the succession: it is uncertain which two of his seams D, E and F should be identified as the Coxtool Coals. All three are between 46 cm and 61 cm thick, and it may well be that the average thicknesses of the Coxtool seams in the Largo area lie in this range, rather than at the higher figures given in the Lundin generalised section. The Lower Coxtool Coal has *Anthracosia spp.* in its roof. In the generalised section over 9 m of mudstone and siltstone lie between the Upper Coxtool and Barncraig coals, but in the coast section only some 3·7 m of strata separate these seams.

Barncraig Coal. Several bores at Lundin Links cut the Barncraig Coal, which is generally about 145 cm thick and is represented in Landale's section (1837, p. 277) by 152 cm of coal. Two of the bores recorded wastes at this horizon, and it seems probable that this was the most extensively worked coal in the Lundin Coalfield. The top part of the seam is still visible in the shore section, where *Anthracosia spp.* were found in its roof.

About 6 m above the Barncraig Coal there is a rather variable seam, recorded as being from 8 to 53 cm thick, which also has *Anthracosia spp.* in the overlying mudstone. This is probably the Wall Coal. It is succeeded by about 45 m of strata coloured variously red, yellow and purple, which consist mainly of sandstone with some siltstone and mudstone. Skipsey's Marine Band has not been recognised in the coast section and no borehole cuts this part of the sequence. Comparison with the succession further west suggests that it should lie about 40 m above the supposed Wall Coal and its position on the coast has accordingly been taken immediately to the east of the vari-coloured strata mentioned above, at a gap in the section about 90 m west of the mouth of the Keil Burn. The strata to the east of this are therefore assigned to the Upper Coal Measures.

UPPER COAL MEASURES

In the Lundin Links–Largo coast section the lowest 60 m of strata assigned to the Upper Coal Measures are incompletely exposed but appear to consist largely of red and purple mudstones with red, green and purple siltstones and some red and yellow cross-bedded sandstone. About 15 m above the base of this part of the sequence there is a band of shale 1·5 m thick, with ostracods, fish debris and non-marine bivalves (p. 134). The remaining 90 m of strata exposed on the coast consist almost entirely of red cross-bedded sandstone.
I.H.F.

REFERENCES

EWING, C. J. C. and FRANCIS, E. H. 1960. No. 3 Off-Shore Boring in the Firth of Forth (1956–1957). *Bull. geol. Surv. Gt Br.*, No. 16, 48–68.
KNOX, J. 1954. The economic geology of the Fife Coalfields, Area III. *Mem. geol. Surv. Gt Br.*
LANDALE, D. 1837. Report on the geology of the East of Fife Coalfield. *Trans. Highl. agric. Soc. Scotl.*, **11**, (vol. 5, new series), 265–348.
MACGREGOR, A. G. 1960. Divisions of the Carboniferous on Geological Survey maps of Scotland. *Bull. geol. Surv. Gt Br.*, No. 16, 127–130.

Chapter 9

CARBONIFEROUS PALAEONTOLOGY

INTRODUCTION

THE extensive exposures of Carboniferous sediments, almost entirely of Lower Carboniferous age, on the coast of east Fife have long attracted the attentions of geologists. In relation to the great thickness of rocks exposed however, relatively few palaeontologists have done detailed work on the contained fossils. This results from a number of factors. Many of the exposures are below high water mark and access is dependent on the state of the tide. Few of the sections lend themselves to straightforward interpretation as folding, faulting and igneous activity have disturbed the sequence in many places. Beds of shale, from which fossils are most easily obtained, tend to be eroded and their outcrops become covered by mud, sand or shingle. Some of the carbonate beds in the Calciferous Sandstone Measures are seen to be packed with shells which are partially exposed by weathering but the matrix is very hard and extraction of specimens is extremely difficult. There are also a few exposures inland but they are disconnected and no great thickness of strata is seen.

The first comprehensive study of the Lower Carboniferous succession including palaeontological details was by Brown (1861). It was the monumental work of James W. Kirkby, however, which laid the foundation of our knowledge of the succession and of the fossil contents of the rocks. The principal results of his researches (Kirkby 1880; *in* Geikie 1902) were the consequence of about thirty years of careful examination for and collection of fossils from the coastal sections.

A comprehensive bibliography was compiled by Tait (*in* Geikie 1902, pp. 406–412) which covers palaeontological work published prior to 1902. Since then, Carboniferous faunas have been published in conjunction with stratigraphical studies by Tait and Wright (1923) from the St Monance area, by Kirk (1925) from the St Andrews area, by Wattison (1962) from the Crail–Anstruther area and by Anderson (*in* Knox 1954) from the Largo area. Detailed studies of fossils from the district have been made by Leitch (1942) and Bennison (1960, 1961) on non-marine bivalves and by Traquair (1906) on fish faunas. Mention must also be made of James Wright and his lifetime of work on the Carboniferous Crinoidea. Many of the calices which he collected and described came from Fife but to the west of the present district. He did record some species however, from the St Monance, Ardross and St Andrews areas. A full bibliography of his works is to be found in Wright (1950, pp. xx–xxi).

The most comprehensive collection of Carboniferous fossils from the district was probably that made by Kirkby. It appears, however, that his collection was dispersed after his death. What is presumably a small part of it is preserved in the Hancock Museum, Newcastle-upon-Tyne, but the specimens, like those in other old collections consulted, are poorly localised. The present study is mainly based on collections made by Messrs. P. J. Brand and D. K. Graham from the whole district, supplemented by work by W. G. E. Graham in the Largo area and on the faunas from several boreholes (Forsyth and Chisholm 1968).

The fossils present in the Carboniferous rocks of the district are very varied and reflect the numerous changes in the depositional environment during the accumulation of the sediments. Bands containing varied marine assemblages are known to occur at intervals throughout the succession except in the Coal Measures but exposures of that formation are extremely limited. These marine bands provide evidence that the sea invaded the district many times. The acme of this activity was reached during the deposition of the upper part of the Calciferous Sandstone Measures (Pathhead Beds) and the Lower Limestone Group, judging by the number of such bands and the richness of the contained faunas in this part of the succession.

The remains of animals that are assumed to have lived in water which was not marine also abound in the succession, particularly in the Calciferous Sandstone Measures. These include non-marine bivalves, the gastropod *Naticopsis? scotoburdigalensis* (Etheridge jun.), certain ostracods which occur in profusion and fish remains which are normally found in dark shales. These various types of fossils tend to occur to the exclusion of other forms, suggesting that each was suited to a particular environmental niche. The water ranged presumably from fresh to brackish, perhaps even hypersaline, the normal sea water being excluded wholly or partially by bars forming lagoons or by distance over estuarine flats where the admixture of fresh water from drainage led to brackish conditions. In the absence of incoming fresh water, evaporation could have led to hypersaline conditions.

The following account presents the sequence of faunas, in ascending stratigraphical order, found in the Carboniferous rocks of the district. The lists are compiled mainly from identifications of specimens collected by Geological Survey staff in recent years, together with those found in boreholes (Forsyth and Chisholm 1968). All previously unpublished identifications are listed in Appendix 1 (pp. 257–263) where the authors of the species are given, together with the ranges of the species in relation to the main stratigraphical subdivisions. The locality numbers given below refer to the list of localities in Appendix 1 (pp. 263–268).

Calciferous Sandstone Measures

The present study has produced much new detailed information concerning the distribution of fossils in the Calciferous Sandstone Measures, especially in the strata exposed on the coast. It is unfortunate that the succession is disturbed by faults, mostly of unknown throw, so that the full sequence is not yet known. It is possible, however, to present a general picture of the distribution of the faunas in these measures and to note those species which appear to have stratigraphical significance.

The distribution of the non-marine faunas is only known in general terms. As yet no detailed work has been done on the ostracods and fish. Non-marine bivalves occur in abundance at many horizons, except in the Fife Ness Beds which so far have proved barren. Two species which have been assigned to *Carbonicola*, namely *C. antiqua* and *C. elegans*, are found only in the lower part of the succession (Anstruther and Pittenweem beds). *Naiadites obesus* is the most common form present and ranges throughout most of the Calciferous Sandstone Measures but is very rare or absent in the Sandy Craig and Pathhead beds where *Curvirimula scotica* is the dominant non-marine species. A similar vertical distribution of non-marine bivalves has been previously noted in the Lower Carboniferous in the Lothians (Wilson *in* Mitchell and Mykura 1962, p. 99).

Except in the Fife Ness Beds, marine bands occur throughout the succession at irregular intervals. In some, rich marine assemblages are present whereas others have yielded only a few marine fossils. The exact stratigraphical relationship between some of the bands is obscure and there may be other bands still to be discovered. The following account presents the current information available on the marine faunas.

Anstruther Beds

The earliest marine fossils in the Carboniferous of the area appear to be those in the sequences exposed on the shore near Randerston and Wormistone. They are probably also the earliest in central Scotland. Several marine bands are present, occurring in or immediately above the limestones, with the Randerston Nos. 5, 6, and 7 limestones (p. 32) yielding the richest faunas. The combined fossil list from localities 120–122 and the Anstruther Borehole, which penetrated similar beds, is as follows: algal nodules, trepostomatous and other bryozoa, *Spirorbis sp.*, *Lingula mytilloides*, rhynchonelloid indet., *Bellerophon randerstonensis*, *Donaldina sp.*, *Hypergonia kirkbyi*, *Meekospira?*, *Naticopsis sp.*, *Aviculopecten* aff. *tabulatus*, *A. sp.*, *Edmondia unioniformis?*, *Leiopteria hendersoni*, *Naiadites crassus?*, *Sanguinolites subplicatus*, *Schizodus pentlandicus*, orthocone and coiled nautiloid fragments.

In the upper part of the Anstruther Beds there are several poorly developed marine bands but the exact position of some of these in the succession is uncertain. Their faunas are meagre and are as follows:

Anstruther Wester Marine Band. Locality 1, *Girtyspira sp.*, *Lithophaga lingualis*.

Billow Ness Marine Band. Locality 6, *Lingula mytilloides*, *Naiadites* cf. *crassus*, *Schizodus pentlandicus*.

Barns Mill Marine Band. Locality 45a, *Donaldina sp.*, *Retispira* cf. *undata*, *Leiopteria hendersoni*, *Naiadites* cf. *crassus*, *Polidevcia attenuata*, *Sanguinolites clavatus*, orthocone nautiloid.

Chain Road Marine Band. Locality 85, *Bellerophon?*, *Retispira undata*, *Sanguinolites* cf. *clavatus*, *S. sp. nov.*

Craig Hartle North Marine Band. Locality 39, *Naiadites* cf. *crassus*, *Schizodus pentlandicus*, crinoid columnals, *Archaeocidaris urii*.

Craig Hartle South Marine Band. Locality 40, gastropod indet., *?Lithophaga lingualis*, crinoid columnals.

Goats Marine Band. Locality 77, orthocone and nautiloid fragments.

Kenly Mouth Marine Band. Locality 69, small high-spired gastropod indet., *Modiolus latus?*, nuculoid?, orthocone fragment.

Most of these bands yielded molluscs only but the crinoid remains and *Archaeocidaris* in the Craig Hartle bands are the lowest records of these forms in the local succession.

Pittenweem Beds

The next marine horizons to be dealt with mark the first incursions of rich faunas into the district. Over a century ago Brown (1861, p. 387) equated a bed of abundant crinoid remains at Pittenweem, now called the Pittenweem Marine Band, with similar beds at Crail and St Andrews. This enabled him to work out a general structure of the Lower Carboniferous rocks of the east Fife coast. The band which he correlated was named the Encrinite-bed or Encrinite Limestone by subsequent authors and is here called the Witch Lake Marine Band at St Andrews, the Pittenweem Marine Band at Pittenweem and the Crail Harbour Marine Band at Crail. Early investigators also found another band overlying it which was called the *Myalina* Bed or Limestone (Geikie 1902, p. 135), and is now named the St Andrews Castle Marine Band at St Andrews (p. 38). During the resurvey it was discovered that there are three marine bands in the sequence at St. Andrews, two major marine bands at Crail and two major marine bands at Pittenweem. The faunas of the marine horizons at St Andrews in ascending order are as follows:

West Sands Marine Band. Localities 8, 127, *Fenestella sp., Lingula mytilloides, L. squamiformis, Orbiculoidea sp., Productus?, Donaldina* cf. *grantonensis, Euphemites sp., Retispira sp., Aviculopecten subconoideus, Edmondia unioniformis, Limipecten?, Myalina sp., Pteronites sp., Sanguinolites clavatus, S. tricostatus?, Schizodus pentlandicus, Sedgwickia gigantea, Streblochondria sp., Streblopteria? redesdalensis, Wilkingia elliptica, Catastroboceras?*, orthocone, goniatite?, crinoid columnals.

Witch Lake Marine Band. Localities 86, 101, 102, 123–127, algal nodules, '*Conularia*' sp., *Fenestella sp., Buxtonia?, Hustedia?, Lingula mytilloides, L. squamiformis, Orbiculoidea nitida*, orthotetoid, *Productus redesdalensis?, Pugnax* cf. *pugnus?*, rhynchonelloid, *Euphemites* cf. *urii, Retispira* cf. *concinna, Dentalium* s.l., *Aviculopecten subconoideus, Edmondia* cf. *maccoyi, E.* cf. *unioniformis, Limipecten* cf. *dissimilis, Modiolus* cf. *latus, Myalina sublamellosa, Palaeoneilo* cf. *laevirostrum, Pernopecten?, Polidevcia attenuata, Prothyris sp., Pteronites angustatus, Sanguinolites clavatus, S. striatus, Schizodus sp., Sedgwickia sp., Streblochondria concentricolineata?, Streblopteria? redesdalensis*, orthocone, crinoid columnals [abundant], *Archaeocidaris urii*.

St Andrews Castle Marine Band. Localities 88, 100, 102, 127, *Fenestella sp., Serpuloides sp., Lingula mytilloides, L. squamiformis*, orthotetoid, *Donaldina?, Euphemites sp., Pseudozygopleura sp., Aviculopecten subconoideus, A. sp., Edmondia unioniformis. Limipecten?, Myalina sublamellosa, Naiadites* cf. *crassus, Polidevcia attenuata, P, attenuata traquairi, Pteronites* cf. *angustatus, Sanguinolites clavatus, S.* cf. *costellatus, S. striatus, Schizodus sp., Sedgwickia gigantea, Streblochondria?*, orthocone, '*Estheria*' sp. [at top of band], crinoid columnals.

The faunas of the two major bands at Crail in ascending order are:

Pans Marine Band. Locality 45, *Fenestella sp.*, bryozoa indet., *Euphemites sp., Retispira?, Aviculopecten subconoideus?, Edmondia* cf. *unioniformis, Naiadites* cf. *crassus, Sanguinolites plicatus, S. tricostatus, Schizodus sp., Streblochondria sp., Wilkingia elliptica*, crinoid columnals.

Crail Harbour Marine Band. Locality 45, *Fenestella sp.*, bryozoa indet., *Composita* cf. *ambigua, Lingula mytilloides, Orbiculoidea sp.*, orthotetoid, *Productus* cf. *redesdalensis, Pugnax* cf. *pugnus?, Punctospirifer?*, rhynchonelloid, *Retispira?, Aviculopecten* cf. *subconoideus, Edmondia sp., Leiopteria sp., Limipecten sp., Palaeolima sp., Polidevcia attenuata, Streblochondria elliptica?, Wilkingia?*, orthocone, crinoid columnals, *Archaeocidaris urii*.

The faunas from the two major bands in the Pittenweem area in ascending order are:

Cuniger Rock Marine Band. Locality 117, sponge?, *Fenestella sp.*, trepostomatous bryozoan, *Donaldina sp., Aviculopecten sp., Edmondia* cf. *senilis, Leiopteria sp., Naiadites* cf. *crassus,* *Pteronites sp.,* *Sanguinolites sp.,* *Schizodus* cf. *pentlandicus,* *Streblochondria sp.*

Pittenweem Marine Band. Locality 117, *Fenestella sp.,* bryozoa indet., *Orbiculoidea nitida,* orthotetoid, *Productus* cf. *redesdalensis, Punctospirifer* cf. *scabricosta, Spirifer sp. crassus* group, *Bellerophon* aff. *costatus, Euphemites* cf. *urii, Retispira* cf. *decussata, Dentalium* s.l., *Aviculopecten subconoideus, Aviculopinna mutica, Limipecten* cf. *dissimilis, Myalina sublamellosa?,?Naiadites* cf. *crassus, Nuculopsis gibbosa, Polidevcia attenuata, Sanguinolites striatus, Schizodus* cf. *pentlandicus, Sedgwickia sp.,* *Wilkingia elliptica,* orthocone, crinoid columnals [common], *Archaeocidaris urii.*

The correlations of the bands in the three areas suggested by the faunas and the relative abundance of various species are as follows:

St Andrews	Crail	Pittenweem
St Andrews Castle M.B.		
	Fault	*Fault*
Witch Lake M.B. =	Crail Harbour M.B. =	Pittenweem M.B.
West Sands M.B. =	Pans M.B. =	Cuniger Rock M.B.

The faunas as a whole mark the first incoming of large numbers of bryozoa and brachiopods into the district. The almost exclusive occurrence of abundant crinoid debris, orthotetoids, *Productus* cf. *redesdalensis, Pugnax* cf. *pugnus,* rhynchonelloids and *Dentalium* s.l. in the Witch Lake, Crail Harbour and Pittenweem marine bands suggests the correlation of these bands. The equation of the Pans and Cuniger Rock marine bands is supported by the fact that they both lie about 180 m below the Crail Harbour and Pittenweem marine bands respectively and that about halfway between the marine bands in each sequence there is a bed containing poorly preserved bivalves including *Sanguinolites* (**Kirklatch** and **Westland Skelly marine bands**).

In addition to the foregoing coastal exposures of these marine bands, there are two isolated inland localities at which representatives of the bands are present.

Chesters Marine Band. Locality 56, *Fenestella sp., Lingula sp., Productus sp., Pugilis* cf. *pugilis,* rhynchonelloid, *Spirifer sp., Euphemites sp., Aviculopecten* cf. *subconoideus, Edmondia* cf. *unioniformis, Polidevcia attenuata, Pteronites angustatus, Sedgwickia?, Streblochondria sp., Wilkingia elliptica?,* orthocone nautiloid, crinoid columnals [common], *Archaeocidaris sp.* This fauna bears a close resemblance to those present in the Witch Lake Marine Band and its correlatives.

Brigton Marine Band. Locality 7, *Retispira?, Aviculopecten subconoideus?, Edmondia?, Naiadites* cf. *crassus, Schizodus pentlandicus,* crinoid columnals. This relatively poor assemblage most closely resembles that present in the St Andrews Castle Marine Band but this suggested correlation is not a firm one.

Sandy Craig Beds

In the Anstruther–Pathhead coast section of the Sandy Craig Beds faults are present and the full sequence is not known. Elsewhere in the district, little evidence is available regarding these beds as there is no borehole information and natural sections are faulted. One marine band has been found, by D. K. Graham, on the coast at Pittenweem and the contained fossils are as follows:

Boat Harbour Marine Band. Locality 118, megaspores, *Fenestella sp., Lingula squamiformis, Orbiculoidea nitida,* orthotetoid, *Productus redesdalensis?, Pleuropugnoides sp.,*

Euphemites sp., Dunbarella sp., Leiopteria hendersoni, Palaeolima?, Sanguinolites cf. *subplicatus, Streblochondria sp., Streblopteria? redesdalensis*; orthocone nautiloid?, goniatite indet., crinoid columnals.

This fauna has close affinities with those in the underlying Pittenweem Beds, and it is possible that the Boat Harbour Marine Band may be part of that sequence. Because of its field relation with the Pittenweem Marine Band, the Boat Harbour Marine Band may be the equivalent of the St Andrews Castle Marine Band of the St Andrews area. The nature of the succession at Pittenweem between the Pittenweem Marine Band and the Boat Harbour Marine Band is not fully known as a fault of unknown throw is present in the intervening measures. There are stratigraphical arguments against the correlation suggested above (p. 25) and further evidence is required before this problem can be solved.

Three marine bands occur in isolated or faulted short sequences, in the St Andrews area, which are difficult to correlate with known sections. Some of these bands may represent horizons in the Sandy Craig Beds. The details of the faunas are as follows:

St Nicholas Marine Band. Localities 128, 129, sponge?, coral indet., *Fenestella sp.*, bryozoa indet., *Echinoconchus punctatus, Lingula mytilloides, L. squamiformis*, orthotetoid, *Productus sp., Spirifer?, Aviculopecten subconoideus?, A.* aff. *interstitialis, Leiopteria sp., Limipecten* cf. *dissimilis, Lithophaga lingualis, Streblochondria sp.* The presence of corals and *Echinoconchus punctatus* in this band, forms not recorded in the Pittenweem Beds or lower, suggests that this fauna is from a higher position in the Calciferous Sandstone Measures.

New Mill Marine Band. Localities 82, 83, sponge?, *Fenestella sp., Penniretepora sp.*, bryozoa indet., *Serpuloides sp., Echinoconchus punctatus, Lingula mytilloides, Linoprotonia?, Orbiculoidea nitida, Phricodothyris* cf. *lineata, Pleuropugnoides sp., Productus redesdalensis?, Pugilis* cf. *pugilis, Schellwienella sp., Spirifer sp., Spiriferellina* cf. *insculpta, Lithophaga lingualis, Palaeolima* cf. *simplex, Streblochondria* cf. *elliptica*. This fauna is sufficiently similar to that of the St Nicholas Marine Band to suggest that the two bands can be correlated with some degree of certainty.

Denbrae House Marine Band. Locality 25, *Orbiculoidea sp., Productus sp.*, rhynchonelloid, pectinoid indet. This is a poor fauna and it does not suggest any definite horizon.

PATHHEAD BEDS

The sequence of faunas in the Pathhead Beds is reasonably well known and the assemblages from the marine bands are given below in ascending order.

West Braes Marine Band. The West Braes Marine Band is named from its occurrence in the shore section near Pathhead, locality 113, where the fossils found were: *Lingula squamiformis, Productus sp., Retispira striata?, Actinopteria persulcata, Modiolus sp., Palaeolima?*. About 15 km to the north-north-west, in the Claremont Burn area, west of St Andrews, there are five records of a marine band, the **Denbrae Farm Marine Band**, which may be equivalent to the West Braes Marine Band. The combined fauna from them, localities 25, 27, 29, 51, 103, is: *Crurithyris sp., Lingula mytilloides, Orbiculoidea sp., Productus sp., Euphemites sp., Retispira striata?, Actinopteria persulcata, Myalina sp., Polidevcia attenuata, Posidonia* cf. *becheri, Sanguinolites clavatus, S.* cf. *plicatus, S. striatus?, Schizodus sp., Streblopteria ornata?*, goniatite indet., *Dithyocaris sp.*, crinoid columnals, *Planolites sp.*

Ardross Limestones. The Ardross Limestones are named from outcrops on the shore at Ardross. Localities 4, 105–107, 113. *Hexaphyllia sp.*, corals indet., *Fenestella sp.*, trepostomatous bryozoa, *Serpuloides sp.*, *Avonia youngiana*, *Buxtonia sp.*, *Crurithyris urii*, *Echinoconchus* cf. *punctatus*, *Eomarginifera* cf. *praecursor*, *E.* cf. *setosa*, *Lingula mytilloides*, *L. squamiformis*, *Linoprotonia sp.*, *Orbiculoidea nitida*, *Productus redesdalensis?*, *Pugilis?*, *Euphemites sp.*, *Glabrocingulum beggi?*, *Pseudozygopleura* cf. *rugifera*, *Retispira* cf. *densistriata*, *Dentalium* s.l., *Actinopteria persulcata*, *Aviculopecten subconoideus?*, *Cardiomorpha hindi?*, *Cypricardella* cf. *acuticarinata*, *C.* cf. *rectangularis*, *Dunbarella sp.*, *Edmondia* cf. *senilis*, *E. sulcata?*, *E.* cf. *unioniformis*, *Leiopteria* cf. *thompsoni*, *Limipecten dissimilis*, *Lithophaga lingualis*, *Modiolus sp.*, *Myalina sp.*, *Palaeolima* cf. *simplex*, *Palaeoneilo laevirostrum*, *P. luciniformis*, *P. mansoni*, *Parallelodon* cf. *semicostatus*, *Pernopecten sp.*, *Polidevcia attenuata*, *Posidonia becheri*, *Prothyris* cf. *scotica*, *Pterinopectinella granosa?*, *Sanguinolites* cf. *abdenensis*, *S. clavatus*, *S.* cf. *plicatus*, *S.* cf. *striatolamellosus*, *S.* cf. *striatus*, *Schizodus salteri*, *Sedgwickia sp.*, *Solemya* cf. *primaeva*, *Solenomorpha minor*, *Streblochondria* cf. *anisota*, *Streblopteria ornata*, *Sulcatopinna flabelliformis*, orthocone, *Beyrichoceratoides sp.*, goniatite indet., crinoid columnals, *Archaeocidaris urii*. This is the richest assemblage in the Calciferous Sandstone Measures in the district and marks the first appearance of many brachiopod and molluscan species which range up into the Lower Limestone Group and some into the Upper Carboniferous.

From boreholes and surface exposures in the area to the west of St Andrews similar faunas have been recorded from the **Newbigging, Claremont Cottage** and **Blebo Hole marine bands.** These bands are correlated with parts of the sequence containing the Ardross Limestones and the composite fauna from localities 19–23, 30, 103 is: corals indet., *Fenestella sp.*, trepostomatous bryozoa, *Serpuloides sp.*, *Buxtonia sp.*, *Crurithyris urii*, *Dielasma?*, *Echinoconchus punctatus?*, *Eomarginifera* cf. *longispina*, *Lingula mytilloides*, *Productus sp.*, *Schizophoria resupinata?*, *Euphemites urii?*, *Glabrocingulum sp.*, *Meekospira?*, *Naticopsis* cf. *variata*, *Platyceras?*, *Pseudozygopleura* cf. *rugifera*, *Retispira decussata*, *R.* cf. *densistriata*, *R. striata*, *Straparollus (Euomphalus) carbonarius?*, *Actinopteria persulcata*, *Aviculopecten subconoideus*, *Ctenodonta pentonensis*, *Cypricardella* cf. *rectangularis*, *Edmondia sp.*, *Limipecten dissimilis*, *Lithophaga lingualis*, *Modiolus?*, *Nuculopsis gibbosa*, *Palaeolima* cf. *simplex*, *Palaeoneilo laevirostrum*, *P. mansoni*, *Paleyoldia macgregori*, *Parallelodon* cf. *semicostatus*, *Pernopecten sp.*, *Polidevcia attenuata*, *Prothyris?*, *Pterinopectinella sp.*, *Sanguinolites costellatus*, *S.* cf. *plicatus*, *S.* sp. *variabilis* group, *Schizodus sp.*, *Solenomorpha minor*, *Streblochondria* cf. *elliptica*, *Streblopteria ornata*, *Wilkingia elliptica*, *W. maxima*, orthocone, goniatite indet., trilobite fragments, crinoid columnals, *Archaeocidaris urii*.

Pathhead Marine Bands. The Pathhead Marine Bands are named from the coast section near St Monance where the combined fauna from the Lower and Upper bands at locality 113 is: Caniniid, Clisiophyllid, *Lithostrotion junceum*, *Fenestella sp.*, bryozoa indet., *Brachythyris?*, *Composita?*, *Echinoconchus elegans*, *E.* cf. *punctatus*, *Girtyella saccula*, *Lingula mytilloides*, *L. squamiformis*, *Linoprotonia?*, *Orbiculoidea nitida*, *Productus redesdalensis?*, *Pleuropugnoides sp.*, *Pugilis pugilis*, *Schizophoria?*, *Donaldina sp.*, *Retispira striata*, *Actinopteria persulcata*, *Aviculopecten sp.*, *Edmondia* cf. *maccoyi*, *E.* cf. *senilis*, *Leiopteria sp.*, *Limipecten* cf. *dissimilis*, *Lithophaga lingualis*, *Myalina sp.*, *Palaeolima* cf. *simplex*, *Palaeoneilo luciniformis*, *Parallelodon sp.*, *Pernopecten sp.*, *Polidevcia attenuata*, *Sanguinolites* cf. *abdenensis*, *S. costellatus*, *S. striatolamellosus*, *Schizodus sp.*, *Streblochondria sp.*, orthocone, goniatite indet., '*Estheria*' *sp.*, crinoid columnals.

From boreholes in the St Andrews area, localities 30, 51, and the Drumcarro Borehole, beds correlated with the Pathhead Marine Bands on stratigraphical evidence contained the following composite fauna: *Hyalostelia sp.*, *Crurithyris urii*, *Dielasma* cf. *hastatum*, *Lingula mytilloides*, *L. squamiformis*, *Orbiculoidea sp.*, *Phricodothyris?*, *Productus sp.*, *Donaldina sp.*, *Euphemites* cf. *urii*, *Retispira striata*, *Actinopteria persulcata*, *Aviculopecten* cf. *subconoideus*, *Edmondia* cf. *senilis*, *Palaeolima* cf. *simplex*, *Palaeoneilo mansoni*, *Pernopecten sowerbii*, *Pterinopectinella?*, *Sanguinolites* cf. *plicatus*,

Schizodus sp., Solemya sp., orthocone, goniatite indet., crinoid columnals, *Archaeocidaris urii*.

Carnbee Marine Band (from inland exposures). Localities 15, 63, '*Conularia*' *sp., Chaetetes?, Lithostrotion junceum, Syringopora?, Fenestella sp.*, bryozoa indet., *Serpuloides sp., Composita sp., Lingula mytilloides, Orbiculoidea sp., Productus concinnus?, Pugnax* cf. *pugnus*, rhynchonelloid, *Schellwienella sp., Spiriferellina* cf. *perplicata, Retispira sp., Actinopteria persulcata, Edmondia sp., Leiopteria sp., Limipecten sp., Polidevcia attenuata, Sanguinolites clavatus, Schizodus sp., Sedgwickia sp., Streblochondria concentricolineata?*, orthocone, goniatite indet., crinoid columnals. A comparison of this fauna with those from known horizons gives a balance of evidence in favour of a position high in the Pathhead Beds.

St Monance White Limestone. The combined fauna from the St Monance White Limestone and the shales immediately above and below it from localities 33, 34, 113, 131, 134 and the Drumcarro and Higham boreholes is: *Diphyphyllum* cf. *lateseptatum, Lithostrotion junceum,* zaphrentid, *Fenestella sp.,* trepostomatous bryozoan, *Serpuloides sp., Avonia youngiana, Composita* cf. *ambigua, Dielasma* cf. *hastatum, Eomarginifera?, Lingula squamiformis, Orbiculoidea nitida, Phricodothyris sp., Pleuropugnoides sp., Productus* cf. *redesdalensis, Pugilis* cf. *pugilis, Rhynchopora?, Schizophoria* cf. *resupinata, Spirifer sp. crassus* group, *Euphemites sp., Limipecten?, Lithophaga lingualis, Modiolus sp., Myalina sp., Naiadites crassus, Pernopecten sp., Polidevcia attenuata, Sanguinolites abdenensis?, S.* cf. *plicatus, S.* aff. *striatogranulatus, S. striatus?, S. sp. variabilis* group, *Schizodus sp., Streblopteria ornata*, crinoid columnals, *Archaeocidaris urii*.

It can be seen from the foregoing that, in general terms, the faunal assemblages become richer in the variety of species present towards the top of the Calciferous Sandstone Measures and that the character of the faunas changes. In the lower bands, molluscan species are the dominant element. Articulate brachiopods, which are rare or absent in the lower bands, are represented by a few forms in the Pittenweem Beds but it is not until the Ardross Limestones are reached that they form a major part of the fauna in terms of species and numbers. Corals display a similar late arrival in the district as their first appearance in beds whose position in the sequence is not in question is in the Ardross Limestones.

The incidence of certain species appears to be of stratigraphical value in the district. *Bellerophon randerstonensis* has been recorded only from the Randerston Limestones. *Punctospirifer* cf. *scabricosta, Pteronites angustatus* and *Streblopteria? redesdalensis* have been identified only from the Pittenweem Beds and the Boat Harbour Marine Band. The West Braes Marine Band and the overlying Ardross Limestones mark the first appearance of several species in the succession. These include *Avonia youngiana, Echinoconchus* cf. *punctatus, Eomarginifera spp., Actinopteria persulcata, Cypricardella spp., Palaeoneilo luciniformis, P. mansoni, Paleyoldia macgregori, Parallelodon* cf. *semicostatus, Posidonia becheri* and *Streblopteria ornata*. It is also of note that the non-marine genus *Curvirimula* appears to be confined to measures above the Pittenweem Beds.

Correlations based on faunal evidence to areas outside the present district can be made only in general terms because of the high degree of lateral variation displayed by the sediments of the Calciferous Sandstone Measures. Kirkby (*in* Geikie 1902, p. 104) correlated the 'Encrinite-bed' of Pittenweem and St Andrews with the marine strata at Cove Harbour, Berwickshire. This has been supported by a recent study of the Lower Carboniferous marine bands of south-east Scotland (Wilson 1974) where a correlation is made of the marine bands in the Pittenweem Beds of the present account with the Macgregor Marine Bands of East Lothian and Berwickshire which include the marine bands at Cove Harbour (Wilson 1974, p. 47). This is based on the presence of *Punctospirifer* cf. *scabricosta*, *Naiadites* cf. *crassus*, *Pteronites angustatus* and *Streblopteria? redesdalensis* in both areas and these forms are confined to the bands in question. There is also support for this correlation on palynological grounds (Neves and others 1973, fig. 15).

The faunal assemblages present in the uppermost part of the Calciferous Sandstone Measures from the West Braes Marine Band up to the St Monance White Limestone are similar to those found in equivalent measures in many other parts of the Midland Valley but individual beds cannot be correlated for any appreciable distance on palaeontological evidence alone.

At present there is no evidence that the Calciferous Sandstone Measures of east Fife are not all of Viséan age. No identifiable goniatites have been found but the presence of *Posidonia becheri* in the Ardross Limestones suggests a Lower Bollandian (P_1) age for these beds. Currie (1954, p. 532) placed the 'Encrinite-bed' in the Cracoean (B) stage on the tentative correlation of this band with the Cove Lower Marine Band which contains *Beyrichoceratoides redesdalensis* (Hind). Macgregor (1930, p. 476) tentatively assigned a C_1 age to the limestone at Randerston which contains rhynchonelloids. This was based on a comparison of the rhynchonelloids with '*Camarotoechia*' *proava* (Phillips), taken to be a Tournaisian species on evidence of Garwood's zonation of the north of England succession. Whether '*C.*' *proava* is a Tournaisian index fossil or not is irrelevant in this context as all the rhynchonelloids examined from the Randerston Limestones are indeterminate and the better preserved specimens bear little resemblance to '*C.*' *proava*.

Recent palynological research on the Scottish Lower Carboniferous (Neves and others 1973, pp. 44, 46) has shown that the oldest miospore zone identified in east Fife is the TC Zone which is assigned to the Viséan.

Lower Limestone Group

The deposition of the Lower Limestone Group marks the acme of marine influences over the district in Carboniferous times. Thick developments of limestones and shales containing rich marine faunas are present, especially in the lower part of the group. The Lower Limestone Group was approximately equated with the Upper Bollandian (P_2) Stage by Currie (1954, p. 534). The fossils identified from the marine horizons are given in ascending order, the faunas being composite ones from the limestones and the shales immediately above and below them.

St Monance Brecciated Limestone. As the St Monance Brecciated Limestone was extensively quarried along its outcrop, collections have been made from numerous places. Localities 5, 10–12, 37, 38, 50, 60, 62, 65, 89, 93, 104, 110, 119, 130, 131, 134, 137–139, 141, 142, Drumcarro and Higham boreholes, *Hyalostelia parallela*, *Aulophyllum fungites*, *Cladochonus?*, *Lithostrotion junceum*, *Zaphrentites sp.*, *Fenestella sp.*, *Pennireteopora sp.*, *Polypora dendroides*, trepostomatous bryozoa, *Buxtonia sp.*, *Brachythyris triradialis*, *Cleiothyridina fimbriata?*, *Composita* cf. *ambigua*, *Crurithyris urii*,

Dielasma cf. *hastatum, Eomarginifera* cf. *lobata, E. longispina?, E.* cf. *praecursor, E. setosa, Hustedia sp., Lingula mytilloides, L. squamiformis, Orbiculoidea nitida, Phricodothyris sp., Pleuropugnoides sp., Productus concinnus?, P.* cf. *redesdalensis, Promarginifera trearnensis, Pugilis* cf. *pugilis, Rhipidomella* cf. *michelini, Rugosochonetes sp., Schizophoria* cf. *resupinata, Spirifer* cf. *trigonalis, Tornquistia* cf. *polita, Donaldina sp., Euphemites urii, Glabrocingulum sp., Hesperiella thomsoni, Naticopsis* cf. *variata, Pseudozygopleura robroystonensis, P.* cf. *rugifera, Retispira decussata, R. striata, Straparollus (Euomphalus) carbonarius, Dentalium* s.l., *Acanthopecten sp., Actinopteria persulcata, Aviculopecten* aff. *interstitialis, A. semicostatus, Aviculopinna mutica, Ctenodonta pentonensis, Cypricardella* cf. *rectangularis, Dunbarella sp., Edmondia arcuata, E. maccoyi, E. sulcata, E. unioniformis, Leiopteria sp., Limipecten dissimilis, Lithophaga lingualis, Myalina sp., Nuculopsis gibbosa, Palaeolima* cf. *simplex, Palaeoneilo laevirostrum, P. mansoni, Paleyoldia macgregori, Parallelodon* cf. *semicostatus, Pernopecten sowerbii, Polidevcia attenuata, Posidonia corrugata, Prothyris* cf. *scotica, Pterinopectinella sp., Sanguinolites* cf. *abdenensis, S. costellatus, S. plicatus, S. striatolamellosus, S. tricostatus, Schizodus obliquus, S. sp., Sedgwickia gigantea, S. suborbicularis, Solenomorpha minor, Streblochondria* cf. *elliptica, Streblopteria ornata, Wilkingia elliptica?, Catastroboceras sp.,* orthocone, goniatites indet., trilobite fragments, crinoid columnals, *Archaeocidaris urii.*

St Monance Little Limestone. Localities 131, 134, Drumcarro and Higham boreholes, coral indet., *Composita* cf. *ambigua, Crurithyris urii, Eomarginifera* cf. *lobata, Gigantoproductus sp. giganteus* group, *Lingula mytilloides, L. squamiformis, Linoprotonia sp., Orbiculoidea nitida, Productus* cf. *carbonarius, P.* cf. *redesdalensis, Rugosochonetes sp., Spirifer sp., Euphemites urii, Retispira decussata, R. striata, Dentalium* s.l., *Actinopteria persulcata, Aviculopecten sp., Aviculopinna* cf. *mutica, Cypricardella* cf. *rectangularis, Dunbarella?, Edmondia maccoyi?, Myalina sp., Palaeoneilo laevirostrum, P. mansoni, Polidevcia attenuata, Pterinopectinella sp., Sanguinolites plicatus, S. striatolamellosus, Schizodus* cf. *salteri, Solenomorpha minor, Streblochondria sp., Streblopteria ornata, Wilkingia?,* goniatite indet., crinoid columnals. This is the only horizon in the district which yields *Gigantoproductus* of the *giganteus* group.

Charlestown Main Limestone and Neilson Shell Bed. Localities 48, 49, 59, 60, 99, 131, 134, Drumcarro and Higham boreholes, *Zaphrentites sp.,* coral indet., *Fenestella sp.,* trepostomatous bryozoa, *Avonia youngiana, Crurithyris urii, Eomarginifera setosa?, Hustedia sp., Lingula mytilloides,* orthotetoid, *Plicochonetes sp., Productus sp., Rhipidomella* cf. *michelini, Rugosochonetes sp., Schizophoria* cf. *resupinata, Spirifer sp., Spiriferellina* cf. *perplicata, Tornquistia* cf. *polita, Euphemites urii, Glabrocingulum sp., Retispira decussata, R. striata, Straparollus (Euomphalus) carbonarius, Tropidocyclus oldhami, Dentalium* s.l., *Aviculopecten sp., Aviculopinna mutica, Euchondria neilsoni, Leiopteria sp., Nuculopsis gibbosa, Palaeoneilo luciniformis, P. mansoni, Parallelodon sp., Pernopecten sp., Polidevcia attenuata, Posidonia corrugata, P. corrugata gigantea, Pterinopectinella sp., Sanguinolites costellatus, Schizodus sp., Streblochondria sp., Streblopteria ornata* [under limestone], *Wilkingia sp., Catastroboceras sp.,* orthocone, goniatites indet., trilobite fragments, crinoid columnals, *Archaeocidaris urii.* The fauna of the Neilson Shell Bed is typically developed in the Central Coalfield area where it contains several characteristic species (Wilson 1966, pp. 111–113). Of these, *Straparollus (Euomphalus) carbonarius, Tropidocyclus oldhami, Euchondria neilsoni* and *Posidonia corrugata gigantea* occur in the shale above the Charlestown Main Limestone in the present district.

Mill Hill Marine Band. Drumcarro and Muircambus boreholes, *Fenestella sp., Polypora dendroides, Brachythyris?, Eomarginifera setosa?, Productus sp., Rugosochonetes sp., Spirifer bisulcatus?, Euphemites urii, Latischisma* cf. *globosa, Retispira decussata,*

Nuculopsis gibbosa, Palaeoneilo mansoni, Pernopecten sp., Polidevcia attenuata, goniatite indet., crinoid columnals.

Seafield Marine Band. Localities 47, 59, 114, 132, 133, Drumcarro and Muircambus boreholes, *Fenestella sp., Avonia youngiana, Dielasma?, Eomarginifera* cf. *lobata, Lingula squamiformis, Phricodothyris sp., Pleuropugnoides sp., Productus sp., Rugosochonetes sp., Spirifer* cf. *trigonalis, Spiriferellina* cf. *perplicata, Euphemites urii, Hesperiella thomsoni, Platyceras sp., Retispira sp., Dentalium* s.l., *Aviculopecten sp., Aviculopinna mutica?, Caneyella?, Curvirimula* cf. *scotica* [in separate band], *Cypricardella* cf. *rectangularis, Dunbarella?, Edmondia* cf. *maccoyi, E.* cf. *senilis, E. sulcata, Leiopteria thompsoni, Limipecten dissimilis, Modiolus?, Nuculopsis gibbosa, Palaeoneilo laevirostrum, P. mansoni, Paleyoldia macgregori, Parallelodon elegans, P.* cf. *semicostatus, Pernopecten sp., Prothyris* cf. *scotica, Sanguinolites striatolamellosus, Schizodus sp., Sedgwickia* cf. *gigantea, Solemya primaeva, Solenomorpha minor, Streblochondria sp., Streblopteria ornata*, orthocone, trilobite fragments, crinoid columnals.

Lower Kinniny Limestone (horizon of). Localities 32, 59, 115, 132, Drumcarro and Muircambus boreholes, *Lingula squamiformis, Productus sp., Euphemites sp., Edmondia sp., Modiolus?, Myalina sp., Palaeoneilo mansoni, Polidevcia attenuata, Posidonia corrugata?, Prothyris?, Sanguinolites* cf. *abdenensis, S. clavatus?, S. costellatus, S.* cf. *plicatus, S.* cf. *variabilis, Schizodus sp., Solenomorpha minor, Streblochondria sp., Streblopteria ornata*.

Mid Kinniny Limestone. Localities 59, 132, 134, Drumcarro, Dunotter and Muircambus boreholes, *Hyalostelia parallela?, Zaphrentites sp., Fenestella sp.*, trepostomatous bryozoa, *Brachythyris ovalis, Eomarginifera* cf. *lobata, Lingula mytilloides, L. squamiformis, Orbiculoidea nitida, Phricodothyris sp., Productus redesdalensis?*, rhynchonelloid, *Rhynchopora sp., Rugosochonetes sp., Schizophoria* cf. *resupinata, Spirifer sp., Spiriferellina* cf. *perplicata, Bellerophon sp., Euphemites urii, Glabrocingulum sp., Retispira decussata, Dentalium* s.l., *Aviculopecten? semicircularis, Lithophaga lingualis, Nuculopsis gibbosa, Palaeolima sp., Palaeoneilo laevirostrum, P. luciniformis, P. mansoni, Pernopecten sp., Polidevcia attenuata, Posidonia corrugata, Pterinopectinella?, Sanguinolites costellatus, Schizodus sp., Solenomorpha minor, Streblochondria sp., Streblopteria ornata*, orthocone, crinoid columnals.

Upper Kinniny Limestone. Localities 57, 61, Callange, Dunotter and Muircambus boreholes, *Lingula mytilloides, L. squamiformis, Productus* cf. *carbonarius, Pugilis?, Euphemites urii, Retispira striata?, Lithophaga lingualis, Palaeoneilo luciniformis, P. mansoni, Polidevcia attenuata, Posidonia corrugata, Pterinopectinella sp., Sanguinolites abdenensis?, S. costellatus, Schizodus sp., Solenomorpha minor, Streblopteria ornata*, orthocone, crinoid columnals.

LIMESTONE COAL GROUP

The Limestone Coal Group, inferred to be of Pendleian (E_1) age by Currie (1954, p. 534) is the lowest division of the Upper Carboniferous in central Scotland. The group is relatively poorly known in the district and there has been little opportunity to examine the strata for fossils. Scarce occurrences of *Naiadites* and bands containing *Lingula* have been recorded in the lower part of the Group but information regarding their distribution is meagre. The Johnstone Shell Bed and Black Metals Marine Band, which are widely distributed over central Scotland, are both present in the district.

Johnstone Shell Bed. Localities 57, Callange and Dunotter boreholes, *Lingula mytilloides, L. squamiformis, Aviculopecten sp., Myalina sp., Sanguinolites costellatus?, Streblopteria ornata.*

Black Metals Marine Band. Locality, Dunotter Borehole, *Lingula squamiformis, Productus sp.*, rhynchonelloid.

Upper Limestone Group

The Upper Limestone Group, included in the Arnsbergian (E_2) Stage by Currie (1954, p. 535), marks the return of limestones and shales with rich marine faunas subsequent to the essentially non-marine conditions that prevailed during the deposition of the Limestone Coal Group. A regional study of the marine faunas of the Upper Limestone Group (Wilson 1967) showed that the faunas present in east Fife and Midlothian differ appreciably from those to the west in the Kincardine Basin and the Central Coalfield (Wilson 1967, pp. 461–463). The outcrop area of the Group in the present district is very restricted and most of the information concerning the contained faunas is from boreholes.

Index Limestone. The occurrence of the Index Limestone has not been proved in east Fife. A bed of calcareous sandstone with lenses of limestone, which may be at this horizon, is present at one locality, Shell Bay, 745 m W. 24°N. of Kincraig: only brachiopod fragments and crinoid columnals were obtained. The succeeding major marine horizon in the Central Coalfield is the Lyoncross Limestone but no identifiable fauna from this horizon has been observed in the district.

Orchard Beds. The Orchard Beds yield the richest fauna in the Upper Carboniferous in the district. Localities, Balcormo Wood, Dumbarnie and Pilmuir boreholes, *'Conularia' sp.*, bryozoa indet., *Serpuloides sp., Antiquatonia* cf. *muricata, Buxtonia sp., Composita* cf. *ambigua, Eomarginifera* cf. *longispina, Lingula mytilloides, Linoprotonia sp., Orbiculoidea* cf. *nitida*, orthotetoid, *Pleuropugnoides sp., Productus carbonarius?, Pugilis?, Schizophoria* cf. *resupinata, Semiplanus* cf. *latissimus, Spirifer sp., Bellerophon anthracophilus?, Donaldina sp., Euphemites* cf. *urii, Glabrocingulum sp., Retispira decussata, R. striata, Straparollus (Euomphalus) sp., Aviculopecten sp., 'Ctenodonta'* cf. *pentonensis, Cypricardella sp., Edmondia* cf. *maccoyi, E.* cf. *senilis, Limipecten* cf. *dissimilis, Myalina sp., Nuculopsis gibbosa, Palaeolima sp., Palaeoneilo laevirostrum, P. luciniformis, P. mansoni, Parallelodon* cf. *semicostatus, Polidevcia attenuata, Prothyris scotica, Sanguinolites* cf. *striatolamellosus, Schizodus sp., Sedgwickia sp., Solenomorpha?, Streblochondria* cf. *elliptica, Wilkingia elliptica*, goniatite indet., orthocone, crinoid columnals.

Calmy Limestone. Localities, Balcormo Wood, Dumbarnie and Pilmuir boreholes, *Fenestella sp., Planolites sp., Serpuloides sp., Buxtonia sp., Crurithyris?, Lingula mytilloides, Orbiculoidea nitida*, orthotetoid, *Productus* cf. *carbonarius, Pugilis* cf. *pugilis, Schizophoria sp., Spirifer sp., Donaldina sp., Euphemites sp., Glabrocingulum armstrongi* E. G. Thomas, *Naticopsis* cf. *variata, Retispira* cf. *decussata, Dentalium* s.l., *Actinopteria regularis* (Etheridge jun.), *Aviculopecten sp., Cardiomorpha hindi, Edmondia punctatella* (Jones), *Leiopteria sp., Limipecten sp., Modiolus sp., Myalina sp., Nuculopsis gibbosa, Palaeoneilo* cf. *laevirostrum, P. mansoni, Paleyoldia macgregori, Pernopecten sp., Polidevcia attenuata, Prothyris scotica, Sanguinolites* aff. *abdenensis, S.* cf. *clavatus, S. striatus?, S. sp. variabilis* group, *Solemya sp., Solenomorpha minor, Streblochondria sp., Wilkingia elliptica*, orthocone, goniatite indet., crinoid columnals.

Plean Limestones. The Plean Limestones are minor marine horizons between the Calmy and Castlecary limestones. The relationships of the fossiliferous horizons have not been observed in a continuous sequence. The fauna given below is a composite one. Several of the localities were previously ascribed to the Upper Coal Measures by Anderson (*in* Knox 1954, p. 77, beds a and b). The correlation of the isolated exposures in the Keil Burn has been reinterpreted and further collections have been made from the beds in question. They are now placed in the upper part of the Upper Limestone Group and interpreted as representatives of the Plean Limestones. Localities 66, 71–76 and Balcormo Wood and Pilmuir boreholes, corals indet., *Lingula mytilloides, Orbiculoidea* cf. *nitida*, orthotetoid, *Productus carbonarius?, Pugilis* cf. *pugilis, Schizophoria?, Bellerophon anthracophilus, Euphemites* cf. *urii, Glabrocingulum sp., Latischisma globosa, Retispira sp., Strobeus sp., Dentalium* s.l., *Aviculopecten sp., Modiolus sp., Myalina?, Palaeolima?, Palaeoneilo mansoni, Pernopecten sp., Polidevcia attenuata, ?Sanguinolites clavatus, S. sp. variabilis* group, *Schizodus sp., Solenomorpha minor, Streblochondria sp.*, orthocone, goniatite indet., crinoid columnals.

Castlecary Limestone. The uppermost bed of the Upper Limestone Group, the Castlecary Limestone, has only been identified at two localities in the district. The first is locality 67 where *Clisiophyllum sp.*, gastropod fragments and crinoid columnals have been collected. The second is at the top of the Dumbarnie Borehole where a thick limestone, separated by a fault from the underlying succession, yielded *Semiplanus* cf. *latissimus* and crinoid columnals. This limestone, if its correlation with the Castlecary Limestone is correct, gives the highest record of *S.* cf. *latissimus* in Scotland.

PASSAGE GROUP

Fossils have been obtained from the Passage Group only in the Largo area. In the Largo Burn and from the shore nearby, localities 91, 91a, the following composite fauna was collected: '*Conularia*' *sp., Serpuloides sp., Spirorbis sp., Lingula mytilloides, Linoprotonia?*, orthotetoid, *Productus* cf. *carbonarius, Pugilis?*, murchisoniid, *Edmondia?*, orthocone. Comparison with other districts in central Scotland suggests that the beds containing these fossils are in the lower part of the Passage Group.

COAL MEASURES

The area of outcrop of Coal Measures in the district is very small and the only place where a small part of the measures can be examined is on the coast section at Lower Largo where, on the evidence of the non-marine faunas, Middle and Upper Coal Measures are present.

Middle Coal Measures. At Locality 98 three musselbands are present within a sequence of about nine metres. The composite fauna is *Anthraconaia sp., Anthracosia* cf. *acutella, A. atra, A. aquilinoides?, A.* cf. *concinna, A.* cf. *fulva, Anthracosphaerium sp.* and *Naiadites sp.*, which points to the measures being near the top of the Middle Coal Measures.

Upper Coal Measures. A band at Locality 97 yielded *Anthraconauta* cf. *phillipsii, A.* cf. *tenuis* and *Carbonita sp.* These fossils indicate a position in the Upper Coal Measures. Previously this band had been tentatively taken to represent a marine horizon above Skipsey's Marine Band (Anderson *in* Knox 1954, p. 77, band c) but the present interpretation is mainly based on better preserved specimens found during a re-examination of the locality.   R.B.W.

REFERENCES

BENNISON, G. M. 1960. Some Lower Carboniferous non-marine lamellibranchs from east Fife, Scotland. *Palaeontology*, 3, 137–152.
—— 1961. Small *Naiadites obesus* from the Calciferous Sandstone Series (Lower Carboniferous) of Fife. *Palaeontology*, 4, 300–311.
BROWN, T. 1861. Notes on the Mountain Limestone and Lower Carboniferous rocks of the Fifeshire coast from Burntisland to St Andrews. *Trans. R. Soc. Edinb.*, 22, 385–404.
CURRIE, ETHEL D. 1954. Scottish Carboniferous goniatites. *Trans. R. Soc. Edinb.*, 62, 527–602.
FORSYTH, I. H. and CHISHOLM, J. I. 1968. Geological Survey boreholes in the Carboniferous of east Fife. *Bull. geol. Surv. Gt Br.*, No. 28, 61–101.
GEIKIE, A. 1902. The geology of eastern Fife. *Mem. geol. Surv. Gt Br.*
KIRK, S. R. 1925. The geology of the coast between Kinkell Ness and Kingask, Fifeshire. *Trans. Edinb. geol. Soc.*, 11, 366–382.
KIRKBY, J. W. 1880. On the marine fossils in the Calciferous Sandstone Series of Fife. *Q. Jnl geol. Soc. Lond.*, 36, 559–590.
KNOX, J. 1954. The economic geology of the Fife coalfields, Area III. *Mem. geol. Surv. Gt Br.*
LEITCH, D. 1942. *Naiadites* from the Lower Carboniferous of Scotland: a variation study. *Trans. geol. Soc. Glasg.*, 20, 208–222.
MACGREGOR, M. 1930. Scottish Carboniferous stratigraphy: an introduction to the study of the Carboniferous rocks of Scotland. *Trans. geol. Soc. Glasg.*, 18, 442–558.
MITCHELL, G. H. and MYKURA, W. 1962. The geology of the neighbourhood of Edinburgh. 3rd edit. *Mem. geol. Surv. Gt Br.*
NEVES, R., GUEINN, K. J., CLAYTON, G., IOANNIDES, N. S., NEVILLE, R. S. W. and KRUSZEWSKA, K. 1973. Palynological correlations within the Lower Carboniferous of Scotland and northern England. *Trans. R. Soc. Edinb.*, 69, 23–70.
TAIT, D. and WRIGHT, J. 1923. Notes on the structure, character and relationship of the Lower Carboniferous Limestones of St Monans, Fife. *Trans. Edinb. geol. Soc.*, 11, 165–184.
TRAQUAIR, R. H. 1906. Notes on the Lower Carboniferous fishes of eastern Fifeshire. *Proc. R. Phys. Soc. Edinb.*, 16, 80–86.
WATTISON, A. 1962. Temporary exposures in the Calciferous Sandstone Measures of east Fife. *Trans. Edinb. geol. Soc.*, 19, 133–138.
WILSON, R. B. 1966. A study of the Neilson Shell Bed, a Scottish Lower Carboniferous marine shale. *Bull. geol. Surv. Gt Br.*, No. 24, 105–130.
—— 1967. A study of some Namurian marine faunas of central Scotland. *Trans. R. Soc. Edinb.*, 66, 445–490.
—— 1974. A study of the Dinantian marine faunas of south-east Scotland. *Bull. geol. Surv. Gt Br.*, No. 46, 35–65.
WRIGHT, J. 1950–54. A monograph on the British Carboniferous Crinoidea. *Palaeont. Soc.*

Chapter 10

OLIVINE-DOLERITE SILL-COMPLEX

INTRODUCTION

NUMEROUS sheet-like bodies of olivine-dolerite and allied rock types crop out in the western part of the district, forming a sill-complex of considerable extent (see Fig. 16). The area in which these intrusions occur extends from the Firth of Forth between Lundin Links and St Monance northwards to the Ceres–Maiden Rock Fault Zone and within this area more than thirty sill-like bodies of varying size have been recorded. Isolated members occur at Spalefield and Kingask, about three kilometres east of the main group of intrusions, and the geographically isolated Isle of May sill may also belong to the same suite. The sills range in thickness from a few metres to a maximum recorded figure of 114·3 m at the Balhousie Water Borehole [NO 43210600]; several others are known to exceed 55 m. Some of the sills have fairly extensive outcrops, the largest which can with reasonable certainty be attributed to a single intrusion being about four kilometres long and up to one and a half kilometres across.

The area invaded by the sills is one of considerable geological complexity, and because the presence of the sills has inhibited exploratory drilling for coal, little is known about its structure, particularly in the area of maximum density of intrusion. In consequence the relationship of the intrusions to the folding and faulting has not been established. A few of the sills, e.g. the Higham and Cameron Burn intrusions (Fig. 16), appear to maintain their horizon across the axis of a fold, but this does not necessarily imply that such intrusions have been affected by the folding. In some areas of poor exposure and conjectural structure, sills are shown on the geological maps as crossing fault-lines, but no definite proof of this has been obtained. Other sills have been mapped as terminating against faults. This relationship too has nowhere been demonstrated with certainty although the circumstantial evidence in several cases appears to be fairly strong. In any case, such a relationship may imply either that the fault was later and cuts the sill or that the sill was later and used the fault-plane to change its horizon of intrusion. On the available field evidence, therefore, it is not possible to determine whether the episode of sill emplacement pre-dates or post-dates the folding and faulting. Indeed the two may well have overlapped.

The geographical ranges of the olivine-dolerites and quartz-dolerites overlap, but no evidence of age-relationships has been obtained in the field. There is, however, clear evidence that some of the volcanic rocks are later than some of the olivine-dolerite sills. The Kinaldy sill, for instance, is cut by two necks, one of which contains blocks of dolerite most probably derived from the sill. Near the other neck the dolerite has been invaded by tuffisite. The Drumcarro sill also has been penetrated by a neck, at the western end of its outcrop, and the Dunicher Law sill has been cut by a monchiquitic dyke believed to be related to the neck intrusions.

The youngest strata cut by members of the olivine-dolerite sill-complex belong to the Upper Limestone Group (E_2 Stage of the Namurian). No sill of this type is known to cut Westphalian strata anywhere in Fife, though petrographically similar sills do so in the Glasgow area and in Ayrshire. Several of the east Fife sills have been dated radiometrically by the Isotope Geology Unit of the Institute of Geological Sciences, using the 'whole-rock' potassium–argon method (Forsyth and Rundle, in press). The results are summarised in Table 6. Forsyth and Rundle stress that these results are all minimum ages. The significantly lower figures for five of the samples—including two from the same sill which gave minimum ages of 235 and 288 m.y. respectively—are attributed to argon loss. The highest figures were obtained from the Kilbrackmont–Baldutho sill, which is fresh and very fine-grained. Forsyth and Rundle consider therefore

TABLE 6

Radiometric age-determinations of members of the olivine–dolerite sill-complex

Locality	Petrographic type	Minimum age (m.y.)
Greigston Waterhole	ophitic olivine–dolerite	262 ± 6
Newbigging of Craighall	ophitic olivine–dolerite	282 ± 6
Drumcarrow Craig	non-ophitic olivine–dolerite	235 ± 6
Ladeddie	non-ophitic olivine–dolerite	288 ± 6
Baldutho Quarry	olivine–basalt of Dalmeny type	298 ± 6
Kilbrackmont Quarry	olivine–basalt of Dalmeny type	304 ± 6
Dunicher Law	teschenitic olivine–dolerite	274 ± 8
Cassindonald	analcime–basanite	283 ± 6

Constants used: $\lambda_e = 0.584.10^{-10}$ yr.$^{-1}$; $\lambda_\beta = 4.72.10^{-10}$ yr.$^{-1}$; $^{40}K = 0.0119$ atom %

The quoted errors in the table and in text are combined standard deviations and take into account uncertainties in the mass spectrometric determinations, in the volume of spike (^{38}Ar isotopic tracer) and the potassium content (based on replicate determinations). The effects of error magnification due to the correction for atmospheric argon are also included.

that this sill is likely to give the best minimum estimate for the time of intrusion of the sill-complex, which they place at 304 ± 12 m.y. (95 per cent. confidence level). From this and from what evidence there is on field relationships they conclude that the intrusion of the sill-complex was most probably associated with the Namurian to early Westphalian period of magmatic activity. It is possible that some members of the sill-complex may have functioned as high-level magma reservoirs related to the volcanicity (cf. Francis 1968) during this period.

With a few exceptions, notably the Isle of May sill, the members of the sill-complex are to a considerable extent concealed by drift deposits. Those which show evidence of differentiation are, unfortunately, poorly exposed, and nothing comparable to the way in which the Braefoot sill (Campbell and others 1932, 1934) is displayed on the coast farther west has been found. Contacts with the country rocks are rarely visible and while some of the sills develop good scarp features which make them easily traceable, many of them are known only from isolated, often rather poor, exposures. Consequently their outcrops are difficult to define, and although boreholes have in places provided useful subsurface information, little is known about the form of some of the intrusions. The Dunotter sill, for example, is not exposed at all, being known only from the Dunotter Borehole.

Previous research

The olivine-dolerites and allied rocks of the east Fife sills have hitherto attracted remarkably little attention, despite their abundance, freshness and diversity. Geikie's (1902) account is brief and does not distinguish between the quartz-dolerite and olivine-dolerite suites, although the presence of olivine-dolerites is indicated, for instance at Baldutho, Denork, Wilkieston and Drumcarro, and in the Cameron Burn. The Spalefield and Isle of May sills are the two whose petrography is described in greatest detail therein (Geikie 1902, pp. 392–394).

Balsillie (1922, pp. 442–452) was the first to distinguish the quartz-dolerite and olivine-dolerite sills as separate groups. He recognised and briefly described three principal types in the latter group, namely the Gathercauld type (the ophitic olivine-dolerites of the present account), the Kilbrackmont and Baldutho type (the olivine-basalts of Dalmeny type of this account) and the Kingask–Spalefield type, which here has been subdivided. He noted that the Gathercauld sill is remarkably fresh and that the Gathercauld type of ophitic olivine-dolerite is not common in east Fife. Balsillie also emphasised the freshness of the rocks which comprise his Kilbrackmont and Baldutho type, among which he included also the non-ophitic and the teschenitic olivine-dolerites of the present account. He distinguished the analcime-basanite sills as the 'rocks of Crossgates, Radernie and Lathones' and regarded them as altered members of the Kilbrackmont and Baldutho type, with much secondary analcime. Balsillie also noted that the Spalefield sill contains fresh olivine and that the Kingask sill has affinities with both teschenites and picrites.

Walker and Irving (1928) described only the more north-westerly members of the olivine-dolerite sill-complex and mapped the ophitic olivine-dolerites in greater detail than had previously been attempted. They further discussed Balsillie's 'rocks of Crossgates, Radernie and Lathones', which they classed as augite-teschenites. Walker (1936) described the Isle of May sill, which he called an augite-teschenite, and distinguished two varieties, one with, and the other without, a dark mesostasis; the variety without the mesostasis has more zeolite and locally has fresh olivine. He provided chemical and micrometric analyses of both varieties.

Forsyth (*in* Forsyth and Chisholm 1968, pp. 88–92) described those members of the olivine-dolerite sill-complex that were encountered in the Geological Survey boreholes in east Fife, in particular the Higham sill, which consists largely of picroteschenite, and the hitherto unknown multiple sill in the Dunotter Borehole. The latter has an olivine-basalt component and also a differentiated component which includes orthoclase-rich and picroteschenitic variants, respectively above and below a central zone of analcime-bearing olivine-dolerite.

Francis (1968) has suggested that the alkaline dolerite sill-complexes in the Midland Valley of Scotland served for long periods as high-level magma reservoirs from which the magma found surface expression through vents; he pointed out that the occurrence of necks which cut sills is not necessarily incompatible with this theory. He also specifically disclaimed the suggestion that all the necks had their roots in the sills.

Field Characters

The rocks of the sill-complex include a variety of petrographical types, ranging from olivine-dolerite and olivine-basalt to more alkalic types such as teschenite, essexite and basanite. Most of them are distinguished by symbol, though not by colour, on the new editions of 'one-inch' Geological sheets 41 and 49. For convenience in description the sills have been grouped according to their petrographical characters, and the areal distribution of the various groups is indicated on Fig. 16. The distribution pattern that emerges seems to suggest that there is a rude zonal arrangement which is independent of geological structure. Ophitic and non-ophitic olivine-dolerite sills form well-defined groups in the north-western and northern parts of the sill-complex respectively. Sills of olivine-basalt of Dalmeny type are largely confined to the south and south-east. More alkalic types crop out in the intervening central zone, in a roughly triangular area between Balhousie, Cassindonald and Lochty; analcime-basanites are concentrated on the northern flank of this area, with teschenitic olivine-dolerite and essexite to the south and south-east. The significance of this pattern is not apparent.

A few sills do not readily fall into any of the main petrographical groups represented. These include the Baldastard olivine-basalt sill whose outcrop, about a kilometre in length, is inferred from scattered exposures in the ground south and south-west of Baldastard [NO 421069]; a sill or boss of rather altered olivine-dolerite exposed on the shore of Wood Haven [NT 497997], Elie, where it is intruded into the uppermost Calciferous Sandstone Measures and is itself cut by veins of carbonate and tuffisite (see p. 189); and the West Coates sill, a 6-m sheet of olivine-basalt of Craiglockhart type which cuts tuffs in a stream 64 m west of West Coates [NO 447041].

Ophitic Olivine-Dolerite

The ophitic olivine-dolerite sills occur mainly in the north-western part of the outcrop of the sill-complex, in an area nearly five kilometres long and over one and a half kilometres broad. These rocks correspond to Balsillie's (1922, p. 444) Gathercauld type, which was also recognised by Walker and Irving (1928, p. 7). The fresh members of this group are compact, almost black rocks: where the olivine is altered they have a greenish tinge.

Details

The Gathercauld sill, which probably reaches 30 m in thickness, is best exposed in the cliffs, nine metres high, on the west side of Gathercauld Craig [NO 423101]. From there it can be followed southwards in good exposures past Gathercauld Farm to the cliffs on the north side of the Craighall Burn, where it is thought to terminate along the line of the Cassingray Fault. The sill can also be traced northwards from Gathercauld Craig for nearly a kilometre, making the total length of almost continuous exposures about one and a half kilometres. The northern limit of exposures is near the eastward extension of the fault which runs through the Newcraighall limestone quarries

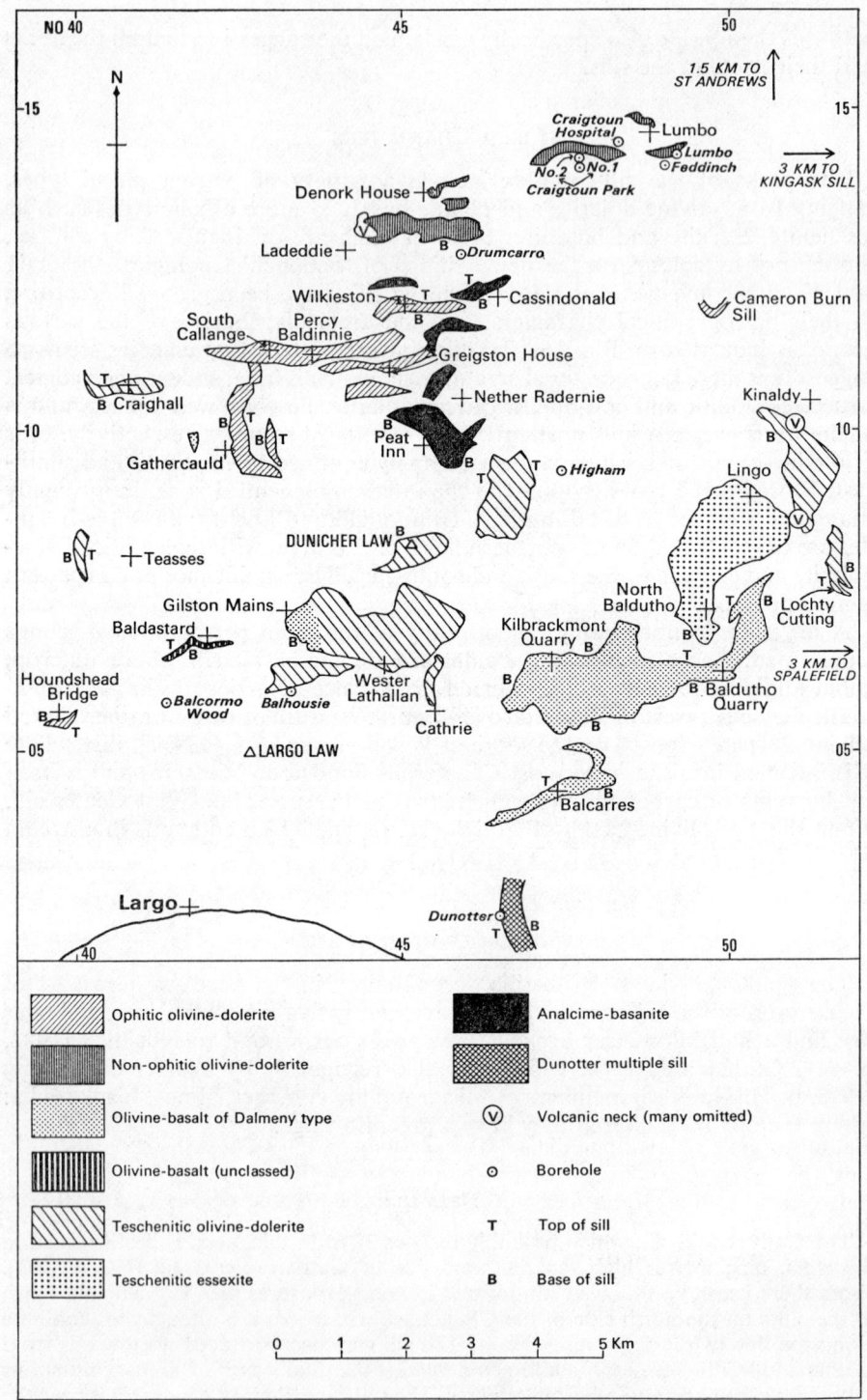

Fig. 16. *Sketch-map of the olivine-dolerite sill-complex showing areal distribution of the various rock-types*

and it may be that the outcrop of the sill stops at this fault, possibly reappearing north of it, as shown on Geological Sheet 41, in an inferred outcrop south of South Callange [NO 427113]. Two small exposures of ophitic olivine-dolerite about 800 m east of Gathercauld Craig are taken to belong to a separate small sill with northerly strike.

Ophitic olivine-dolerite is poorly exposed in an old quarry 180 m north of South Callange. In another disused quarry 90 m north-west of Percy Baldinnie [NO 435113] 3·7 m of this material can be seen and there are several exposures west and north-west of Northbank [NO 447115]. The structure of this area is inadequately known, but it is suggested that these exposures all belong to one sill with easterly strike. Another east-trending ophitic olivine-dolerite sill can be traced eastwards from exposures near Greigston Mains [NO 443111], past Greigston House, close to which three metres of dolerite are exposed in an old quarry, to exposures in fields on both sides of the Peat Inn–St Andrews road. About 400 m S.S.W. of Greigston House, dolerite exposed in an old quarry [NO 44491075]—the Greigston Waterhole of Walker and Irving (1928, p. 7)—is thought to belong to another sill with easterly strike, which is probably separated from the other two by the Lathones Fault. The analysed specimen (Anal. I, Table 7) was obtained from this quarry, the only known exposure of the sill.

North-east of the Radernie Fault only one outcrop of ophitic olivine-dolerite has been traced. East of Wilkieston [NO 450121] it is well exposed in cliffs which show the sill to be over nine metres thick, and it continues eastwards across the Peat Inn–St Andrews road to exposures in old quarries [NO 456121] on the east side.

More than three kilometres east of the main group of ophitic olivine-dolerite outcrops there are two small exposures [NO 49761187; NO 50081199] in the Cameron Burn of a dolerite sill at least three metres thick, which is similar to the other ophitic olivine-dolerites except that it contains less olivine. The disposition of the dips in the adjacent strata suggests that these exposures belong to the same sill outcropping on the two limbs of a small syncline with northerly trend.

The only other occurrence of ophitic olivine-dolerite is in the south-western part of the district where the Balcormo Wood Borehole [NO 41320575] cut a sill 10·7 m thick between depths of 227·4 m and 238·0 m. This sill has been intruded into Upper Limestone Group strata, whereas all the other sills of this type in east Fife lie in the lower part of the Limestone Coal Group or the Lower Limestone Group, more than 300 m lower in the sequence. No surface exposure of this body is known within the district under review, but it is well displayed in the Hatton Burn, 135 to 180 m south of Houndshead Bridge (see Fig. 16), and only 90 m beyond the western limit of the district.

NON-OPHITIC OLIVINE-DOLERITE

The sills of non-ophitic olivine-dolerite, of which the Drumcarro intrusion is the most prominent, crop out only in the northern part of the sill-complex area where they occur in a belt extending from Ladeddie by Drumcarro and Denork to Lumbo, a distance of over five kilometres. Balsillie (1922, p. 446) regarded this type as a coarse variety of his porphyritic Kilbrackmont and Baldutho type (the olivine-basalts of Dalmeny type of the present account), but the textural and chemical differences and the geographical separation between the two groups appear to justify their being regarded as separate types. Walker and Irving (1928, p. 8) distinguished the non-ophitic dolerites on textural grounds from the otherwise very similar ophitic type.   I.H.F.

Details

The Drumcarro intrusion is an irregular body which exceeds 30 m in thickness and is elongated in an easterly direction. It forms the high ground of Ladeddie Hill (180 m above O.D.) and Drumcarrow Craig (210 m above O.D.) on both of which it is well exposed. The outcrop of the intrusion is just over one and a half kilometres in length, and varies from 150 m to 450 m in width. The rock is best exposed in an old quarry [NO 45971313] at the south-eastern end of Drumcarrow Craig where about 12 m of columnar dolerite are seen. This is the locality from which the analysed specimen (Anal. II, Table 7) was obtained. The limits of outcrop of the body are well known from old records and maps relating to former coal-mining activities, particularly along its southern margin which is shown to be steeply transgressive towards the country rocks. On its northern side the body is separated from the nearby Denork Craig intrusion by a narrow strip of sedimentary rocks but a geomagnetic traverse across this area suggests that dolerite is present at shallow depth beneath these sediments. It is quite possible therefore that these two intrusions are connected beneath the surface, forming part of a single irregularly transgressive, northward-dipping, sheet. At its eastern end the Drumcarro intrusion appears to plunge at a fairly steep angle beneath the strata on the western limb of the Denhead Syncline: Denhead No. 4 Borehole [NO 46381339] encountered 'whinstone' at a depth of 52·4 m, probably just above the horizon of the Charlestown Main Limestone. At its western end, on Ladeddie Hill, the intrusion is cut by basalt and agglomerate of the Ladeddie neck (see Fig. 16). The dolerite of the Drumcarro intrusion maintains its coarse-grained character right up to its contact with the fine-grained basalt, suggesting that the neck is younger than the dolerite.

The Denork Craig intrusion is exposed on Denork Craig [NO 45651370], a steep-sided knoll of dolerite in which columnar jointing is well developed. On its eastern, southern and western sides there is evidence of old coal workings, and on its northern side the intrusion is thought to end against an east–west fault, one of a series of sub-parallel faults in the Ceres–Maiden Rock Fault Zone. Geophysical evidence (see above) suggests that the intrusion may be linked to the Drumcarro body. It may also be connected to the nearby Denork House intrusion, a roughly horizontal sill-like body about 7·6 m thick, exposed in a small gully [NO 45491383] 60 m south-east of Denork House. This body appears to rest transgressively on coal-bearing sediments which dip at high angles to the east.

The Mount Melville sill is exposed in a line of old quarries excavated along the northern slopes of a smooth ridge which runs westwards for about 900 m from Mount Melville [NO 483145]. Three shallow boreholes drilled by the Geological Survey in 1966 provide evidence that the sill is intruded more or less conformably into southward-dipping sediments near the base of the Limestone Coal Group. Craigtoun Park No. 2 Borehole [NO 47721442] was drilled just to the south of the dolerite outcrops, and encountered dolerite at a depth of 14·9 m, beneath sediments which include a *Lingula* band and two coal seams (p. 100). Sediments probably lying at the same stratigraphical horizon were encountered near the bottom of Craigtoun Park No. 1 Borehole [NO 47751426], which lies a few hundred metres to the south of No. 2. The results of these two boreholes suggest that both dolerite and sediments dip in a roughly southerly direction. The Craigtoun Hospital Borehole [NO 48281464] was sited north of the dolerite outcrop and encountered coal-bearing sediments only slightly lower in the sequence than those in the Craigtoun Park boreholes. Calculations based on the range of dips available from the boreholes suggest that the thickness of the sill lies between 76 and 137 m.

The upper part of the Lumbo sill is exposed in the sides of Lumbo Den and in the bed of the Lumbo Burn about 180 m north-west of Lumbo Farm. The base of the sill is not exposed, so that the visible thickness of 7·6 m may not represent the full thickness which, according to an old borehole record, reaches 12·2 m nearby. The lateral extent of the sill is not known. In a small cliff-like exposure on the east side of the den [NO 48821490], about a metre of altered sediments rests on the fine-grained top of the sill; the conformable contact can be traced for several metres. In Mount Melville No. 1 Borehole (1855) [NO 48711488] 'whinstone', probably the Lumbo sill, was encountered at a depth of 22·6 m; it is at least 6 m thick and lies at a horizon just below the Radernie Duffie Coal (p. 74).

The Catcraig intrusion is exposed about 450 m south-east of Lumbo Farm at Catcraig Quarry [NO 492145] where a few metres of fine-grained xenolithic dolerite are visible. A petrographically similar but non-xenolithic rock was encountered at rockhead about 180 m east of the quarry, in the Geological Survey Lumbo Borehole (1966) [NO 49361454]. The bore was stopped after penetrating 3·7 m of the intrusion, but without reaching the base. The Feddinch Borehole (1966) [NO 49151434], drilled about 90 m south of Catcraig Quarry, encountered sediments belonging to the upper part of the Calciferous Sandstone Measures, but no dolerite. The Catcraig dolerite is interpreted as a concordant sheet which dips northwards and is truncated by the Maiden Rock Fault on its northern side.

J.I.C.

OLIVINE-BASALT OF DALMENY TYPE

Olivine-basalts of the Dalmeny type described by MacGregor (1928) are prominently displayed at Balcarres Craig, and also at Kilbrackmont and Baldutho—from which two localities they acquired the composite type-name given to them by Balsillie (1922, p. 446). They mostly occur in the southern and south-eastern parts of the sill-complex. Individual sills are known to exceed 30 m in thickness and some probably exceed 60 m. These olivine-basalts tend to be very fresh, dark-grey uniform rocks, and contain at Baldutho the only working roadstone quarry in east Fife.

Details

The outcrop of the main body of olivine-basalt of Dalmeny type in east Fife, the Kilbrackmont–Baldutho sill, extends for four kilometres from Kilbrackmont Craig [NO 473064] by Baldutho Craig to beyond Craighead Quarry [NO 50540685]. Like several of the other members of the sill-complex it appears to crop out in the core of a syncline, in this case the Baldutho Syncline. The surface trace of the base of the sill is therefore much more extensive than that of the top, and can be followed with reasonable accuracy round Kilbrackmont Craig, past Lathallan Mill and Balniel, eastwards to a streamlet 550 m north-west of Gibliston where the base is actually exposed [NO 49250602], and on to Baldutho Craig and Craighead Quarry. The sill is not seen north of the latter and is presumed to die out, possibly near Knights Ward. The top of the sill is nowhere exposed but its surface trace is thought to extend westwards from Baldutho Craig for about a kilometre and thence northwards to the Cassingray Fault; the line of this fault on Baldutho Craig is marked by a gully, probably indicating that there was some movement of the fault after the intrusion of the sill. 'Whinstone' encountered at depths of about 60 m in several bores 400 m south-east of East Cassingray [NO 491070] probably belongs to this sill. North of the Cassingray Fault it does not appear at the surface on the north-west limb of the syncline, but the occurrence of a 21-m body of 'whin' at 76·2 m in Cassingray No. 1 Borehole (1910) [NO 48930726] suggests that it persists some distance beyond the axial zone, and a small sill of similar material found in several bores situated between 400 and 800 m west of South Cassingray [NO 486074] may be an offshoot from the main body.

The Kilbrackmont–Baldutho sill is quite well displayed at the surface, the best exposures being in the disused quarry [NO 47260636] at Kilbrackmont and the working one [NO 499063] at Baldutho Craig; in both quarries at least 18 m of fresh, dark-grey uniform basalt can be seen. The analysed specimens (Anals. III, IV, Table 7) were obtained from these quarries. Other good exposures are found in Craighead Quarry [NO 50540685], Clachreid Ha' Wood Quarry [NO 48550580] and around Balniel [NO 474056]. Over most of its length the sill is intruded into the lower part of the Limestone Coal Group or the uppermost part of the Lower Limestone Group, but at its north-eastern end it is thought to cross the Lathones Fault into Calciferous Sandstone Measures.

The thickness of the sill must be considerable and probably exceeds 60 m. Baldutho No. 5 Borehole (1902) [NO 49870704] cut a body of 'whin', at least 68·6 m thick, between 22·8 m and the bottom of the bore at 91·4 m. This 'whin' almost certainly belongs to the Kilbrackmont–Baldutho sill. The greatest thickness of basalt which is definitely known to belong to this sill is 31·7 m in a bore at Kilbrackmont Craig farm steading, which stopped without reaching the base of the sill.

The Balcarres sill forms Balcarres Craig [NO 478045] where up to 30 m of uniform basalt are prominently displayed. Thence it can be traced both westwards at least as far as an old quarry [NO 467041], and northwards by means of exposures [NO 47950475; NO 47650505] in the Den Burn, apparently round the nose of a westward-plunging syncline. The sill is emplaced in little-known strata of Limestone Coal Group age. The multiple Dunotter sill, known only from the Dunotter Borehole [NO 46500250], consists partly of olivine-basalt of Dalmeny type, the rest being teschenitic olivine-dolerite with picroteschenitic and essexitic differentiates. The olivine-basalt appears to be the earlier component and forms the highest part of the sill (about 2·06 m thick) that was cut by the borehole and most of the lowest part (16·8 m thick). The presence of olivine-basalt of Dalmeny type in exposures south and east of Gilston Mains [NO 432072], within an outcrop mainly composed of teschenitic olivine-dolerite (see p. 145), suggests that here also the two types may be components of a multiple intrusion. A small sill of the same type of basalt is exposed in the Craighall Burn 320 m S.S.E. of Newbigging of Craighall [NO 416102]. Another sill of this type, 10·6 m thick, was encountered in the Drumcarro Borehole [NO 45911290], intruded near the base of the Lower Limestone Group: no surface exposure of this body is known.

TESCHENITIC OLIVINE-DOLERITE

The sills of teschenitic olivine-dolerite are geographically more scattered than those of the other main groups (see Fig. 16), though they tend to occur mainly in the central and eastern parts of the sill-complex and some of them, e.g. the Higham and Dunotter sills (Forsyth *in* Forsyth and Chisholm 1968, pp. 89–92), show marked vertical differentiation, from teschenite or essexite through teschenitic olivine-dolerite to picroteschenite. Indeed it is quite likely that this feature is generally associated with the teschenitic olivine-dolerites. Differentiation of this kind is suspected in the Lochty and Kinaldy sills, and it may be that only the scattered and inadequate nature of the exposures prevents its detection elsewhere. The greatest known thickness for a member of the olivine-dolerite suite was recorded in the Balhousie Water Borehole [NO 43210600], which proved 114·3 m of teschenitic olivine-dolerite. No marked vertical differentiation was however noted in this sill.

DETAILS

The Craighall sill is the most north-westerly of the teschenitic olivine-dolerites (see Fig. 16). It is well exposed on the ridge north of Craighall [NO 407106], where it appears to be over 15 m thick, and is also well displayed in an old quarry [NO 401107] beside the Craighall Burn, where fully 7·6 m of columnar dolerite can be seen. In Teasses Den about 800 m west of Teasses House [NO 407081] olivine-dolerite related to teschenite has recently been discovered among the leaves of a quartz-dolerite sill. The outcrop extends for nearly 800 m in the woods on the east side of the den, and exposures are locally very good. The sill is probably about 15 m thick. To avoid confusion with the Teasses sill of quartz-dolerite, this body is here referred to as the Woodtop sill, after a nearby cottage.

The main concentration of teschenitic olivine-dolerite sills is in the area between Balhousie [NO 425063] and Higham [NO 469094]. A water bore [NO 43210600] at Balhousie proved a sill of teschenitic olivine-dolerite, 114·3 m thick, the top of which lies only about a metre below rockhead. This sill appears to have a considerable area of outcrop to the north of the bore, but exposures are few and poor. In consequence the shape of this outcrop is ill-defined. Another body forms a more extensive outcrop a short distance to the north-east, which can be traced by means of scattered, rather poor, exposures of weathered dolerite. It is best seen in two old quarries [NO 437075; NO 449064]. It is believed that the intrusion is an irregular sheet lying in a syncline with the top nowhere preserved, but the structure of the country rocks is too little known for this to be established. Near its western end teschenitic olivine-dolerite appears to merge into an area of olivine-basalt of Dalmeny type (see p. 144), suggesting that the intrusion may be a multiple one similar to the Dunotter sill (see pp. 144, 146). Exposures of dolerite in the railway cutting at Cairn and in a small wood 275 m north of Cathrie [NO 455056] are thought to belong to a small sill of olivine-dolerite of teschenitic affinities, here named the Cathrie sill.

The best single exposure of teschenitic olivine-dolerite is in the old quarry [NO 451083] on Dunicher Law from which the analysed specimen (Anal. V, Table 7) was obtained. The dolerite, which is cut by a basanitic dyke 1·5 to 2·4 m wide (cf. Walker and Irving 1928, p. 2) is at least 15 m thick here. The extent of the outcrop of this intrusion is ill-defined, as is its form, which may well be irregular. Another poorly-defined area of teschenitic olivine-dolerite lies around and south of Higham [NO 469094]. The only good exposures are in an old quarry 450 m south of Higham farm steading. The northern part of the outcrop lies just below the top of the Calciferous Sandstone Measures. It appears to be in the axial region of an anticline, and this impression is strengthened by the presence of the sill at the same stratigraphical position in the Higham Borehole [NO 47460941]; here it is 28·7 m thick, excluding an intercalation of sediments 1·2 m thick near its base. This borehole provides the only good section through the Higham sill and shows that the bulk of this intrusion (20·4 m) consists of picroteschenite. Southwards the sill appears to transgress the Lathones Fault, changing horizon upwards as it does so, but maintaining its position in the axial zone of the anticline. South of the Cadger's Bridge Fault only the lower part of the sill is apparently preserved. It does not occur in any of the numerous bores in the Largoward area south of its outcrop.

Over three kilometres to the east of the main group of sills of teschenitic olivine-dolerite there are two other intrusions—the Kinaldy and Lochty sills—which have this rock as one of their principal components. Both include picroteschenitic layers and the latter also contains material best classed as teschenite. The Kinaldy sill is best exposed in Kinaldy Den [NO 509100] as a coarse-grained black dolerite locally invaded by tuff associated with the adjacent neck; the latter therefore post-dates the sill. There are scattered exposures in the fields on either side of the den, and the sill can be traced southwards for about one and a half kilometres to the Dunino Burn near Lingo Burnside [NO 511087]; here also it has been cut by a later vent, in which blocks of the dolerite are now incorporated. The geological structure of the area is little known, but it appears possible that the Kinaldy sill lies in a syncline, with its top nowhere exposed. The Lochty sill may be a continuation, at a different horizon, of the Kinaldy sill. Its northern limit is probably about 400 m east of the latter's southern extremity, but exposures are poor. Indeed the Lochty sill is now visible only beside and in the railway cutting [NO 516075–517075] 600 to 700 m south-west of Lochty goods station, where it is split into two leaves. The lower one is estimated to be about 15 m thick and contains a layer of picroteschenite; the upper is about six metres thick. Both leaves dip south-westwards with the sediments at between 20° and 27°.

The Dunotter Borehole proved at rockhead a previously undetected multiple sill of basalt and dolerite at least 58 m thick. The major component, nearly 40 m in thickness, shows marked differentiation, with essexitic variants some 20·4 m thick in the upper part and a picroteschenite layer 16·2 m thick in the lower part (Forsyth *in* Forsyth and Chisholm 1968, pp. 88–90). In between there is a zone of teschenitic olivine-dolerite, almost two metres thick, the composition of which is similar to the average for the whole body.

TESCHENITE

The Isle of May, which lies about 15 km beyond the limits of the sill-complex, is made up entirely of teschenite. The configuration of the island clearly points to the intrusion being a sill although no actual contacts are visible. The island is about 1·8 km long, with a precipitous scarp slope, 45 m high, on its south-western side and a gently dipping north-eastern side. The thickness of the sill was estimated to be more than 50 m by Geikie (1902, p. 196) and at least 60 m by Walker (1936, p. 227). Exposures are abundant throughout the island, but the cliffs on the south-west side are in most places difficult or impossible to reach.   I.H.F.

PICROTESCHENITE

The Kingask sill, whose only visible component is picroteschenite, has been intruded into Calciferous Sandstone Measures strata about four kilometres E.S.E. of St Andrews. It is exposed at Kingask Quarry [NO 550146] where up to 4·5 m of coarse-grained black dolerite can be seen. The analysed specimen (Anal. VIII, Table 7) was obtained at this locality. A mass of baked sandstone and shale stands up in the middle of the quarry; the dolerite is chilled against this mass and includes irregular veins of tuffisite. The form of the intrusion is not known, but it is assumed to be sill-like. The presence of picroteschenite layers in the Higham, Dunotter, Kinaldy and Lochty sills has already been mentioned.
                                                                                                                                                                                            J.I.C., I.H.F.

TESCHENITIC ESSEXITE

In the eastern part of the area of the sill-complex there is a sill of teschenitic essexite between Lingo and North Baldutho. It is sufficiently coarse-grained to show its more felsic nature in hand specimen by a grey and white or pink mottling. Similar material is known to occur in the Dunotter sill in association with teschenitic olivine-dolerite and picroteschenite.

DETAILS

The Lingo sill extends from the railway cutting 450 m south of North Baldutho [NO 496072] to Lingo House [NO 50350895]. Exposures are scattered and generally poor, but together with some data from nearby boreholes, they are sufficient to establish that the outcrop limit in the south and west is in fact the base of the sill. It is thought most likely that the top is nowhere preserved and that the sill lies in a postulated north-easterly extension of the Baldutho Syncline. The best exposures are at the southern end of the outcrop, in the old quarry 320 m west of North Baldutho, from which the analysed specimen came (Anal. X, Table 7), and beside the Lingo Burn up to 800 m W.S.W. of Lingo Burnside [NO 511087].

ANALCIME-BASANITE

In the north-central part of the outcrop area of the sill-complex there are several sills which have been classed as analcime-basanites. They are the 'rocks of Crossgates, Radernie and Lathones' of Balsillie (1922, p. 448) which Walker and Irving (1928, pp. 8–9) termed augite-teschenites. In hand specimen these rocks are grey, fine-grained and porphyritic, usually with a distinctive dull appearance.

Details

The most northerly sill of analcime-basanite known in east Fife was encountered in the Drumcarro Borehole between 141·7 m and 149·7 m from the surface; the sill, eight metres thick, lies just below the St Monance White Limestone. Nearly 800 m to the south another sill of analcime-basanite is exposed in and around an old quarry [NO 46151236] near Cassindonald, from which the analysed specimen (Anal. XI, Table 7) was taken. The stratigraphical level of intrusion here is rather higher, between the St Monance Brecciated and Charlestown Main limestones. Above the latter there is another sill of this type exposed beside and in the Kinninmonth Burn 400 m N.N.W. of Wilkieston [NO 450121]. Scattered exposures west and south-west of Threefords [NO 460118] may all belong to a single intrusion of analcime-basanite emplaced near the top of the Calciferous Sandstone Measures. Its outcrop probably ends to the south-west at the Radernie Fault. South-west of this fracture exposures of analcime-basanite at Wester Radernie [NO 455107] are presumed to indicate the presence of a sill lying near the horizon of the Charlestown Main Limestone. Exposures in an old quarry [NO 46001005] 850 m to the south-east are thought to belong to a small basanite sill intruded below the Radernie coals.

The largest outcrop of analcime-basanite in east Fife is that of the southernmost sill, north and east of Peat Inn [NO 453099], on the south-west side of the Lathones Fault. The extent of this outcrop is however largely inferred from a few borehole and shaft records, one of which started in 'whin' and continued in it for 15 m, thus giving a minimum figure for the thickness of the intrusion. Only in the northern part of the outcrop, on both sides of the Nether Radernie farm-road, are there good exposures, the best being in an old quarry [NO 45281039] on the south side, in which three metres of columnar basanite are seen. The only other exposures are about 685 m E.S.E. of Peat Inn.

BIOTITE-BASANITE

The only known biotite-basanite in the district is the Spalefield sill, which lies fully three kilometres to the east of the main outcrop of the sill-complex. This rock was described by Geikie (1902, p. 196) and Balsillie (1922, p. 450) as a teschenite, but the ratio of plagioclase to analcime suggests that it should be classed with the basanites, among which it can be distinguished by its abundance of biotite.

Details

The sill was formerly seen to be about 3·7 m thick in Whinnyhall Quarry [NO 554064], but unfortunately both this quarry and the smaller one on the other side of the road have now been filled in. There is, however, a small exposure of similar basanite in a streamlet 400 m S.S.W. of Whinnyhall Quarry, and this probably marks an attenuated southward extension of the sill.

Petrography

The sill-complex includes rocks which range from olivine-dolerites with 48 per cent. of silica to picroteschenites with only 42 per cent. The olivine-dolerites contain very little analcime or orthoclase whereas the essexites have 20 per cent. of orthoclase and the analcime-basanites over 15 per cent. of analcime. The olivine content of the picroteschenites is over 30 per cent. in some cases but the sill-complex has no known picrites. The analcime-deficient olivine-dolerites are divided into two types, one showing pronounced ophitic texture and the other non-ophitic; the former corresponds to the Gathercauld type of Balsillie (1922). These two types are distinguished from the olivine-basalts of Dalmeny type (Balsillie's Kilbrackmont and Baldutho type) mainly by their coarser grain-size. On the whole the rocks of these three types are little altered. The teschenitic olivine-dolerites form a more variable group and are more altered. With the exception of those which are transitional to the picroteschenites, they contain more analcime and alkali-feldspar and less olivine than the rocks of the other three groups. Their affinity to the teschenites is confirmed by their association, in certain differentiated bodies, with teschenite, teschenitic essexite and picroteschenite. Further reduction in the amount of olivine and increase in that of analcime characterise the teschenites, which commonly contain some fibrous zeolite. The picroteschenites on the other hand show great increase in the amount of olivine (to fully 30 per cent. in some cases), some reduction in the amount of augite and a good deal less plagioclase. The teschenitic essexites are rendered distinctive by their comparative richness in orthoclase, combined with a relative deficiency in plagioclase, and, especially, in olivine.

The basanite members of the olivine-dolerite suite fall into two groups. The first are Balsillie's (1922) 'rocks of Crossgates, Radernie and Lathones', which he regarded as porphyritic olivine-dolerites subsequently analcimised. Walker and Irving (1928, pp. 9–10) noted the richness of these rocks in augite and analcime and classed them as augite-teschenites. They thought that the plagioclase was altered very soon after crystallisation, concomitant with the development of interstitial analcime as a late-stage primary mineral, and this view is accepted here. Because of the abundance of primary analcime and the deficiency in plagioclase these rocks are probably best regarded as analcime-basanites. The second type of basanite is represented only by the Spalefield sill, which was originally called a teschenite by Geikie (1902). It is now considered, however, that this name should not be used for a rock with only some 20 per cent. of modal feldspar, and that it would be more appropriately classed as a biotite-basanite.

Chemistry

Eleven chemical analyses have been carried out on members of the olivine-dolerite sill-complex. They are reproduced in Table 7, together with other analyses for comparison, mainly from the Midland Valley of Scotland. There is not very much variation between these eleven analyses, and a good deal less if the one of picroteschenite is excluded from consideration, yet considerable mineralogical variation occurs, as demonstrated by the modal analyses (Table 8). The extent to which two chemically similar rocks can differ in mode is illustrated by the olivine-dolerite of Dunicher (Anal. V, Table 7; Anal. 8, Table 8) and the analcime-basanite of Cassindonald (Anal. XI, Table 7; Anal. 17, Table 8).

In the former the silica deficiency is shown in the large amount of olivine, in the latter by the richness in analcime. Chemically similar rocks may also differ in other respects. There is for instance a marked textural difference between the coarse Dunicher rock and the much finer-grained olivine-basalts from Baldutho and Kilbrackmont, all of which have similar analyses. Here the difference in grain-size does not arise from difference in size of the intrusions, for the Kilbrackmont–Baldutho sill is one of the thickest in east Fife. The ophitic and non-ophitic olivine-dolerites are similar both chemically and mineralogically, but the pyroxene ratio (pyroxene × 100)/(pyroxene + plagioclase) of the non-ophitic ones (32·6) is higher than that for the ophitic ones (26·6). This agrees qualitatively with the result obtained by Tomkeieff (1945) and Clark (1952) for Carboniferous basalts from Berwickshire and Arthur's Seat respectively.

The analysis of the ophitic olivine-dolerite from Greigston (Anal. I, Table 7) compares so closely with Walker's (1934) analysis from nearby Gathercauld that the latter need not be reproduced; it is also very similar to that of the non-ophitic Drumcarro intrusion (Anal. II), but no other closely comparable analysis has been traced among the sills in the Midland Valley. An olivine-analcime-dolerite from Nottinghamshire (Anal. A, Table 7) is fairly similar, but the combination of comparatively high silica, low alumina and high iron, lime and magnesia found in the olivine-dolerites in east Fife is rather unusual. The two analyses of olivine-basalt of Dalmeny type (Anals. III and IV) resemble each other and also that of the teschenitic olivine-dolerite of Dunicher (Anal. V). They are quite similar to the analysis of the type rock from Dalmeny (Anal. C), the comparative richness in soda and deficiency in lime of the east Fife rocks being shown in the composition of the plagioclase and the presence of an appreciable amount of analcime. An analysis of a basalt of Dalmeny type from Ayrshire (Anal. D) is also similar, as is that of the 'olivine-diabase' member (Anal. B) of the Stankards sill (Flett 1932).

The analysis of the Kingask picroteschenite (Anal. VIII) shows that it closely resembles the picroteschenite layers in the Stankards and Braefoot sills (Anals. K and L respectively) and contrasts with the picrite layer in the Stankards sill (Flett 1932) and the peridotite one in the Lugar sill (Tyrrell 1917, p. 114). In the last two alumina and lime are much scarcer, whereas iron and magnesia are much more abundant. Comparison of all the relevant analyses demonstrates that whereas the Inchcolm picrite (Campbell and Stenhouse 1908) is rightly so called, several other rocks previously classed as picrites either fall within the range of picroteschenites, e.g. the one from Burntisland (Anal. M), or are between the picroteschenites and the olivine-dolerites, e.g. the 'picrite' layer in the Lugar sill (Anal. N). (The 'picroteschenite' layer in the last-named body resembles chemically the olivine-dolerites of east Fife rather than the picroteschenites.) It is suggested in this connection that in general rocks termed 'picrite' should not have a silica content higher than 40 per cent or a magnesia content lower than 20 per cent. Picroteschenites should not have more than 45 per cent of silica or less than 13 per cent of magnesia. The analysis of the Kinaldy sill (Anal. IX) provides another example of a rock-type intermediate between the picroteschenites and the olivine-dolerites. Others comparable to it include the Leckstone 'picrite' (Anal. O), the Carcraig 'teschenite' (Anal. P) and the Galliston type of olivine-dolerite (Anal. Q).

The analysis of the Lochty teschenite (Anal. VI) is not easily matched in the Midland Valley, a teschenite from near Linlithgow (Anal. F) probably being

TABLE 7
Analyses of members of the olivine-dolerite sill-complex

	Olivine-Dolerites and Olivine-Basalts							Teschenites				
	I	II	A	B	III	IV	C	D	V	E	VI	F	VII	G	H	J
SiO_2	48·23	47·96	47·95	47·66	47·71	47·14	46·24	46·16	46·36	46·91	43·90	43·55	46·47	46·31	45·26	46·99
Al_2O_3	14·56	14·21	14·42	14·03	14·16	13·92	13·83	13·60	14·29	12·54	14·87	14·84	15·60	16·91	15·28	15·92
Fe_2O_3	2·68	2·65	2·49	3·13	1·97	1·88	1·92	3·19	2·69	4·51	3·98	3·09	3·30	3·25	4·53	1·57
FeO	8·28	8·28	8·97	6·04	9·34	10·13	9·36	9·39	8·66	8·17	5·46	7·44	6·40	6·46	8·12	7·20
MgO	8·75	8·06	7·96	9·65	8·43	10·20	9·51	8·84	8·65	9·57	7·38	7·41	6·49	6·01	5·15	7·17
CaO	9·36	8·98	7·97	7·74	8·39	8·03	10·06	9·41	8·83	8·53	8·31	7·66	8·48	9·33	9·57	6·82
Na_2O	2·74	3·35	3·86	2·93	3·56	3·50	2·53	2·81	3·39	2·87	3·70	3·82	4·51	3·08	2·98	4·30
K_2O	0·58	0·88	0·78	0·84	1·19	1·16	1·71	1·36	1·24	0·96	1·83	1·90	1·62	1·33	1·18	1·38
H_2O+	2·44	2·17	2·33	3·06	1·53	1·74	0·71	1·52	2·87	1·89	4·38	5·08	3·35	4·31	1·28	4·67
H_2O-	0·69	0·65	1·18	1·62	0·82	0·45	0·32	0·37	0·59	0·47	1·49	1·79	0·75			0·69
TiO_2	1·19	1·84	1·77	1·62	1·87	1·24	2·54	2·36	1·76	3·02	1·83	2·48	2·01	1·82	3·46	2·20
P_2O_5	0·16	0·34	0·26	0·27	0·36	0·38	0·47	0·47	0·11	0·53	0·37	0·63	0·63	0·45	0·59	0·22
MnO	0·16	0·17	0·16	0·16	0·18	0·19	0·15	0·15	0·19	0·38	0·14	0·23	0·14	0·17	0·40	0·10
CO_2	0·07	0·11	0·08	0·00?	0·15	0·07	0·63	0·12	0·28	0·02	2·04	0·53	0·14	0·36	2·90	0·12
Total S as FeS_2	0·07	0·11	—	0·10	—	—	0·15	0·08	0·15	0·00	—	0·13	—	0·22	—	0·41
Allow for minor constituents	0·17	0·20	0·09	0·12	0·21	0·22	0·02	0·05	0·23	0·04	0·19	0·07	0·18		0·08	
Other constituents	—	—	—	—	—	—	—	—	—	—	—	—	—	—	—	—
Totals	100·13	99·95	100·28	100·16	99·87	100·25	100·15	99·88	100·29	100·41	99·87	100·65	99·96	100·01	100·78	99·76
	p.p.m.	p.p.m.			p.p.m.	p.p.m.			p.p.m.		p.p.m.		p.p.m.			
*Ba	90	170			140	180			340		400		220			
*Co	33	35			35	32			33		15		10			
*Cr	210	190			150	240			170		130		100			
*Cu	23	30			40	32			50		25		10			
*Ga	16	14			17	13			16		10		12			
*Li	20	10			16	14			<10		18		18			
*Ni	160	150			190	210			150		54		43			
Sr	200	360			250	320			380		270		400			
V	230	220			210	190			190		180		160			
*Zr	90	160			200	140			200		150		120			
*B	13	12			8	2			15		3		22			
F	440	440			400	380			500		350		500			
S	—	—			300	250			—		140		100			

*Spectrographical determination.   p.p.m.   parts per million.

Analyses of members of the olivine-dolerite sill-complex

	PICROTESCHENITES AND OLIVINE-RICH DOLERITES										ESSEXITES				BASANITES			
	VIII	K	L	M	N	IX	O	P	Q	X	R	S	T	XI	U	W	Y
SiO_2	42.10	43.82	41.92	44.62	44.47	44.70	44.73	43.52	44.50	45.94	45.71	47.03	45.03	45.72	44.20	46.13	45.33
Al_2O_3	11.16	10.60	11.08	10.80	7.59	13.09	11.89	13.95	14.48	15.82	15.23	15.36	14.82	13.91	15.04	16.01	14.99
Fe_2O_3	4.28	4.36	1.83	5.00	6.25	3.40	4.85	5.01	3.13	2.90	2.84	3.38	2.77	3.44	1.56	3.88	3.38
FeO	7.24	6.88	9.02	8.26	9.57	7.12	6.61	6.70	9.01	6.52	6.93	7.35	8.81	8.02	9.52	7.40	8.75
MgO	15.46	12.77	14.96	15.51	11.93	11.40	10.77	10.84	12.26	7.65	8.11	5.10	7.79	7.69	10.26	7.77	6.93
CaO	5.40	6.75	5.91	6.29	10.24	6.18	7.69	8.74	9.55	5.78	7.34	8.47	9.83	8.32	10.13	8.59	8.06
Na_2O	1.42	3.14	2.35	1.50	4.27	2.49	2.77	2.72	1.77	3.55	3.96	4.32	4.33	3.24	3.46	2.78	2.94
K_2O	1.32	1.52	0.70	0.23	1.46	0.99	0.89	1.29	0.79	2.05	1.31	3.00	1.51	1.38	1.15	2.23	1.00
H_2O+	6.24	4.59	4.97	} 5.50	0.73	4.40	4.15	2.20	1.80	5.15	4.70	2.60	1.71	4.02	1.85	} 2.26	4.74
H_2O-	3.03	2.66	4.40		0.48	3.96	3.49	2.62	0.60	1.72	1.54	0.22	0.24	1.08	0.20		0.92
TiO_2	1.45	2.34	1.62	1.64	2.73	1.54	1.53	1.69	1.90	1.68	1.64	2.64	2.30	1.98	1.95	2.38	2.23
P_2O_5	0.29	0.48	0.19	0.29	0.54	0.33	0.46	0.40	0.18	0.60	0.47	0.73	0.58	0.40	0.32	0.74	0.28
MnO	0.16	0.17	0.20	0.20	0.49	0.12	0.41	0.22	0.20	0.13	0.54	0.14	0.37	0.21	0.14	0.19	0.20
CO_2	0.15	tr.	0.29	0.04	tr.	0.06	0.00	0.05	nil	0.07	—	0.00	0.22	0.15	tr.	0.08	0.15
Total S as FeS_2	0.09	0.22	0.52	—	—	—	—	0.39	—	0.08	—	—	0.15	0.24	—	0.21	0.18
Allow for minor constituents	0.21	—	—	—	—	0.20	0.19	—	—	0.20	—	0.33	0.05	0.27	—	—	—
Other constituents	—	0.11	—	0.25	—	—	—	—	—	—	0.19	—	—	—	—	—	—
Totals	100.00	100.00	99.96	100.13	100.75	99.98	100.43	100.34	100.17	99.84	100.51	100.67	100.51	100.07	99.78	100.65	100.08
	p.p.m.					p.p.m.				p.p.m.				p.p.m.			
*Ba	180					160				420				390			
*Co	55					32				15				31			
*Cr	370					300				38				190			
*Cu	17					15				14				33			
*Ga	13					11				15				12			
*Li	10					14				<10				24			
*Ni	310					170				30				160			
*Sr	160					300				360				430			
*V	150					150				200				220			
*Zr	120					150				200				200			
B	16					6				7				25			
F	420					320				620				750			
S	—					10g				—				—			

*Spectrographic determination.   p.p.m. parts per million.   tr. trace.

Details of Analysed Rocks of Table 7

I. Ophitic olivine-dolerite; sill. Old quarry at Greigston Waterhole. [NO 44491075]. S 50480. Lab. No. 1995. Anal. J. M. Nunan and G. A. Sergeant, spectrographic work by C. Park. *A. Rep. Inst. geol. Sci. for 1966*, 1967, p. 96.

II. Olivine-dolerite; intrusion. Drumcarro Quarry. [NO 45991314]. S 50486. Lab. No. 2000. Anal. J. M. Nunan and W. H. Evans, spectrographic work by C. Park. *A. Rep. Inst. geol. Sci. for 1966*, 1967, p. 96.

A. Olivine-analcime-dolerite; sill. Bulcote Borehole, 8 km N.E. of Nottingham. E 23936. Lab. No. 1573. Anal. W. F. Waters and K. L. H. Murray. Guppy 1956, pp. 31–32.

B. Olivine-diabase; sill. Stankards No. 2 Borehole, West Lothian, at a depth of 12·8 m from top of sill. Anal. B. E. Dixon. Flett 1932, p. 150.

III. Olivine-basalt of Dalmeny type; sill. Kilbrackmont Quarry. [NO 47260636]. S 50479. Lab. No. 2015. Anal. J. M. Nunan and W. H. Evans, spectrographic work by C. Park.

IV. Olivine-basalt of Dalmeny type; sill. Baldutho Top Quarry. [NO 49930631]. S 50493. Lab. No. 2004. Anal. J. M. Nunan and W. H. Evans, spectrographic work by C. Park.

C. Olivine-basalt of Dalmeny type; lava. Quarry 210 m south of Dalmeny Church, West Lothian. S 25062. Lab. No. 834. Anal. E. G. Radley. *Summ. Prog. geol. Surv. Gt Br. for 1926*, 1927, p. 77.

D. Olivine-basalt of Dalmeny type; lava. Quarry on left bank of Annick Water, 213 m E.35°S. of Rashillhouse, Ayrshire. S 25065. Lab. No. 835. Anal. B. E. Dixon. *Summ. Prog. geol. Surv. Gt Br. for 1926*, 1927, p. 77.

V. Teschenitic olivine-dolerite; sill. Quarry on Dunicher Law. [NO 45090883]. S 50481. Lab. No. 1996. Anal. J. M. Nunan and G. A. Sergeant, spectrographic work by C. Park. *A. Rep. Inst. geol. Sci. for 1966*, 1967, p. 96.

E. Olivine-basalt, intermediate between Craiglockhart and Dalmeny types; lava. Left bank of Little Calder Water, 365 m N.7°W. of Caldergreen Farm, Avondale, Lanarkshire. S 25757. Lab. No. 866. Anal. E. G. Radley. Guppy 1931, pp. 68–69.

VI. Teschenite; sill. Railway cutting, 640 m S.W. of Lochty Goods Station. [NO 51670752]. S 50492. Lab. No. 2019. Anal. J. M. Nunan and W. H. Evans, spectrographic work by C. Park.

F. Teschenite; sill. Old quarry in Mochrie's Craig, 4 km S.E. of Linlithgow. S 11894. Lab. No. 278. Anal. E. G. Radley. *Summ. Prog. geol. Surv. Gt Br. for 1907*, 1908, p. 54.

VII. Teschenite; sill. Horse Hole, 400 m N.W. of lighthouse, Isle of May, Fife. [NT 65209966]. S 51789. Lab. No. 2050. Anal. J. M. Nunan and R. L. Clements, spectrographic work by C. Park.

G. Teschenite; sill. Links Quarry, Gullane, East Lothian. Anal. T. C. Day. Day 1930, pp. 264–265.

H. Teschenite; sill. Old quarry, Salisbury Crags, Edinburgh. Anal. J. B. Harrison and K. D. Reid. Washington 1917, p. 888.

PETROGRAPHY

J. Teschenite; sill. Quarry near summit of Braefoot promontory, Fife. Anal. T. C. Day. Campbell and others 1932, p. 354.

VIII. Picroteschenite; sill. Kingask Quarry, 730 m N.W. of Winchester. [NO 55021462]. S 50487. Lab. No. 2001. Anal. J. M. Nunan and W. H. Evans, spectrographic work by C. Park.

K. Picroteschenite; sill. Stankards No. 2 Borehole, West Lothian, at a depth of 63·1 m from top of sill. Anal. B. E. Dixon. Flett 1932, p. 150.

L. Picroteschenite; sill. Braefoot Bay, Fife. Anal. T. C. Day. Campbell and others 1934, p. 168.

M. Picroteschenite ('picrite'); sill. Colinswell, Burntisland, Fife. Anal. J. B. Harrison and K. D. Reid. Washington 1917, p. 650.

N. 'Picrite' (transitional to theralite); sill. Glenmuir Water, Lugar, Ayrshire. Anal. A. Scott. Tyrrell 1917, p. 114.

IX. Olivine-rich dolerite; sill. Kinaldy Den, 775 m N.N.W. of South Kinaldy. [NO 50880997]. S 50478. Lab. No. 2014. Anal. J. M. Nunan and W. H. Evans, spectrographic work by C. Park.

O. 'Picrite'; sill. 'Leckstone' quarry, Blackburn, West Lothian. S 11859. Lab. No. 175. Anal. W. Pollard. *Summ. Prog. geol. Surv. Gt Br. for 1905*, 1906, p. 74.

P. 'Teschenite'; sheet. Carcraig, Firth of Forth. Anal. T. C. Day. Day and Stenhouse 1930, p. 241.

Q. Olivine-dolerite ('Galliston type'). Galliston Quarry, Kirkcaldy, Fife. Anal. W. H. Herdsman. Allan 1924, p. 497.

X. Teschenitic essexite; sill. Old quarry, 400 m west of North Baldutho. [NO 49410725]. S 50491. Lab. No. 2003. Anal. J. M. Nunan and W. H. Evans, spectrographic work by C. Park.

R. 'Teschenite'; sill. 'Leckstone' quarry, Blackburn, West Lothian. S 11860. Lab. No. 174. Anal. W. Pollard. *Summ. Prog. geol. Surv. Gt Br. for 1905*, 1906, p. 74.

S. Essexite; intrusion. Craigleith Island, off North Berwick, Firth of Forth. S 11897. Lab. No. 170. Anal. W. Pollard. *Summ. Prog. geol. Surv. Gt Br. for 1905*, 1906, p. 75.

T. Essexite; plug. Hillside, 1½ km north of Lennoxtown, Stirlingshire. S 13311. Lab. No. 270. Anal. E. G. Radley. *Summ. Prog. geol. Surv. Gt Br. for 1907*, 1908, p. 55.

XI. Analcime-basanite; sill. Quarry 365 m N.W. of Cassindonald. [NO 46151236]. S 50483. Lab. No. 1998. Anal. J. M. Nunan and W. H. Evans, spectrographic work by C. Park.

U. Analcime-basanite; lava. River Ayr, 1½ km east of Stair, Ayrshire. Anal. W. H. Herdsman. Tyrrell 1928, pp. 273–274.

W. 'Basalt'; sill. The island of The Lamb, off North Berwick, Firth of Forth. Anal. T. C. Day. Day 1930, p. 265.

Y. Olivine-basalt; sill. Reef A, Braefoot Bay, Fife. Anal. T. C. Day. Campbell and others 1932, p. 349.

the one most similar to it. The analysis of the Isle of May sill (Anal. VII) shows it to be a typical teschenite, similar to those from Links Quarry near Gullane, East Lothian (Anal. G) and Salisbury Crags (Anal. H). It is so similar to Walker's (1936, p. 283) analysis from the Isle of May that the latter need not be reproduced here.

The analysis of the teschenitic essexite from North Baldutho (Anal. X) shows it to have affinities with the teschenites, the Leckstone one in particular. The available analyses of essexites from the Midland Valley form a rather heterogeneous group (e.g. Anals. S and T), on the whole rather more mafic than the North Baldutho rock. The closest resemblance chemically, and probably also mineralogically, is to the Craigleith essexite (Anal. S), which is much finer-grained.

The similarity between the analysis of the analcime-basanite from Cassindonald (Anal. XI) and those of some of the olivine-dolerites has already been noted. This rock is however rich in analcime and this probably accounts for the abundance of retained water indicated by the analysis. Only a few analyses of Midland Valley basanites are available for comparison. A basanite lava from Ayrshire (Anal. U) is fairly similar but more mafic. A closer match is provided by the 'basalt' of The Lamb (Anal. W) and the olivine-basalt near the base of the Braefoot sill (Anal. Y) which, according to Campbell and others (1932, p. 348) varies in mineralogical composition from types resembling the olivine-dolerite of Dunicher to others more like the analcime-basanite of Cassindonald. The two analysed basanites from neck intrusions (Table 10) are generally similar but rather more mafic.

Mineralogy

The rocks of the olivine-dolerite suite consist of varying proportions of plagioclase (mainly labradorite), titanaugite, olivine and analcime, with small amounts of iron ore, orthoclase and apatite usually present. Biotite, zeolite and hornblende occur locally but nepheline is scarce or absent. The amount of alteration differs a good deal from type to type and also to some extent within types, and most rocks contain some secondary material including analcime, carbonate and various viriditic minerals. The olivine-dolerites, except those of teschenitic affinities, and the olivine-basalts are on the whole the freshest, some of them being almost entirely free of alteration whilst others have partly or completely altered olivine and a little interstitial secondary material. The non-ophitic dolerites are on the whole fresher than the ophitic ones.

In the other types the olivine is usually replaced completely by green secondary minerals of the chlorite–serpentine group and the plagioclase is locally altered as well. Partly or wholly fresh olivine, however, occurs sporadically in, for instance, the Woodtop sill, the mafic differentiates in the Dunotter sill, parts of the Isle of May teschenite, the Kingask picroteschenite and the Dunicher Law intrusion. None of the teschenitic essexites has fresh olivine.

In marked contrast to the proneness of the olivine—and, to a lesser degree, of the plagioclase—to alteration, the augite is almost invariably fresh. (The highly altered lower leaf of the Higham sill is the only rock which shows complete alteration of the pyroxene, to carbonate, granules of iron ore and viriditic material.) The augite of the olivine-dolerite suite is a purplish-brown titaniferous variety, many of the crystals being deeper in tint at the margins than in the centres. Its crystal habit is very variable. In the finer-grained rocks (the basalts and basanites) the augite forms abundant subhedral granules. In the non-ophitic rocks the crystals vary from anhedral to euhedral, and in the ophitic ones, which include some of the essexites, large plates of pyroxene enclose or partly enclose the abundant laths of plagioclase. In several of the teschenites, teschenitic olivine-dolerites and essexites, some of the pyroxene takes the form of green aegirine-augite, particularly where it is in contact with analcime.

Orthoclase is in general fresh and is usually anhedral and interstitial, but in the essexites it often forms an investing rim round the plagioclase crystals, which in many cases are more or less altered. Analcime varies in aspect from clear to cloudy in an apparently random manner: it is usually interstitial or secondary to feldspar, but locally it is euhedral against the other zeolites, which form radiating rosettes of fibrous crystals. Natrolite is much the commonest of this group of minerals.

Modal analyses

Numerous modal analyses of members of the olivine-dolerite suite were made, using a point counter. Selected analyses are reproduced in Table 8. Others have already been published in Forsyth and Chisholm (1968, pp. 89–92). The analyses of the more altered rocks are necessarily less accurate than the others, but none the less they give an approximate indication of the quantitative mineralogical composition. In such rocks distinctive pseudomorphs, such as many of those after olivine, were counted as the original mineral.

The ophitic and non-ophitic olivine-dolerites are shown by these analyses (Anals. 1–4, Table 8) to be similar mineralogically as well as chemically. There are, however, some slight differences, as for instance the higher pyroxene ratios of the non-ophitic dolerites (see p. 149). There appears, too, to be an eastward decrease in the amount of olivine in the ophitic rocks and a local enrichment, at Wilkieston (Anal. 2), in analcime (which otherwise is scarce in both types), but on the whole the proportions of the various minerals remain quite consistent within each type. With a few exceptions, the olivine-basalts of Dalmeny type (Anals. 5–7) have less plagioclase and olivine than the olivine-dolerites, but they are in general richer in augite and analcime. This antipathetic relationship between olivine and plagioclase on the one hand, and augite and analcime on the other, is widespread among the rocks of the sill-complex. It is largely responsible for the degree of mineralogical variation shown by this group of chemically very similar rocks. The enrichment in augite and analcime at the expense of olivine and plagioclase is particularly marked in the analcime-basanites, which are on the whole a fairly constant group in their mineralogical composition.

Constancy of composition is also a feature of the picroteschenites (e.g. Anal. 13, Table 8), which are of course characterised by their abundance of olivine, to the extent of 25 to 30 per cent. of the rock. The analysed rock from Kinaldy (Anal. 14, Table 8) is listed as a picroteschenite, although it is really intermediate between the picroteschenites and the olivine-dolerites, in that it is enriched in olivine but retains a large amount of plagioclase. This is achieved at the expense of analcime and, more especially, of augite. Indeed the Kinaldy rock provides one of the very few modal analyses from this suite showing less than 10 per cent. of pyroxene.

The teschenites are noteworthy for their abundance of both plagioclase and analcime: together with the teschenitic essexites they are the most felsic members of the olivine-dolerite suite in east Fife. The amount of analcime present usually

TABLE 8

Mineralogical composition of members of olivine-dolerite sill-complex
(expressed in percentages by volume)

| | Olivine-Dolerites | | | | Olivine-Basalts of Dalmeny Type | | | Teschenitic Olivine-Dolerites | | | Teschenites | | Picroteschenites | | Teschenitic Essexites | | Basanites | |
	Ophitic		Non-Ophitic															
	1	2	3	4	5	6	7	8	9	10	11	12	13	14	15	16	17	18
Plagioclase	53	50	49	44	49·5	42·5	40	45·5	36·5	39	50·5	39·5	30	43·5	38	35	28·5	20
Orthoclase	1	2	3·5	5·5				1	6	4	3·5	2·5	4	2·5	16·5	21·5	37	36
Augite	20	18·5	23	22	24	29	35·5	23	24·5	19	21	24	16·5	9·8	16·5	20	10	9·5
Olivine	19	13·5	13·5	17·5	17·5	16	12·5	16	18	12·5	7	3·5	28·5	25·5	8	4	16·5	18
Analcime	0·5	6	2	1·5	3·5	4·5	2	6·5	3·5	13·5	11	12·5	6	3	11·5	5	5	6
Iron ore	3	3·5	5	5	4	6	6·5	4·5	4·5	4	4	7·5	4	2·5	4	6·5		
Minor constituents	3·5	6·5	3·5	4·5	1·5	2	3·5	3·5	7	8	3	10·5	11*	13·2	5·5	8†	3	10·5‡

*includes 2·5 per cent. of natrolite   †includes 3 per cent. of natrolite   ‡includes 6 per cent. of biotite

1. Greigston   NO 44491075. S 50480
2. Wilkieston   NO 45601210. S 51185
3. Drumcarro   NO 45991314. S 50486
4. Denork   NO 45571373. S 48045
5. Baldutho   NO 49930631. S 50493
6. Kilbrackmont   NO 47260636. S 50479
7. Gilston   NO 43520711. S 50929
8. Dunicher Law   NO 45090883. S 50481
9. Woodtop   NO 39930792. S 49509
10. Lochty   NO 51680752. S 51187
11. Isle of May   NT 65209966. S 51789
12. Lochty   NO 51670752. S 50492
13. Kingask   NO 55021462. S 50487
14. Kinaldy   NO 50880997. S 50478
15. North Baldutho NO 49410725. S 50491a
16. Dunotter Borehole NO 46500250. S 49607a
17. Cassindonald   NO 46151236. S 50483
18. Spalefield   NO 55450655. S 9968 and 37104

exceeds 10 per cent. whereas that of olivine rarely does so. The teschenitic essexites are characterised by the abundance of orthoclase, which occurs to the extent of some 15 to 20 per cent. of the rock.

Details

Ophitic olivine-dolerite. Plagioclase (labradorite) is the dominant mineral (see Table 8) in the ophitic olivine-dolerites, in which it makes up about half the rock. It occurs as fresh laths and rectangular plates up to 1·5 mm long which are in markedly ophitic relationship with augite. The latter is invariably completely unaltered and consists of purplish-brown anhedral plates up to 5 mm long (S 51185). The olivine crystals are up to 3 mm long (S 37061, 37072) and are usually fresh or partly fresh, but in some slices (e.g. S 37084) complete alteration to viriditic material has taken place. Alkali-feldspar and iron ores are the principal accessories and apatite is also present; analcime is scarce, biotite is very rare and fibrous zeolites are absent. A little green interstitial secondary material is usually present. The Cameron Burn sill (S 37084) is unusually rich in apatite and also in alkali-feldspar. The latter occurs in interstitial patches with plagioclase, which contain numerous apatite needles, biotite flakes and iron ore granules. This sill is deficient in olivine compared with the other ophitic olivine-dolerites and the olivine present is completely altered. In view of these compositional differences and of the geographical isolation of the sill (see Fig. 16), it is doubtful whether this rock should be grouped with the ophitic olivine-dolerites or with the teschenitic essexites.

Non-ophitic olivine-dolerite. Labradorite is not quite as plentiful in the non-ophitic olivine-dolerites as it is in the ophitic ones (see Table 8), but it remains much the most abundant mineral. It is fresh in all the slices examined, as is the augite. The latter occurs in abundant, usually purplish-brown, anhedral or subhedral crystals, and locally shows a slight tendency towards ophitic texture. Augite is rather more abundant in this group than it is in the ophitic olivine-dolerites. Olivine, on the other hand, is somewhat less abundant; it forms phenocrysts up to 4 mm long (S 36969), which are usually fresh. The accessory minerals are similar to those found in the ophitic olivine-dolerites. Analcime, though by no means common, is somewhat more abundant, a very little biotite is usually present and a little fibrous natrolite occurs in some slices.

Olivine-basalt of Dalmeny type. The phenocrysts in the basalts of this type consist mainly of olivine; a few of augite and labradorite also occur. Some of the olivine phenocrysts are fresh: others are partly or wholly altered. They are commonly about 1 mm long and many are surrounded by rims of subhedral augite granules. The groundmass is composed largely of those granules and of plagioclase laths; the latter are everywhere the more abundant, but in some slices the difference in amount is not great. Among the olivine-dolerites and allied rocks in east Fife only the analcime-basanites are richer in augite than the basalts of Dalmeny type (see Table 8). Interstitial patches of alkali-feldspar and analcime occur in most of these basalts, but the latter mineral is uncommon except in one from near Gilston Mains (S 49879; see p. 144) and in some of the specimens from the Dunotter sill (e.g. S 50922). Secondary viriditic material and in places a little fibrous natrolite or thomsonite are associated with these minerals. Iron ore granules and apatite needles are the most abundant accessories and a little biotite is locally present.

Other olivine-basalts. The Baldastard sill (S 42194, 49530) consists of abundant laths of fresh labradorite, olivine crystals up to 1·5 mm long, which are mainly unaltered, and abundant granules of fresh augite. Iron ore and apatite occur as accessories and there is a brownish isotropic mesostasis of analcime and glass. The West Coates sill (S 49537) was examined by Mr. R. W. Elliot, who reports that it is an olivine-basalt of Craiglockhart type, containing numerous phenocrysts and microphenocrysts of augite and of pseudomorphs after olivine.

Teschenitic olivine-dolerite. The teschenitic olivine-dolerites are a rather mixed assemblage of rocks from several sills, including some that show definite evidence of marked vertical differentiation and others in which this is suspected. Plagioclase (labradorite) is the most abundant mineral, and usually contributes 35 to 50 per cent. of the rock (see Table 8). The degree of alteration of the plagioclase varies both between one sill and another and also within individual bodies. Augite almost always comes next in order of abundance, forming 15 to 25 per cent. of the total. It is almost everywhere completely fresh and is purplish-brown in colour except in a few cases where green sodic varieties occur at the edges of crystals adjacent to analcime. The crystal habit is variable from anhedral to subhedral or, in a few slices, locally euhedral. Ophitic texture was noted in a few slices (e.g. S 50915–6, 50933). Olivine usually makes up 12 to 18 per cent. of the rock, but both lower and higher amounts are known. It is almost invariably altered to green or brown pseudomorphs, easily recognisable because of their distinctive shape. Analcime is ubiquitous in interstitial patches, within which it may be euhedral against fibrous zeolites; the amount present is very variable, from 1 to over 10 per cent. Interstitial alkali-feldspar, grains of iron ore and apatite needles are ubiquitous accessory minerals, but fibrous zeolites and biotite are rare and local. Most of the slices contain some secondary viriditic material in the groundmass.

Individual sills of teschenitic olivine-dolerite show various special features. The Woodtop sill is noteworthy for the local presence of phenocrysts, up to 2 mm long, of largely fresh olivine (S 49509, 50936). The rock in an old quarry near Wester Lathallan has sub-ophitic plates of augite up to 4 mm across (S 50928). Plagioclase is unusually abundant in the Cathrie sill (S 50931) in which some crystals attain lengths of 4 mm. The amount of olivine (partly fresh) in the Dunicher Law intrusion locally exceeds 20 per cent. (S 34031) although the analysed specimen (S 50481) has only 16 per cent. One slide (S 42193) from the Higham sill shows partial alteration of the pyroxene, which locally has the greenish tinge of aegirine-augite. The middle zone (S 50917) of the coarser component of the Dunotter multiple sill (Forsyth *in* Forsyth and Chisholm 1968, pp. 89–90) is petrographically similar to the teschenitic olivine-dolerites. It is notable for the general freshness of the olivine and the comparative abundance of natrolite (almost 4 per cent. of the rock). The Kinaldy sill is one of the few teschenitic olivine-dolerites which have more olivine (in the form of pseudomorphs) than augite (S 34041, 50478, 50935). The chemical analysis (Anal. IX, Table 7) confirms that part at least of this sill is transitional in type to the picroteschenites.

Teschenite. The Isle of May sill contains abundant plagioclase, in places partly or wholly altered, which in some slices forms phenocrysts up to 4 mm long (S 37095). Walker (1936, pp. 280–281) determined it as labradorite in both the modifications he recognised in this intrusion. The distinction between the two modifications is based on the presence or absence of a brown mesostasis which Walker regarded as probably devitrified glass. He noted that the presence of this mesostasis is associated with decrease in amount and complete alteration of olivine, increase in the amount of iron ore and decrease in the amount of analcime and other zeolites. These features have been confirmed by the present investigation. Walker found that the distribution of the modification with the mesostasis is apparently random.

Augite usually comprises between 20 and 25 per cent. of the Isle of May sill (Anal. 11, Table 8), in which it is fresh throughout. It occurs in crystals up to 4 mm long (S 37098) which vary in shape from subhedral to euhedral. It is in general a purplish-brown titaniferous variety, but locally it develops a greenish tinge where sodic varieties occur adjacent to analcime. In one slice (S 37095) discrete crystals of aegirine were noted.

Some slices with no mesostasis show complete alteration of the olivine, but in others (S 37087, 37090) the mineral is fresh and, for the most part, euhedral (cf. Walker 1936, p. 281); locally it shows a well-marked ophitic relationship with the plagioclase. The amount of olivine present varies from 15 per cent. to less than 5 per cent. of the rock. Analcime is generally abundant and clear. With alkali-feldspar it forms interstitial areas, in some of which (e.g. S 37086) the analcime is euhedral against rosettes of radiating needles of natrolite or thomsonite. Titaniferous iron ore locally forms skeletal crystals. Apatite is ubiquitous and abundant; a little biotite or brown hornblende occurs rarely. Walker (1936, p. 282) also found pegmatitic and aplitic veins, particularly in the upper part of the Isle of May sill. The former he found to be similar in composition to the modification with mesostasis except that they were even poorer in olivine and richer in iron ore. The aplitic veins consist largely of altered soda-orthoclase with some sodic plagioclase. Mafic minerals, diopside and aegirine, comprise only about 15 per cent. of the rock. Reference should be made to Walker's paper for fuller details.

A few of the specimens from the mainland of east Fife may be classed as teschenites. They come from components of differentiated bodies such as the Higham sill (S 50908; Forsyth *in* Forsyth and Chisholm 1968, p. 92) and the Lochty sill (Anal. 12, Table 8; S 49500, 50492) and are distinguished from the olivine-dolerites by their abundance of analcime and smaller amount of olivine.

Picroteschenite. The picroteschenites (see Table 8) are characterised by an abundance of olivine, which makes up 20 to 35 per cent. of the rock. The olivine crystals are up to 3 mm long and most are largely or completely altered to serpentinous material. The principal exception is the picroteschenite layer in the Dunotter sill in which the olivine crystals are altered only along cracks and at their margins (Forsyth *in* Forsyth and Chisholm 1968, p. 90). Augite is in general appreciably less abundant than olivine or plagioclase, and usually forms 15 to 20 per cent. of the rock. It forms phenocrysts which vary from subhedral to euhedral in shape, and locally (S 50919) enter into an ophitic relationship with the plagioclase laths. The freshness of the latter varies from slice to slice. Fresh labradorite crystals reach 4 mm in length (S 50918) in the Dunotter picroteschenite, which is the freshest rock of its type found in east Fife. Analcime is quite abundant, usually forming between 5 and 10 per cent. of the rock. It is clear for the most part and occurs interstitially in association with a little alkali-feldspar and, locally, with rare fibrous natrolite (e.g. S 50487, 50918). Iron ores, apatite and biotite occur as accessories.

Teschenitic essexite. The distinctive features of the teschenitic essexites (see Table 8) are the abundance of alkali-feldspar (usually between 10 and 20 per cent. of the total) and the scarcity of dark minerals, especially olivine, which contributes only some 5 to 10 per cent. of these rocks. Plagioclase (labradorite) is very abundant (35 to 40 per cent.): it forms crystals up to 2 mm long, most of which show partial or complete alteration. The alkali-feldspar was determined as orthoclase in the essexite layer (S 49607) of the Dunotter sill, in which it locally invests plagioclase laths and is euhedral against the interstitial patches of analcime and fibrous natrolite. In another slice (S 50915) from the same sill it includes numerous microlites of apatite, iron ore, sodic pyroxene and sodic amphibole.

Analcime is universally present, but the amount varies a good deal, e.g. in the southern part of the Lingo sill, around North Baldutho, it forms about 10 per cent. of the rock whereas nearer to Lingo itself the amount is reduced to 1 or 2 per cent. The analcime has locally been observed (S 49547, 49848) to be euhedral against fibrous natrolite. The latter is of common occurrence in this group, to the extent of 6·5 per cent. in one slice (S 49547A) and 5 per cent. in another (S 49848). Augite is the main ferromagnesian mineral and usually forms between 15 and 20 per cent. of the total. It is invariably fresh and occurs mainly in the form of purplish-brown ophitic plates up to 5 mm long (S 50930): where it occurs adjacent to analcime, the marginal zone is often composed of a green sodic variety. The olivine has everywhere been found to be completely altered. Iron ore and apatite are ubiquitous accessories. Biotite and brown hornblende occur in places.

Analcime-basanite. The most abundant mineral (see Anal. 17, Table 8) in the analcime-basanites is augite, which occurs in multitudes of fresh subhedral granules forming about one-third of the rock. Olivine forms the only phenocrysts, which are now nearly everywhere completely altered to serpentinous material. The labradorite laths are locally altered, and the abundant interstitial analcime is usually cloudy. The latter is generally associated with a little alkali-feldspar and in places (e.g. S 49652) with rare natrolite. Iron ore grains and apatite needles are the commonest accessories and a little biotite occurs sporadically.

Biotite-basanite. The Spalefield rock (Anal. 18, Table 8; S 9968, 37104) is unique in east Fife. Augite is the most abundant mineral and occurs in numerous subhedral granules and occasional small phenocrysts which are completely fresh and purplish-brown in colour, with deeper tints in the smaller crystals and the marginal zones of the larger ones. The centres of some of the latter show greenish tints. Olivine forms phenocrysts up to 1·5 mm long, most of which are completely altered although locally (e.g. S 9968) alteration is confined to the cracks and margins. Biotite and barkevikite are both present, the former being the more abundant. Cloudy analcime is abundant interstitially. For a more detailed description of this unusual rock reference should be made to Flett (*in* Geikie 1902, pp. 392–393).   I.H.F.

REFERENCES

ALLAN, D. A. 1924. The igneous geology of the Burntisland district. *Trans. R. Soc. Edinb.*, **53,** 479–501.
BALSILLIE, D. 1922. Notes on the dolerite intrusions of east Fife. *Geol. Mag.*, **59,** 442–452.
CAMPBELL, R., DAY, T. C. and STENHOUSE, A. G. 1932. The Braefoot outer sill, Fife: part I. *Trans. Edinb. geol. Soc.*, **12,** 342–375.
—— DAY, T. C. and STENHOUSE, A. G. 1934. The Braefoot outer sill, Fife: part II. *Trans. Edinb. geol. Soc.*, **13,** 148–173.
—— and STENHOUSE, A. G. 1908. The geology of Inchcolm. *Trans. Edinb. geol. Soc.*, **9,** 121–134.
CLARK, R. H. 1952. The significance of flow structure in the microporphyritic ophitic basalts of Arthur's Seat. *Trans. Edinb. geol. Soc.*, **15,** 69–83.
DAY, T. C. 1930. Chemical analyses of thirteen igneous rocks of East Lothian. *Trans. Edinb. geol. Soc.*, **12,** 263–266.
—— and STENHOUSE, A. G. 1930. Notes on the Inchcolm anticline. *Trans. Edinb. geol. Soc.*, **12,** 236–251.
FLETT, J. S. 1932. The Stankards sill. *Summ. Prog. geol. Surv. Gt Br.* for 1931, pt. ii, 141–156.
FORSYTH, I. H. and CHISHOLM, J. I. 1968. Geological Survey boreholes in the Carboniferous of east Fife, 1963–4. *Bull. geol. Surv. Gt Br.*, No. 28, 61–101.
—— and RUNDLE, C. C. *In press.* The age of the volcanic and hypabyssal rocks of east Fife. *Bull. geol. Surv. Gt Br.*, No. 60,
FRANCIS, E. H. 1968. Effect of sedimentation on volcanic processes, including neck–sill relationships, in the British Carboniferous. *23rd Sess. Int. geol. Congr.*, Czechoslovakia 1968, Sect. 2, 163–174.
GEIKIE, A. 1902. The geology of eastern Fife. *Mem. geol. Surv. Gt Br.*
GUPPY, E. M. 1931. Chemical analyses of igneous rocks, metamorphic rocks and minerals. *Mem. geol. Surv. Gt Br.*

GUPPY, E. M. 1956. Chemical analyses of igneous rocks, metamorphic rocks and minerals, 1931–1954. *Mem. geol. Surv. Gt Br.*

MACGREGOR, A. G. 1928. The classification of Scottish Carboniferous olivine-basalts and mugearites. *Trans. geol. Soc. Glasg.*, **18,** 324–360.

TOMKEIEFF, S. I. 1945. Petrology of the Carboniferous igneous rocks of the Tweed basin. *Trans. Edinb. geol. Soc.*, **14,** 53–75.

TYRRELL, G. W. 1917. The picrite–teschenite sill of Lugar (Ayrshire). *Q. Jnl geol. Soc. Lond.*, **72,** 84–131.

—— 1928. A further contribution to the petrography of the late-Palaeozoic igneous suite of the west of Scotland. *Trans. geol. Soc. Glasg.*, **18,** 259–294.

WALKER, F. 1934. The term 'crinanite'. *Geol. Mag.*, **71,** 122–128.

—— 1936. Geology of the Isle of May. *Trans. Edinb. geol. Soc.*, **13,** 275–285.

—— and IRVING, J. 1928. The igneous intrusions between St Andrews and Loch Leven. *Trans. R. Soc. Edinb.*, **56,** 1–17.

WASHINGTON, H. S. 1917. Chemical analyses of igneous rocks. *U.S. geol. Surv. Prof. Pap.*, **99,** 1201 pp.

Chapter 11

QUARTZ-DOLERITE AND THOLEIITE INTRUSIONS

INTRODUCTION

THE quartz-dolerite and tholeiite intrusions that occur extensively throughout central Scotland are well represented in east Fife where, however, they are confined to the western part of the district. They include both dykes and sills, and the two types of intrusion have been found to occupy geographically separate areas, the dykes in the north and the sills in the south. The latter are known from bores to extend underground for some distance south-east of their outcrops. The presence of quartz-dolerite and tholeiite intrusions in east Fife was first indicated by Balsillie (1922). Walker and Irving (1928) listed most of them, provided a map of the outcrops and discussed their petrography.

The quartz-dolerite and tholeiite intrusions in the Midland Valley of Scotland have long been regarded as late-Carboniferous or early Permian in age (Walker 1935, pp. 138–139). Recent determinations by Fitch, Miller and Williams (1970) on a sill and a dyke from this suite, which have yielded ages of 295 to 300 m.y., support this view. Radiometric dates have not been obtained in east Fife, where the quartz-dolerites and tholeiites are not fresh enough to be suitable for dating in this way.

The field evidence regarding the relationship between the episode of quartz-dolerite and tholeiite intrusion and other major igneous or structural events in east Fife is locally conflicting and generally inconclusive. There is no definite indication of the relative age of the quartz-dolerites and olivine-dolerites. Analcimisation of the plagioclase in some of the quartz-dolerites of Fife was attributed by Walker and Irving (1928, pp. 13–15) to emanations from the 'teschenite' (i.e. teschenitic olivine-dolerite) sills, suggesting that the quartz-dolerites are the older of the two groups of intrusions. These authors, however, indicated that the analcime is extremely sporadic in its occurrence, and its presence has not been confirmed in any of the slices examined from the only locality in east Fife that they cited in this connection, namely Kinninmonth Hill.

The youngest strata cut by the quartz-dolerites and tholeiites in the district are of Upper Limestone Group age, but further south, in undersea workings off Leven and Methil, a quartz-dolerite sill cuts Middle Coal Measures strata (Knox 1954). One of the dykes (p. 168) cuts the Kinkell Farm neck near St Andrews, and in the Bruntshiels neck there is a sheet-like body of quartz-dolerite. The latter, however, can be interpreted either as a post-volcanic intrusion or as a very large included block.

Mrs. I. F. Wallace (1916) claimed to have identified blocks of quartz-dolerite in the Viewforth, Ardross and Lundin Links necks: although the blocks from Viewforth and Ardross were decomposed she concluded that some at least of the late-Carboniferous quartz-dolerite sills were intruded before the formation of the Lundin Links neck. Balsillie (1923, p. 541) confirmed the presence of quartz-dolerite blocks in the Viewforth neck: he found that many of them were decomposed but that some were 'fairly fresh and showing interstitial micropegmatite'. Balsillie also reported blocks of quartz-dolerite, in which he claimed to detect fresh hypersthene, in the agglomerate of the St Monance neck. Wallace's material has not been traced but Balsillie's slides (nos. 1952.3.14 and 1952.3.24) are now in the Royal Scottish Museum, Edinburgh. They have been re-examined by Mr. R. W. Elliot, who has confirmed Balsillie's findings and also identified a recently collected block from the Viewforth neck as a decomposed quartz-dolerite in which a little micropegmatite has survived. The Lundin Links neck cuts Middle Coal Measures and contains fine-grained intrusions whose petrographical affinities (see p. 217) suggest that it may belong to an eruptive phase associated with the intrusion of the quartz-dolerite and tholeiite sills. On the other hand, the Viewforth neck (which cuts Passage Group strata and has no neck intrusions) contains, interbedded with the tuffs, sediments which have been reported by Dr. R. Neves (personal communication, 1974) to contain spores indicative of a Lower Coal Measures age. This strengthens Balsillie's contention that some of the quartz-dolerite intrusions in the Midland Valley of Scotland are older than those to which a Permo-Carboniferous age is usually assigned. It does not necessarily follow that any of these older intrusions are exposed in east Fife: no quartz-dolerite or tholeiite is seen near either the Ardross neck or the St Monance neck. If such earlier intrusions exist, and cannot be distinguished petrographically from the later ones, then the discovery of quartz-dolerite blocks is of much less importance as regards the determination of the relative ages of the volcanic necks and the Permo-Carboniferous quartz-dolerite suite.

The relations of the quartz-dolerites and tholeiites to the faults are also rather obscure and probably complex. The Kemback dyke (p. 167) appears to post-date the Dura Den Fault and near Ladeddie a body of quartz-dolerite is thought to have been intruded along the Ceres Fault. Further south, the main sill changes its horizon from the Lower to the Upper Limestone Group at or near the Branxton Fault, but it has not been proved that the sill used the fault-plane to accomplish this transgression.

Geophysical investigations

Numerous geomagnetic traverses have been carried out, mainly by staff of the Geophysical Division of the Institute of Geological Sciences, in an attempt to delimit these intrusions more accurately than surface exposures allow. Areas underlain by dolerite sheets are characterised by a marked variability in the magnetic profiles and are easily distinguished from areas underlain by sedimentary rocks, where magnetic profiles tend to be more constant. Precise demarcation of the limits of outcrop, especially at the tops of these bodies, has, however, proved difficult. Dykes on the other hand, especially in areas of sedimentary rocks, characteristically produce large and very localised magnetic anomalies and this has facilitated the mapping of such bodies in areas of poor exposure.

J.I.C., I.H.F.

Sills

The quartz-dolerite sill-complex which crosses west and central Fife extends into east Fife and crops out around Teasses, Ceres and Blebo. The outcrop is, however, broken up into many parts, of variable size and shape, whereas in the rest of the county it usually forms one or two extensive broad bands. Indeed the shape of some of the outcrops is difficult to reconcile with sill-form at all, and some of the intrusions may perhaps be better regarded as irregular sheets or bosses. Unlike the dykes, which occur also in Lower Old Red Sandstone strata to the north, the sills in east Fife appear to be confined to the Carboniferous rocks. Their outcrops tend to form higher ground than do those of the adjacent sediments. Members of the sill-complex are now known to occur at depth at least as far south as the Little Pilmuir Borehole [NO 40720392] and a related dolerite of tholeiitic affinity was cut by the Dunotter Borehole [NO 46500250] (Forsyth and Chisholm 1968). In the Teasses area the main sill is known to be fully 60 m thick and a bore near Wester Pitscottie [NO 405127], which started in the sill, cut 78 m of dolerite. Even greater thicknesses may be attained by the sills in the Little Pilmuir and Balcormo Wood [NO 41320575] boreholes. J.I.C., I.H.F.

Details

Teasses area. The more southerly part of the quartz-dolerite sill-complex extends eastwards into the district under review in the vicinity of Teuchats [NO 402074], where 6 m of dolerite are exposed in an old quarry 90 m north of the farm steading. This sill lies just above the St Monance Brecciated Limestone. On the east side of the fault which passes 90 m east of Teuchats, the sill is apparently in two leaves. The upper, here termed the Teasses sill, is well exposed, particularly around Teasses west lodge [NO 403079] and up to 800 m east and north-east of Teasses House [NO 407081], e.g. at the disused quarry [NO 41450833] in which 4·5 m of dolerite are exposed. The basal part of the Teasses sill is exposed in places on a scarp slope which extends from Backbraes [NO 401085] to Hall Teasses [NO 413090], and also in an old quarry [NO 407086] 400 m north-east of Windygates. The St Monance Brecciated Limestone outcrops on both sides of the sill, which in this area must therefore transgress from one side of the limestone to the other (cf. Geikie 1902, fig. 24, p. 192). The lower leaf, to which the name Fleecefaulds sill is here given, is visible at several places between Fleecefaulds [NO 401088] and Bankhead [NO 407092]: the best exposures are in the rather inappropriately named Teasses Quarry [NO 405090] in which 4·5 m of dolerite can be seen. The Teasses and Fleecefauld sills appear to join about 400 m E.N.E. of Bankhead. The outcrop of the combined sill terminates towards the north, probably at the Cassingray Fault; this fracture may have been exploited by the sill in changing its horizon of intrusion.

From its outcrop, the Teasses sill extends south-eastwards beneath a cover of Lower Limestone Group strata. In this area the Teasses Moss Borehole (1902) [NO 42120864] passed through fully 60 m of 'whin' with its base at a depth of about 150 m. (The exact thickness of igneous rock is not known because the items adjacent to the 'whin' were logged as 'calm' which may represent heat-altered sediments, bleached basalt or a mixture of both.) This 'whin' sill must be presumed to be quartz-dolerite: the bore lies only 400 m down-dip from the outcrop of the sill, the horizon of intrusion is similar and no large mass of olivine-dolerite is known in the immediate vicinity. The Baldastard Borehole (1896) [NO 42240747], which lies about 1370 m south-east of the outcrop of the sill, was stopped at a depth of 184 m, having penetrated 46 m of 'whinstone' which must also be presumed to be the Teasses quartz-dolerite sill, here intruded at a slightly higher level, above the Charlestown Main Limestone.

Farther south, borehole evidence suggests that at or near the Branxton Fault the sill rises abruptly southwards, probably by more than 300 m, but the exact location and manner of the transgression are not known.

Balcormo Wood, Little Pilmuir and Dunotter boreholes. The Balcormo Wood Borehole entered a quartz-dolerite sill about 18 m below the base of the Orchard Beds at a depth of 284·1 m and proved 25·6 m of this material before stopping, still in the sill, at 309·7 m. The Little Pilmuir Borehole cut just over 30 m of quartz-dolerite without reaching the base of the sill. The sill here is intruded in tuffs and volcanic detritus which occur in the lower part of the Upper Limestone Group and possibly extend down into the Limestone Coal Group. It is therefore at a lower horizon in this bore than it is in the Balcormo Wood Borehole (Forsyth and Chisholm 1968). Robertson (*in* Robertson and Haldane 1937, pp. 102–103, fig. 18) found that in the quartz-dolerite sills in the Kilsyth area, near Glasgow, coarse-grained material 'usually extends down to near the bottom of the top third of the sill.' In the Little Pilmuir Borehole, the base of the coarse-grained part appeared to have been reached at about 23 m below the top of the sill, suggesting a total thickness for the sill there of about 70 m. The base of the coarse-grained part was not reached in the Balcormo Wood Borehole, thus suggesting a total thickness for the sill of more than 75 m. The Dunotter Borehole cut a sill 14·6 m thick, with its base at 196·9 m (Forsyth *in* Forsyth and Chisholm 1968, p. 69), which appears to be a tholeiitic dolerite.

Gilston sill. The presence of another body of quartz-dolerite, the Gilston sill, has recently been detected among the olivine-dolerite sills in the Gilston district. It is known to be present in the Boghall Burn about 450 m south-east of Baldastard [NO 421069] and on Gils Law [NO 427069]. From there its base has been traced north-eastwards, by means of a topographic feature, to exposures on the south side of the Berryside Burn, about 400 m south-west of Berryside [NO 437079]. Thence the base can be followed by another surface feature north-westwards to an old quarry at New Gilston [NO 429081]. Backmuir of New Gilston is believed to be built on the outcrop of the sill, which appears to continue north-eastwards to exposures 275 m north of Patieshill [NO 489083]. Here the sill probably splits into two leaves; one apparently dies out north-east of Patieshill, while the other continues at least as far as exposures in a small stream 400 m west of Falfield House [NO 445088].

Bankhead area. Between the Cassingray and Lathones faults two bodies of quartz-dolerite appear to be present, but not a great deal is known about them. Walker and Irving (1928, fig. 4, p. 6) showed them as one intrusion and although it now seems preferable to regard them as separate bodies, this cannot be proved. One of them has been traced from an old quarry 275 m west of Larennie [NO 444098] to an exposure in a stream 135 m farther west. The best exposures of the other are in an old quarry [NO 436107] 400 m north-west of Bankhead. Other dolerite exposures in that vicinity, and also some in the ground west and north-west of Bruntshiels [NO 435102], are thought to belong to this intrusion, the form and extent of which are ill-defined. Quartz-dolerite is also known to occur as an elongate body 320 m long, apparently lying within the Bruntshiels neck. The age-relation between the quartz-dolerite and the agglomerate is not, however, determinable.

Farther west, beyond the inferred limit of the Lathones Fault, there are two quartz-dolerite sills exposed in Craighall Den [NO 405106], one a little above and the other immediately beneath the St Monance White Limestone.

Between the Lathones and Ceres faults several occurrences of quartz-dolerite are known. The most westerly is over 150 m below the surface, at the bottom of the Callange Borehole [NO 41711191], which at its base penetrated about one metre into an intrusion of this material, a short distance below the Largoward Splint Coal. About 180 m north of North Callange [NO 420122] there are a few exposures of quartz-dolerite but the ill-defined outcrop of the body to which they belong must be of limited extent. Abundant exposures of quartz-dolerite around Lower Baldinnie [NO 429119] indicate the presence of another body of quartz-dolerite; scattered exposures between 800 m and 1200 m to the east may belong to the same intrusion. I.H.F.

Ceres intrusion. The Ceres intrusion comes to crop near the western limit of the district under review, on the hillside north of Ceres. It is exposed only in a small quarry

[NO 40501274] behind the farm steading at Wester Pitscottie, but a borehole at the same farm proved 78 m of dolerite resting on sedimentary rocks. To the north-east the extent of the outcrop is conjectural, but magnetic traverses in the ground north-east of Wester Pitscottie suggest that dolerite is present only at depth in that area. To the north-west the intrusion ends in the region of the Dura Den Fault, but the age-relationship between these two features is not known.

Intrusions north and east of Pitscottie. A roughly triangular outcrop of quartz-dolerite between Pitscottie [NO 417131], Craiglumphart Quarry [NO 424139] and a point [NO 42781329] about 275 m south-west of Newbigging of Blebo is interpreted as an irregular and transgressive sill-like body. It appears to pinch out at its three extremities. At Craiglumphart Quarry about 1·5 m of baked sediments are seen resting concordantly on about 4·5 m of dolerite; the base of the sill and the sediments below it are exposed in the roadsides nearby [NO 42351397]. The dolerite is also well exposed in a quarry [NO 42331328] near Blebo Hole and in another at Pitscottie [NO 417131]. To the east of the former locality the sill is clearly transgressive towards the sedimentary rocks exposed in the banks of the nearby stream.

To the north-west of this intrusion, in Pitscottie Vale, the top contact of a roughly horizontal sill is exposed in the banks of the Ceres Burn [NO 41731368]. It is overlain concordantly by a few metres of baked sediments with several thin sills. This intrusion seems to lie at a lower level than the larger one just described. It is faulted off at its northern end [NO 41761386] near Charlie's Rock; its base is not exposed.

Intrusions round Blebo House. A few hundred metres north of Charlie's Rock a large sill-like mass of dolerite is exposed at Blebo Skellies [NO 41581422], a waterfall in the Ceres Burn. This dolerite can be traced northwards around the end of a small plunging syncline and back again into the banks of the Ceres Burn, where it can be seen [NO 41581438] lying concordantly among the country rocks which are dipping to the south-east. The intrusion appears to thicken north-eastwards, and 275 m south-west of Blebo House [NO 42171467] it forms a small ridge where a thin band of baked sediments can be detected within it. An old shaft [NO 41951437], probably the more westerly of two lead mines referred to in the Old Statistical Account (McDonald 1795, p. 306), has been excavated into the sediments. North of Blebo House a small hill is capped with dolerite, which may originally have been part of the same intrusion.

Intrusions between Kinninmonth and Denork. A large quartz-dolerite intrusion of uncertain shape forms Kinninmonth Hill, about 2 km east of Pitscottie, where it lies among strongly deformed sedimentary rocks on the south side of the Ceres Fault. The dolerite is well seen in quarries and crags along the southern side of Kinninmonth Hill. The steep northern face of the hill marks the line of the Ceres Fault but the contact between the fault and the dolerite is not exposed, so that their age-relationship remains conjectural.

About 1½ km north-east of Kinninmonth Hill a second, smaller, quartz-dolerite intrusion lies obliquely across the line of the Ceres Fault. Its shape is very irregular and at its eastern and western ends [NO 44971383, NO 44561368] the field exposures suggest that it locally assumes a dyke-like form. It is interpreted as a fault-intrusion.

Upper Magus sill. North of the Ceres Fault a large mass of quartz-dolerite forms a mainly sill-like intrusion capping the hills between Blebo Mains [NO 429140], Broomside [NO 43611522], Upper Magus [NO 448148] and Backfield of Ladeddie [NO 440138]. Magnetic traverses across the low ground along the south side of this intrusion suggest that there is no subsurface connection between it and the intrusions along and to the south of the Ceres Fault. The topography suggests that the sill steps down on its northern side to a lower level around Greenside [NO 437149]. At St Andrews Wells an old quarry in the St Monance Brecciated Limestone shows 6 m of dolerite in its western face [NO 43721455] but this thins eastwards to 3 m; at the eastern end of the quarry the dolerite lies in the sediments above the limestone. The relationship between these sedimentary rocks and the main mass of the intrusion is not entirely clear but it seems that they form large lenticular inclusions near its base. At Morton Wood [NO 43001453] the sill is in contact with agglomerate of a volcanic neck, and a dyke-like apophysis projects from the sill in a west-north-westerly direction towards Blebo House. Magnetic traverses across the low ground to the north and west suggest that this dyke-like body does not continue very far, and does not join up with any of the intrusions around Blebo House. The age-relationship between the dolerite and the vent agglomerate cannot be demonstrated, owing to lack of critical exposures. A short distance to the east of the last locality, and about 135 m south-east of Morton of Blebo, a small excavation [NO 43311472] probably marks the site of a small lead mine worked in the eighteenth century (see p. 254; McDonald 1795, pp. 305–306; Anderson 1841, pp. 420–421).

J.I.C.

Dykes

Quartz-dolerite dykes are known only in the northern part of the district under review. Here three broad dykes of considerable linear extent are present: the Kemback, Knockhill–St Andrews and Brownhills dykes. They have been intruded into Lower Carboniferous sediments, and one of them also cuts the agglomerate of a volcanic neck. The dykes all follow the general east–west orientation which is characteristic of quartz-dolerite and tholeiite dykes throughout the Midland Valley. The three intrusions are probably part of a group of dykes—the Callander–Auchterarder group of south Perthshire (Francis and others 1970, p. 227)—which can be traced across Scotland from Loch Long to north Fife.

Details

Kemback dyke. The Kemback dyke is exposed on a small knoll [NO 42111578] at the foot of Kemback Hill, about 400 m south of Kinnaird. It appears to be about 30 m wide at this point and is intruded into Upper Old Red Sandstone strata. It runs eastwards up the face of Kemback Hill, where it crosses the line of the Dura Den Fault, and is exposed in a small quarry [NO 42541574] in Kemback Wood, cutting sediments of the Calciferous Sandstone Measures. There appears to be a downthrow of about 15 m to the south across the line of the dyke, which must therefore lie in a small fracture here. The dyke cannot be traced eastwards beyond this last exposure and probably dies out.

Knock Hill–St Andrews dyke. The Knock Hill–St Andrews dyke extends for a distance of about 8 km from a point [NO 42841652] on Kemback Hill eastwards to Grange Road [NO 51581566] on the south-east side of St Andrews. It is intruded into sedimentary rocks of the Calciferous Sandstone Measures throughout its length. From its western end to a point [NO 44351635] 180 m south-east of Knockhill Farm the course of the dyke is known accurately from temporary exposures, an old quarry, and magnetic anomalies, and it is clear that this part of the intrusion consists of two lengths of dyke *en échelon*. West of Knockhill Farm the two lengths overlap for about 365 m and are about 135 m apart.

From the Knockhill Farm area to New Mill at the west side of St Andrews the course of the dyke is inferred from magnetic traverses, but from New Mill to its eastern end the dyke is known from exposures as well. Just to the south of New Mill it was worked in a narrow quarry but the dyke can now be seen only in the bed of the Kinness Burn nearby [NO 49761594], where it appears to be about 15 m in width. Farther east, the dyke is poorly exposed in a small stream gully [NO 51311570] about 450 m northeast of New Grange House. Magnetic evidence suggests that the dyke ends about 275 m east of this last locality.

Brownhills dyke. The Brownhills dyke is probably an extension of the Knockhill–St Andrews dyke, whose continuity, at least at the present level of erosion, is broken across the line of the Maiden Rock Fault. At its western end the dyke has been located accurately by closely-spaced magnetic traverses, and it curves from an initial northwesterly trend [at NO 51521507] to an east–west trend [NO 51771498] near Easter Grange. The dyke was at one time visible in a small quarry [NO 52391501] by the Anstruther road near Brownhills, and its course can be traced for a short distance in this area by the presence of the characteristic orange-brown weathering products in the soil. About a kilometre farther east, in a roadside exposure [NO 53441490] near Kinkell, the dyke was formerly exposed (Geikie 1902, p. 198) cutting agglomerate of the Kinkell Farm neck. This exposure is now degraded, although the position of the dyke can still be located by the presence of blocks of spheroidally-weathered dolerite in the soil. In the nearby coast sections no large dolerite dyke is present, and the only possible representative there of the Brownhills dyke is a pair of very narrow east–west dykes of altered basalt [NO 54851520]. Geikie (1902, p. 198) recorded the dyke traversing 'the sandstones at the south end of Kittock's Den near Broomhills' but it is not now certain to which locality this record refers. No dolerite is now exposed in Kittock's Den.
J.I.C.

Petrography

The petrography of the Scottish Permo-Carboniferous suite of quartz-dolerite sills and dykes has been described in detail by Walker (1935) who has compiled a comprehensive list of references. Most members of the suite in east and northeast Fife have in addition been included in Walker and Irving's (1928) account. They have been found to be typical of the suite as a whole and are therefore discussed only briefly here. Flett and Seymour (*in* Geikie 1902, pp. 390–392) provided brief petrographic descriptions of a few slices from some of the sills. Most of the east Fife tholeiites[1] appear to belong to Walker's (1935, pp. 145–146) Corsiehill type. The tholeiitic dolerite in the Dunotter Borehole (p. 165) was described by Mr. R. W. Elliot (*in* Forsyth and Chisholm 1968, pp. 90–91) who stated that it is similar to the tholeiite of Dalmahoy type (cf. Campbell and Lunn 1927). It has no known analogues in east Fife, and contains a bright green, isotropic mineral, possibly chlorophaeite.

Details

The most abundant mineral is plagioclase which usually occurs in tabular or acicular crystals. These consist largely of labradorite but in many cases are zoned outwards to andesine or oligoclase. The degree of alteration is variable from negligible to complete; it tends to be most marked in sills which are probably cut by volcanic necks. Replacement of the plagioclase by analcime was reported by Walker and Irving (1928, pp. 13–15) and attributed by them to liquors emanating from the teschenite (i.e. teschenitic olivine-dolerite) sills. The only locality in east Fife where these authors

---

[1] The term 'tholeiite' is used in this account for those members of the quartz-dolerite suite which have an intersertal mesostasis of glass or devitrified glass.

found evidence of such analcimisation is Kinninmonth Hill, but specimens since obtained from there (S 37069, 37082, 48048) show mainly fresh plagioclase. Walker and Irving indicated, however, that the occurrence of the analcime is extremely sporadic, even in a single intrusion.

The principal ferromagnesian mineral is augite, which is usually anhedral or, rarely, ophitic to the plagioclase. It is usually but not invariably fresh and locally survives unaltered in slices where the plagioclase has been affected. Fresh hypersthene is of very rare occurrence. Biotite and brown hornblende occur in places as rare accessories associated with the augite and the iron ore. The latter is usually skeletal in habit and locally altered to leucoxene, from which it is deduced that ilmenite is the main constituent. Pyrite also occurs in places. Some slices show occasional viriditic pseudomorphs, the shape of which suggests that they may be after olivine, with rhombic pyroxene as a possible alternative. Apatite is universally present as the most abundant accessory mineral.

Most of the slices contain a small amount of anhedral or interstitial quartz, but it is not usually clear whether the quartz is to be regarded as a late-stage primary mineral or a secondary one. A little alkali-feldspar also occurs interstitially in some slices.

The distinction between the quartz-dolerites and the tholeiites was made originally on the nature of the mesostasis, which is glass or its devitrification products in the tholeiites and a micropegmatitic intergrowth of quartz and alkali-feldspar in the quartz-dolerites. Walker (1935, p. 145) added as an additional criterion the universal alteration of 'the earlier ferromagnesian minerals' in the tholeiites, by which it is assumed that he meant principally rhombic pyroxene because he found (Walker 1935, p. 142) that olivine is invariably altered in both quartz-dolerites and tholeiites. Francis and others (1970, pp. 235–236) also recognised textural differences between the quartz-dolerites and the tholeiites of the Stirling district, e.g. a greater tendency towards ophitic texture in the quartz-dolerites and stellate arrangements of plagioclase laths and granular augite in the tholeiites.

In east Fife the mesostasis has frequently been found to be altered, with the introduction of secondary quartz and viriditic material, and also carbonate. It appears, however, that tholeiites may be more common in east Fife than Walker (1935, p. 148) suggested: he regarded them as rare, except near the margins of the intrusions. Micropegmatite is uncommon (S 37064, 48075, 49881), whereas many slices contain brown intersertal material which is abundant in some examples (S 36972, 37053–4, 37070, 49552). In almost all cases it is now birefringent and locally occurs in the form of radiating microlites showing radial extinction. There is one example (S 48049) in which the brown mesostasis is in places fairly clear and almost isotropic, suggesting incomplete devitrification of the original glass. I.H.F.

References

ANDERSON, J. 1841. On the geology of Fifeshire. *Trans. Highl. agric. Soc. Scotl.*, **7**, (new series), 376–431.

BALSILLIE, D. 1922. Notes on the dolerite intrusions of east Fife. *Geol. Mag.*, **59**, 442–452.

—— 1923. Further observations on the volcanic geology of east Fife. *Geol. Mag.*, **60**, 530–542.

CAMPBELL, R. and LUNN, J. W. 1927. The tholeiites and dolerites of the Dalmahoy Syncline. *Trans. R. Soc. Edinb.*, **55**, 489–505.

FITCH, F. J., MILLER, J. A. and WILLIAMS, S. C. 1970. Isotopic ages of British Carboniferous rocks. *C.R. 6th Int. Congr. Carb. Strat. Geol.*, **2**, 771–789.

FORSYTH, I. H. and CHISHOLM, J. I. 1968. Geological Survey boreholes in the Carboniferous of east Fife, 1963–4. *Bull. geol. Surv. Gt Br.*, No. 28, 61–101.

FRANCIS, E. H., FORSYTH, I. H., READ, W. A. and ARMSTRONG, M. 1970. The geology of the Stirling district. *Mem. geol. Surv. Gt Br.*
GEIKIE, A. 1902. The geology of eastern Fife. *Mem. geol. Surv. Gt Br.*
KNOX, J. 1954. The economic geology of the Fife Coalfields, Area III. *Mem. geol. Surv. Gt Br.*
MCDONALD, J. 1795. Parish of Kemback. *The [Old] Statistical Account of Scotland*, **14**, 297–310. Edinburgh.
ROBERTSON, T. and HALDANE, D. 1937. The economic geology of the Central Coalfield of Scotland, Area I. *Mem. geol. Surv. Gt Br.*
WALKER, F. 1935. The late Palaeozoic quartz-dolerites and tholeiites of Scotland. *Mineralog. Mag.*, **24**, 131–159.
—— and IRVING, J. 1928. The igneous intrusions between St Andrews and Loch Leven. *Trans. R. Soc. Edinb.*, **56**, 1–17.
WALLACE, I. F. 1916. Notes on the petrology of the agglomerates and hypabyssal intrusions between Largo and St Monans. *Trans. Edinb. geol. Soc.*, **10**, 348–362.

Chapter 12

VOLCANIC NECKS

Introduction

MORE than a hundred volcanic necks have been located at outcrop or in borings in east Fife (see Fig. 19). They are approximately vertical pipe-like bodies ranging in diameter from less than a hundred metres to more than one and a half kilometres. Consisting mainly of pyroclastic rocks, they pierce the Carboniferous sediments and mark the sites of former subterranean channels through which eruptive materials passed. Because of erosion the present exposures lie at various depths below the land surface on which the volcanoes once stood. Except on the coast, and locally in streams, the pyroclastic rocks are not well exposed. The term 'vent' has long been used as a synonym for 'neck' in Scottish geological maps and literature. It is, however, a usage which should be discontinued both on the grounds that 'vent' properly relates to an orifice of an active or recently extinct volcano and because the shape and content of necks owe as much to post-volcanic processes as to eruptive mechanisms (see below).

History of research

The necks were first recognised and later described at length in classic accounts by Geikie (1879, 1902). His notes on the petrography of some of the agglomerates and related minor intrusions were amplified by Mrs. Wallace (1916), while his detailed observations of exposures were extended by subsequent workers in Kinninmonth Den (Craig 1912) and along the shore east of St Andrews (Balsillie 1911a, 1911b, 1919, 1920a, 1920b; Kirk 1925). Although Geikie believed that all the east Fife necks were Permian in age, Balsillie (1923, 1927) and Knox (1954) cited stratigraphical evidence suggesting that some, at least, are contemporaneous with Carboniferous sedimentation.

Along the shore between Earlsferry and St Monance, Cumming (1928, 1936) located further necks and allied structures and defined the major Ardross Fault for the first time. Large-scale re-mapping led Francis and Hopgood (1970) to suggest that the fault is the final expression of a tectono-volcanic structure which was active throughout Upper Carboniferous and possibly during Permian times. The bedding of tuffs and agglomerates in the necks of the same coast section has formed the basis for a discussion of post-volcanic subsidence (Francis 1970).

Xenocrysts in the necks were mentioned first by Heddle (1901) and subsequently by Geikie (1902) and Balsillie (1923, 1927): they include pyrope garnets (Heddle's 'Elie rubies') which have been reinvestigated by Colvine (1968). Ultrabasic inclusions comprise lherzolitic and pyroxenitic types which, according to Chapman (1974), appear to represent two different modes of origin.

Excursion guides to some of the best coastal exposures have been published by Francis (*in* Mitchell and others 1960; Francis 1969) and MacGregor (1968).

Structure and Lithology

In plan the necks are generally sub-circular to sub-oval and they are so mapped in the poorly exposed hinterland, though where seen in the coast sections the neck margins show both minor and major irregularities. In section the necks tend to taper downwards in funnel shape, most margins being essentially ring-fractures inclined inwards at angles of 45° to 80° and accompanied, at most localities, by narrow crush-zones. Further sub-parallel outer ring-fractures are common. They enclose disorientated or wholly brecciated sediments which include large discrete masses in which the bedding is inclined inwards at high angles towards the necks. The sedimentary wall rocks adjacent to the neck-marginal ring-fractures are turned down at progressively higher angles towards the necks, particularly where they strike parallel or sub-parallel to the fractures.

The necks are formed of fragmental material for which the terms 'tuff' and 'agglomerate'—according to coarseness—are here used in a general sense. The material consists of two main components. One is derived by comminution of the Carboniferous sedimentary rocks pierced by the necks and ranges in size from large blocks down to matrix sand grains or dust from coal and mudstone. The other is derived from the parent magma—almost invariably one of the varieties of alkali-basalt including basanite and monchiquite: it consists either of ragged-edged, bleached or otherwise highly altered fragments detached in molten form and chilled during eruption (juvenile basalt) or of angular to rounded crystalline blocks. Proportions of the two components range widely from almost entirely sediment-derived to almost entirely basaltic and colours correspondingly range from grey to green.

The tuffs and agglomerates occur either in unbedded or bedded forms and some necks include both. The unbedded rocks are of various compositions and range in texture from well-mixed to chaotic. The well-mixed rocks generally occur at or near the centres of necks, with the chaotic near the margins, where large rafts of tuff-permeated sediments are common. The long axes of particles are commonly flow-foliated adjacent to neck margins or to large xenoliths. Most of the bedded rocks contain a high proportion of basalt and include layers ranging from dust grade to coarse lapilli-tuffs and agglomerates. The fine layers locally show crude grading and wedge- and cross-bedding. Some of these layers indicate sub-aqueous sedimentation, though Lorenz (1973, p. 199) has also recognised foreset bedding with radial flow and other features indicative of deposition from base-surges. Particle shapes are approximately equant while elongated forms and achneliths are rare. The lithologies and textures of the bedded rocks conform almost entirely to the criteria laid down by Walker and Croasdale (1972) for deposits of Surtseyan type, formed by basaltic eruptions in shallow water or in other situations where water has free access to vents. Modern deposits of this type are additionally characterised by extensive palagonitisation. It is reasonable to suppose that the Fife rocks were also originally palagonitised, for palagonite is known to be too unstable to have survived from Carboniferous or Permian times to the present without alteration. Thus, Geikie (1879, p. 515) may have been intuitively correct in his determination of palagonite even though his material is more likely to have been a chlorite in terms of modern mineralogical nomenclature (see also Sabine *in* Francis and Ewing 1961).

Where bedding in the necks is relatively undisturbed it has a centroclinal disposition (Plate VB), usually with high dips at and parallel to neck margins decreasing inwards towards the centres, where the bedded tuffs and agglomerates commonly give way to basaltic breccias. Near the margins of the Craigforth and Elie Harbour necks (Fig. 22a and b) the centroclinal arrangement is locally reversed across anticlinal axes, so that beds are inclined outwards, though still parallel to the margins. Locally the bedding gives place along the strike to collapse structures. One such structure is represented by a breccia in which individual blocks of bedded tuff have been reorientated, approximately in situ, and cemented by fine-grained material derived from the same tuffs: the structure implies that the bedded tuffs were indurated to some extent before collapse. More commonly a lack of pre-collapse induration is indicated by original bedding which has not only been reorientated, but has also buckled and flowed so that the margins of individually bedded masses pass imperceptibly into rocks of heterogeneous texture showing evidence of autointrusion. Such structures are best seen at the eastern margin of the Elie Ness neck and are the only traces of original bedding still visible in the Ardross, St Monance and inner Coalyard Hill necks. In these and many other necks, the bedded or collapse-bedded tuffs adjacent to the margins contain abundant carbonate veins aligned parallel or sub-parallel to those margins.

Related Intrusions and Cryptovolcanic Structures

Both basaltic and pyroclastic intrusions are associated with the necks. The main basaltic intrusions are plugs, the largest of which, at Craig Rock, Rires Craigs and Ruddons Point, are about 300 m across. They tend to be better exposed inland than the tuffs and agglomerates and range in petrographic type from mafic olivine-basalt to monchiquite. Two of the largest necks, at Largo Law and Rires, contain more than one such type. Apart from a sill in bedded tuffs at Ruddons Point and another extending outside a neck at Chapel Ness, most of the other basaltic intrusions are dyke-like. Exceptionally, as at Elie Harbour and St Monance, the grey crystalline dyke rocks cross the neck margins into the adjacent sediments and pass into altered rocks (Plate IIIb) known in the Scottish literature as 'white trap', though the term 'bleached basalt' is preferred here. Such bleached basalt has also been emplaced along the ring-fracture delineating the eastern margin of the Wadeslea neck.

Although the term 'tuffisite' was devised by Cloos (1941) for tuff-permeated breccias it has become a synonym for intrusive tuff by general usage. In this account, therefore, discrete intrusions of tuff are called 'tuffisite' while the breccias are qualified as 'tuffisitic'. Tuffisites are represented mainly by dykes, though a sill has also been mapped near Newark Castle. The dykes are generally thinner than the basalt dykes and cut both necks and adjacent sediments. They have also been emplaced along the planes of some of the marginal ring-faults. Most tuffisites contain only a small proportion of basaltic debris, as do most of the veins in the tuffisitic breccias margining the necks or contained within them as rafts. The dykes cutting the bedded tuffs in the necks, however, are composed of fragments derived from those tuffs and some are multiple intrusions. Tuffisite dykes of both types contain carbonates or pass into carbonate veins. Both types also show flow-foliation, especially at the margins.

Areas of breccia, commonly oval in plan and ranging from three to two hundred metres in diameter, occur sporadically among the sediments cropping out along the whole of the intertidal zone between Lundin Links and St Andrews. They are in general located near the necks, most notably between the Elie Ness and Wadeslea necks (Cumming 1936, fig. 2), though they have also been mapped in areas where no necks are known, between St Monance and Balcomie. Collectively, they have been described as cryptovolcanic ring-structures and are assumed to represent incipient necks (Francis 1960, 1962). Some of the smallest are marked by little more than minor updoming and sediment disorientation in situ, with or without veins of tuffisite. In larger structures there are sharp vertical contacts between the breccias and surrounding sediments, and peripherally the flat surfaces and long axes of the blocks within the breccias are orientated parallel to the contacts. With further increase in size there is, broadly, a correspondingly larger proportion of tuffisite matrix, sometimes accompanied by tongues of bleached basalt. In places, moreover, dykes of bleached basalt link ring-structures either to one another or to neighbouring necks.

Mechanism of Neck Emplacement

Magmas are generally assumed to begin their ascent from the depths as dykes along fissures. When water gains access, steam is generated under high pressure and temperature and from this, gas and ash streams are formed. Either by fluidisation (Reynolds 1954) or turbulent convection (McBirney 1959), these successively fracture, penetrate and brecciate the surrounding (country) rocks until those rocks are absorbed by stoping and comminution to form part of the gas–ash streams (Cloos 1941). The stoping effect is accentuated by collapse of the side walls and roof as pressure decreases after each eruptive pulse, so that first eruption chambers, then cylindrical conduits extending up to the surface, form above the dyke-like magma columns (Lorenz 1973). The transition from fissure to neck is well documented in kimberlites (Dawson 1971) and though it is not demonstrable in Fife, it is probably exemplified by the necks aligned along major north-easterly disturbances such as the Ardross Fault (p. 228) and, more obviously, the Ceres–Maiden Rock Fault Zone (Fig. 19). Other necks which appear to be distributed at random may have emanated from sills. Evidence from Nottinghamshire and western Fife cited by Francis (1968b) in support of such an origin is matched by the stratigraphical and geographical associations of some necks and sills in east Fife. In particular, the idea that both sills and necks are limited to areas of thick Carboniferous sedimentation is supported by the decrease in size and abundance of the east Fife necks northwards in the direction of an inferred basin margin and by the absence of both sills and necks among the outcrops of pre-Carboniferous rocks still farther to the north. Support is provided also by the correspondence in the ages of sills and necks given by radiometric methods (pp. 137, 179) and their common derivation from alkali-basalt magma, though the neck intrusions are, in general, more mafic and more basic than the sill rocks.

Whether emanating from dykes or sills the early gas and tuff streams contained a high proportion of country-rock debris and this is reflected in the composition of the necks if the volcanoes were short-lived. The outer Coalyard Hill neck provides an example of this. With continuing activity, however, progressively higher proportions of magmatic debris enter the stream and evidence of the early phases remains only in the tuffisites, tuffisitic breccias and tuff-permeated sediments near neck margins (Francis 1960; Plate VI). Where magma reaches the surface, lavas are extruded and the necks retain evidence of this activity in the form of dykes or plugs of crystalline rocks. Blocks of such crystalline rocks in many of the bedded layers of the Fife necks testify to periodic quiescence followed by fragmentation and expulsion of plug material during later eruptions.

The predominance of pyroclastic rocks and some of their characteristic lithologies (p. 172) indicate that in addition to initiating eruptions, water continued to exert a major influence on the pattern of volcanism in east Fife. Consideration of the palaeoenvironment suggests two sources of water. In a region undergoing continuous shallow-water sedimentation the upper layers of sediment must have been water-laden, of low density and only partly lithified. This water gave rise to phreatomagmatic eruption and the sediments were readily comminuted into the sand, mud and coal dust which are major matrix constituents of many of the Fife tuffs. Surface waters, in lagoons or shallow seas, provided the other source and gave rise to phreatic activity of Surtseyan type (Walker and Croasdale 1972) which built wide-diameter low-altitude tuff-rings at the surface. As long as water continued to gain access either from the sediment-pile or at the surface, by breaching the tuff-rings, Surtseyan activity continued. Breaches during periods of quiescence, or after activity finally ceased, would have allowed the temporary establishment of modified regional sedimentation as exemplified by the tuffaceous cyclic sediments at the centre of the Viewforth neck. Rarely, however, the volcanoes may have become sufficiently mature and established subaerially to have passed into a Strombolian phase giving rise to lavas, as at Largo Law, and possibly also scoria-spatter cones within the outer Surtseyan tuff-rings (Walker and Croasdale 1972, p. 313).

The structures of the necks cannot be explained entirely by explosive interaction of magma with water and water-laden sediments, for it is clear that subsidence must have been a major additional factor. Geikie (1902) first recognised this on the evidence that bedded tuffs and agglomerates, originally deposited at the surface, now occupy lower positions in the necks and have thus become juxtaposed with older country-rocks which are themselves dragged down against the neck margins. The amount of subsidence may be quantified crudely by reference to Fig. 18 in which several of the east Fife necks are shown relative to the stratigraphical position of the sediments surrounding them at the same level of erosion. It can be seen that if the bedded pyroclasts in the Elie Ness, Wadeslea, Coalyard Hill and St Monance necks are correlated with even the oldest of the known tuffs interbedded with the sediments (i.e. tuffs of Namurian age) subsidence of 500 m may be inferred. Radiometric age-determinations (p. 179), which have become available since this estimate was made by Francis (1970), suggest that some of those necks may be late Westphalian or Stephanian rather than Namurian in age. If so the amount of subsidence must be in the order of 2000 m unless the volcanism was locally preceded by substantial late Carboniferous erosion.

Geikie (1902) believed that the subsidence was entirely post-volcanic and caused by the withdrawal of magma at depth: he assumed that bedded material accumulated on the inner slopes and bottoms of craters which increased their capacity as the volcanoes were heightened by the addition of fresh ejecta. On this basis, for instance, he estimated the height of the original Kincraig volcano to have been more than 300 m. Francis (1970) followed Geikie in supposing the subsidence to have been entirely post-volcanic and suggested that the magma may have been still molten in the conduits at the time of the subsidence: this would give rise to reactivation of gas–tuff streaming and would explain the carbonate veining at neck margins and the tuffisitic and basaltic intrusions which are emplaced in the bedded tuffs along the ring-fractures and beyond.

The postulation of entirely post-volcanic subsidence can now be seen to conflict with some of the evidence in east Fife. If the volcanoes were 300 m or more in height, they would have been elevated well above the level of the shallow Carboniferous seas and activity would thus have been subaerial and Strombolian in type. Strombolian cinder cones, however, do not display inner-slope layering of the kind described here nor are their diameters generally wide enough to accommodate the extensive areas of inward-dipping pyroclasts observed. The

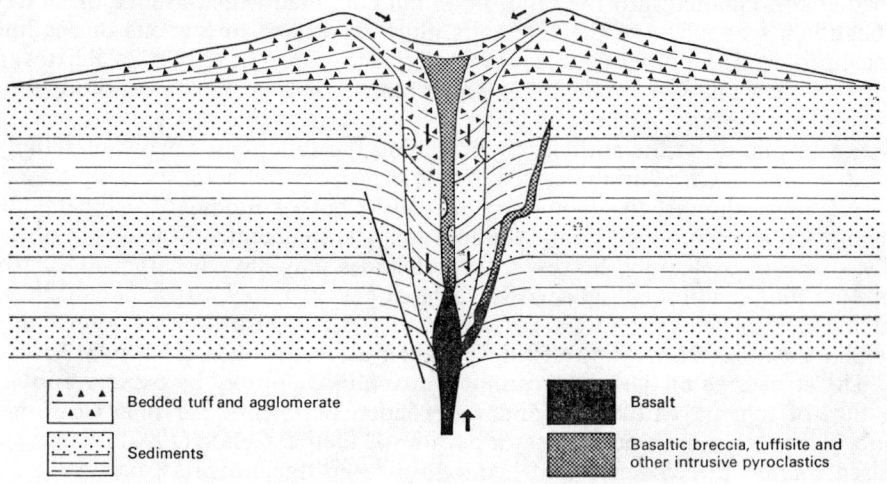

FIG. 17. *Schematic diagram illustrating some of the processes involved in the emplacement of volcanic necks in a sedimentary pile and, particularly, the intermittent subsidence of bedded pyroclastic rocks from the inner flanks of superficial tuff-rings*
(after Lorenz 1973 and Francis 1970)

Surtseyan model (Fig. 17) accords better in having a wide enough diameter and closely comparable lithologies; moreover it allows of subaerial emergence at the rims (thereby accounting for the fragments of wood found in the bedded pyroclasts), though it too fails to accommodate the necessary thickness of bedded material on any theory of subsidence which is wholly post-volcanic.

A solution to the problem of reconciling the thickness of the bedded pyroclasts with continuing activity of Surtseyan type is offered by Lorenz (1973) in a global review of the question. He suggests that ring-fracturing is initiated as early in the volcanic process as the first formation of cylindrical conduits above fissures and that each eruptive pulse thereafter is followed by subsidence. Thus, as the earliest bedded tuffs descend progressively farther down the neck, they are pierced by later conduits filled with basaltic debris (Fig. 17). This proposal accounts for the blocks of early bedded tuffs contained in later pyroclastic layers and fits the evidence of basaltic breccia pipes found in the Kincraig, Elie Harbour, Elie Ness and inner Coalyard Hill necks. It also allows for a final, entirely post-volcanic, phase of subsidence which would have brought tuff-ring material down over the tops of the final conduits, thereby explaining why necks such as Viewforth lack basaltic breccia pipes at present levels of erosion (Fig. 17). Progressive subsidence during the lifetime of the volcanoes also offers a better explanation than purely post-volcanic subsidence for the presence of marginal rafts of sediments spalled off the walls into the bedded tuffs at various levels in the necks. Such periodic spalling would have been facilitated if, as envisaged by Francis (1962, pp. 55–57), the inwardly-inclined ring-fractures were multiple at any given level, though not necessarily continuous from one level to another.

Lorenz (1973) gives due weight to the effects of erosion and inward slumping at the rims of modern tuff-rings and these factors may account for some of the collapsed bedding in the east Fife necks. It is difficult, however, to distinguish slumping from the collapse of poorly cemented layers during subsidence and the latter seems the more likely where, at the margins of such necks as Elie Ness, friction against ring-faults must have been an added factor of disruption.

Age-Relations

The age of the necks may be examined by means of four criteria, namely, the distribution of bedded tuffs within the sedimentary sequence, cross-cutting relationships, identification of blocks in agglomerates, and radiometric dates.

The evidence of the interbedded tuffs (Geikie 1902; Balsillie 1923; Knox 1954; Francis and Ewing 1961; Forsyth and Chisholm 1968; Francis 1968a) establishes east Fife as an area which had an almost unbroken history of volcanic activity at one short-lived centre after another, from early Namurian almost to mid-Westphalian times (Fig. 18). The upper part of the Westphalian B sequence and the lower part of Westphalian C, which includes the youngest strata now preserved in east Fife, are not volcanic. There is thus no stratigraphical evidence of alkaline eruptions in late Westphalian, Stephanian or early Permian times. Only the small Lundin Links neck with its intrusions of unusual petrography (p. 216) is certainly later than Westphalian B.

It is seldom possible to correlate any bedded tuff with a specific neck: it is only safe to infer that a neck is younger than the rocks which it has pierced. In addition to the example of the Lundin Links neck, quoted above, the Ladeddie neck cuts the Drumcarro olivine-dolerite sill, tuffisite associated with the Kinaldy neck cuts the Kinaldy teschenitic olivine-dolerite sill and field evidence suggests that the neck in Balniel Den is later than the Kilbrackmont–Baldutho olivine-basalt sill.

The Lundin Links neck also provides an example of the use of included blocks for the purpose of dating. The recognition by Wallace (1916) of quartz-dolerite blocks in the agglomerate at this locality conforms with the radiometric evidence quoted below and in Fig. 18, for quartz-dolerite sills in Scotland and northern England have consistently given late Westphalian or early Stephanian ages of about 295 m.y. (Fitch and others 1970). Wallace's material was not available for re-examination nor was quartz-dolerite recollected during the resurvey, but her discovery of further quartz-dolerite blocks in the Viewforth neck (also reported by Balsillie 1923, p. 541) has been confirmed (p. 183). Here, however, the evidence is equivocal, for the bedded tuffs and agglomerates of the Viewforth neck have now been shown, by palynological evidence (pp. 119 and 183), to be Westphalian A in age, that is older than any quartz-dolerite so far dated by radiometric methods. This reopens the suggestion by Balsillie (1923) that there

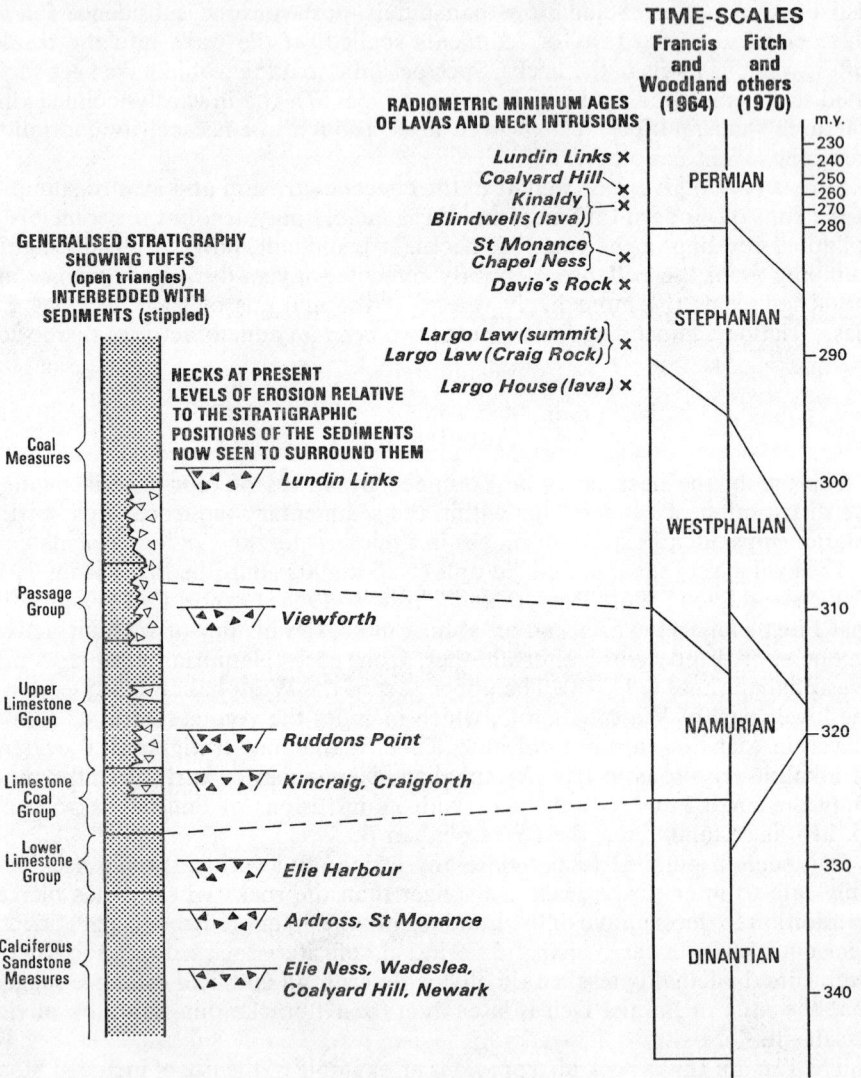

FIG. 18. *Diagram showing the stratigraphical position of the volcanic necks, at present levels of erosion, exposed between Lundin Links and St Monance*

were two suites of quartz-dolerite intrusions, the earlier being pre-Westphalian in age and hitherto undetected, unexposed or not differentiated from the later, radiometrically dated suite.

Potassium–argon dating of ten neck intrusions and lavas by the whole-rock method (Forsyth and Rundle, in press) has given the minimum ages listed in

Table 9. As can be seen from Fig. 18, where these ages are plotted relative to current Carboniferous time-scales (Francis and Woodland 1964; Fitch and others 1970), there is no agreement between the ages determined radiometrically and those derived from the stratigraphic evidence. There are no radiometric ages corresponding to the tuffs locally interbedded with the Namurian and

TABLE 9

Radiometric age-determinations of lavas and neck intrusions in east Fife

Locality	Mode of occurrence	Minimum age (m.y.)
Lundin Links	neck intrusion	239 ± 6
Coalyard Hill	neck intrusion	254 ± 7
Kinaldy	neck intrusion	267 ± 7
Blindwells	lava	267 ± 6
St Monance	neck intrusion	282 ± 6
Chapel Ness	neck intrusion	282 ± 6
Davie's Rock	plug	284 ± 6
Largo Law	neck intrusion	289 ± 6
Largo Law (Craig Rock)	neck intrusion	289 ± 8
Largo House	lava	292 ± 6

Constants used: $\lambda_e = 0.584 \cdot 10^{-10}$ yr.$^{-1}$; $\lambda_\beta = 4.72 \cdot 10^{-10}$ yr.$^{-1}$;

$^{40}K = 0.0119$ atom %

The quoted errors in the table and in text are combined standard deviations and take into account uncertainties in the mass spectrometric determinations, in the volume of spike (^{38}Ar isotopic tracer) and the potassium content (based on replicate determinations). The effects of error magnification due to the correction for atmospheric argon are also included.

early Westphalian sediments: instead all the radiometric ages appear to relate to the period following the deposition of the non-volcanic sediments of the Middle and Upper Coal Measures (see Chapter 8).

Forsyth and Rundle found that the Kinaldy and Coalyard Hill neck intrusions gave radiometric ages that are significantly lower than the five oldest results obtained from neck intrusions, and this they ascribe to argon loss. The five oldest individual dates, ranging from 282 m.y. to 289 m.y., are indistinguishable within the limits of analytical error and Forsyth and Rundle regard the figure of 289 ± 10 m.y. as the best minimum estimate for the time of emplacement. As the five intrusions are very fresh and exceedingly fine-grained they are considered to have lost very little argon and it seems probable therefore that they are Stephanian in age. The possibility exists, however, that all may have lost some argon and in consequence may be appreciably older, i.e. that they may belong to the known Namurian to early Westphalian volcanism.

The radiometric data therefore suggest, but do not prove, the existence, in Stephanian times, of a phase of basanitic volcanism that is wholly missing from the stratigraphic record in east Fife. Necks of Namurian and early Westphalian age must obviously exist in this district, but only the Viewforth neck can with any certainty be attributed (on palynological evidence) to this phase of volcanism. Apparently none of the earlier necks were sampled for radiometric analysis, probably because they lack suitable intrusions. It appears likely that all those which contain basanitic intrusions belong to the same episode of volcanism, tentatively dated on radiometric grounds as Stephanian, but no method has been found of determining the ages of those that lack both intrusions and spore-bearing sediments.

The Lundin Links neck, which gives the youngest minimum age of all (239 ± 6 m.y.), may be considered separately, for the intrusions of the neck do not belong to the alkaline basalt suite (p. 217). Moreover it is the only neck which cuts strata younger than the youngest bedded tuffs, and as both its lithology and structure suggest that the present erosional surface is well below the penecontemporaneous surface of erosion, a very late Carboniferous or Permian age seems acceptable. Whether or not it is linked with the emplacement of the quartz-dolerite sill-complex, some argon loss seems likely.

Field Relations

The areal distribution of the volcanic necks of east Fife is shown on Fig. 19, which also gives the locations of the more important necks cited by name in the descriptions that follow. The necks on the coast between Lundin Links and St Monance are among the best-exposed, and best-known, examples in east Fife and these are described first and, for the most part, in greatest detail. Then follows a brief account of the cryptovolcanic structures exposed at intervals on the coast between St Monance and Kingsbarns. Next comes a description of the necks and cryptovolcanic structures that occur on or near the coast in the Kinkell area, between St Andrews and Kingsbarns. The remaining necks are all inland and are not in general well exposed. Those in the Cairnsmill and Ladeddie areas, south and west of St Andrews, are described first. The final sections are devoted to the concentration of necks exposed on the hilly ground that extends from the Keil Burn eastwards by way of Largo Law and Balniel to Kellie Law.

Coast from Lundin Links to St Monance

Lundin Links neck. The southern part of the Lundin Links neck (Fig. 20a) is exposed in a semicircle measuring approximately 150 × 75 m on the shore [NO 411024] at Lundin Links where it cuts Middle Coal Measures strata some 30 m below the Barncraig Coal. It is formed of unbedded tuff consisting mainly of juvenile basalt lapilli averaging 1 cm in diameter, but occasionally up to 30 cm, in a green fine-grained cement, with impersistent veins of white carbonate. The tuff is well-mixed towards the centre, but marginally it includes fragments of coal and shale with blocks of sandstone up to 10 m long. The tuffs adjacent to these blocks are flow-foliated. Wallace's (1916) report of quartz-dolerite blocks within the tuff was not confirmed during the resurvey. Most of the tuffs are green, except towards the margins of the neck where they are patchily purple. The purple rimming of otherwise green lapilli suggests a process of secondary reddening resulting from circulation of groundwater from the surrounding, partly red sediments. Similar reddening accentuates the flow-banding in a bleached basalt dyke described by Geikie (1902, p. 266). This dyke is one of several intrusions which cut the neck and surrounding sediments and which are somewhat different in petrographic type from other basalts in east Fife (p. 217). Within the neck the intrusions are partly sill-like and partly dyke-like. Locally they pass without any

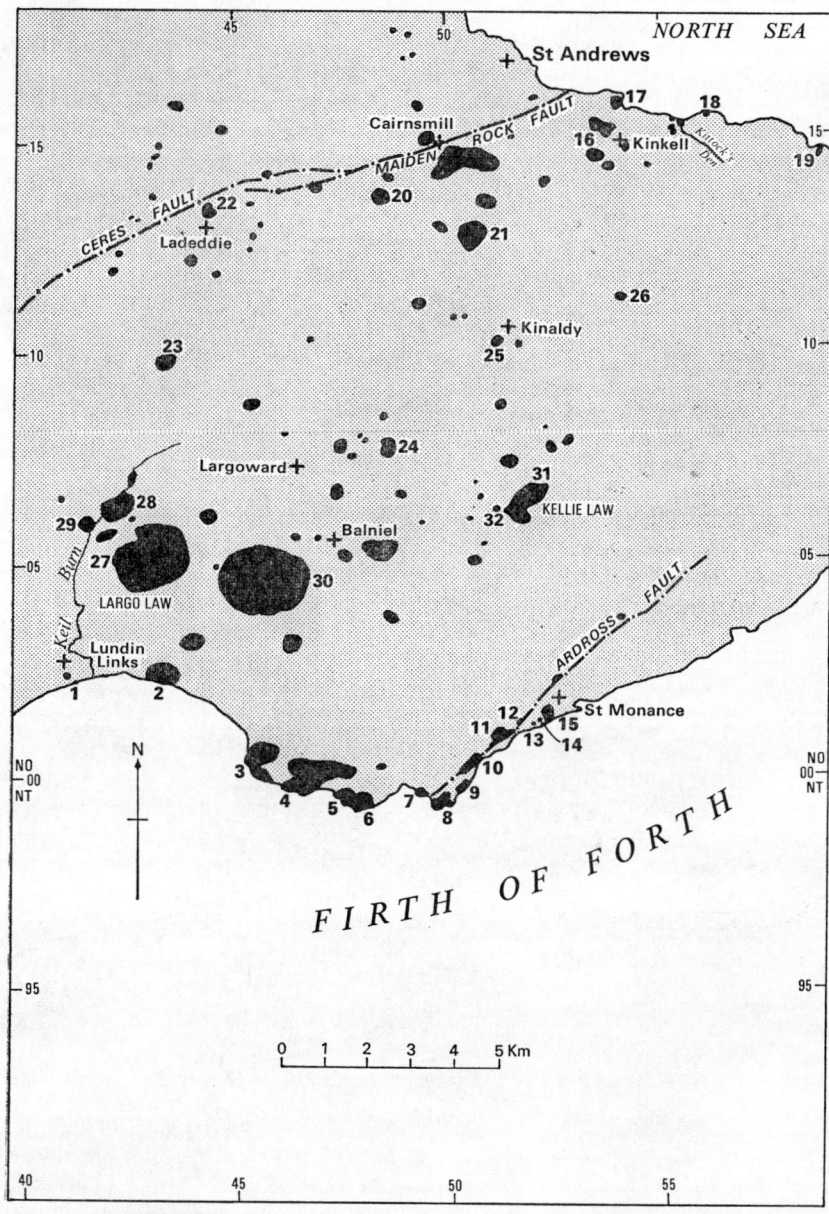

FIG. 19. *Sketch-map showing distribution of volcanic necks in east Fife*

Necks named in text are numbered on map, as follows: 1. Lundin Links. 2. Viewforth. 3. Ruddons Point. 4. Kincraig. 5. Craigforth. 6. Chapel Ness. 7. Elie Harbour. 8. Elie Ness. 9. Wadeslea. 10. Ardross. 11. Coalyard Hill. 12. Newark. 13. Dovecot. 14. Davie's Rock. 15. St Monance. 16. Kinkell Farm. 17. Kinkell Ness. 18. Buddo Ness. 19. Salt Lake. 20. Feddinch. 21. Lambieletham. 22. Ladeddie. 23. Bruntshiels. 24. Pepper Knowe. 25. Kinaldy. 26. Dunino. 27. Largo Law. 28. Balhousie. 29. Balmain. 30. Rires. 31. Kellie Law. 32. Gillings Hill.

182 VOLCANIC NECKS

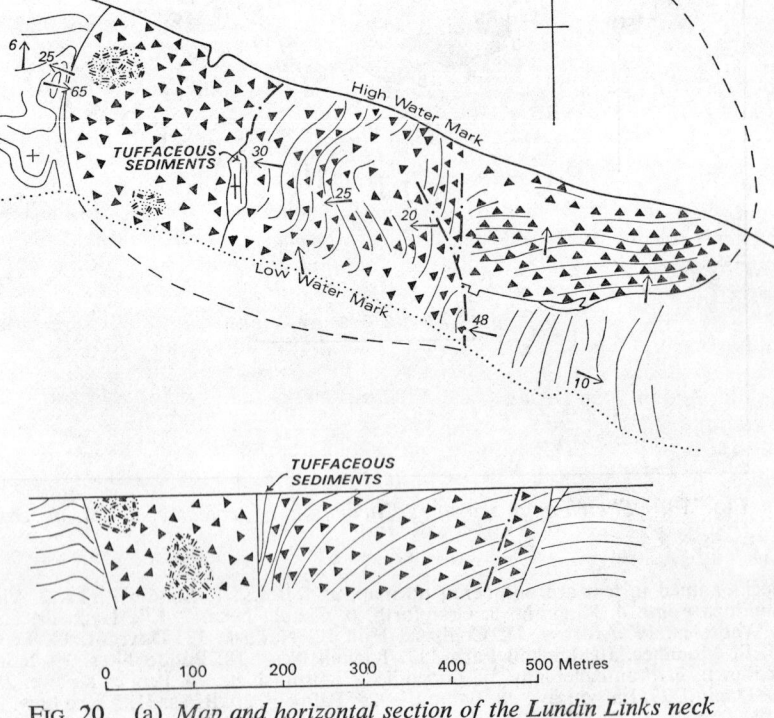

FIG. 20. (a) *Map and horizontal section of the Lundin Links neck*
(b) *Map and horizontal section of the Viewforth neck*

obvious discontinuity from dark crystalline basalt through bleached-basalt breccia to the main lapilli-tuffs of the neck.

The margin of the neck is irregular in outline and the sandstone wall, inclined inwards at angles of 45° to 60°, is breached in places, notably in the south, by tongues of tuff. A sub-parallel ring-fault cuts the sediments about 50 m from the margin in the west and south but in the east it is only 10 m from the margin. About 50 m west of the fault, on the west side of the neck, there is a second sub-parallel ring-fracture linked to the inner ring by a radial fault. Here, and to the south, basalt has been intruded along the planes of both ring and radial fractures. Basalt and tuffisite have also been emplaced patchily along another radial fracture which displaces the eastern margin of the neck near high water mark. Between the ring-faults and the neck the sediments—mainly sandstones—are disturbed and locally penetrated by tuffisite. Sandstones near the eastern margin of the neck are contorted and reconstituted in a fashion which suggests that they were not completely consolidated at the time of the volcanism. Sediments generally are turned down against both ring-fractures and neck margin.

Some 200 m to the east of the neck, beds just below the Barncraig Coal form a cryptovolcanic ring-structure measuring 25×15 m across and elongated parallel to the strike. It is delineated by a narrow zone in which the sediments are vertical, though the bedding inside shows only slight buckling or displacement. There is no evidence of brecciation or of tuffisite intrusion.

Viewforth neck. Much of the central and southern parts of the Viewforth neck (Fig. 20b) are seen, surrounded by Passage Group strata, in the tidal zone near Viewforth [NO 431025]. The neck is at least 700 m across and the outcrop is divided meridionally into two structurally contrasting areas. The eastern area is formed of greenish bedded tuffs and agglomerates in which fine and coarse layers contain a preponderance of green juvenile basaltic lapilli, with subordinate blocks, commonly 15 to 30 cm in diameter, of crystalline basalt, dolerite, mudstone, limestone and earlier bedded tuffs. Mr. R. W. Elliot has examined a thin section of one of the dolerites and reports that 'although it contains pseudomorphs apparently after olivine, the type of mesostasis is very similar to that which occurs in some members of the quartz-dolerite suite'. This confirms the record of quartz-dolerite blocks made by Wallace (1916) and Balsillie (1923). The bedding traces in the eastern sector strike N.E.–S.W., parallel to the southeast margin of the neck, but swing to N.–S. towards the centre.

In the western area the tuffs and agglomerates are similar in content, but are heterogeneous rather than bedded. Subsidence collapse of original bedding is indicated, however, by the presence, especially near the margin, of disorientated rafts of bedded tuff and red country-rock mudstone dipping vertically or at high angles. Further evidence of collapse is seen between the two areas where tuffaceous sediments, including a thin coal, crop out over an area of 85×25 m. They overlie the eastern bedded tuffs conformably, but are separated from the heterogeneous tuffs of the west by an irregular plane which cuts across the bedding and locally gives rise to a step feature on the beach. Towards that plane the dip increases to vertical (Fig. 20b) and is even overturned in places. The sediments are assumed to have been deposited within a breached tuff-ring after activity had ceased but before the final phase of subsidence. Miospores obtained from the sediments (p. 119) establish the date of this event—and by inference, the age of the neck—as early Westphalian.

There are no basaltic intrusions within the neck, but several veins of carbonate and tuff can be seen at high water mark near Viewforth and a tuffisite/carbonate dyke, 30 cm wide and 80 m long, traverses the bedded rocks of the eastern area near low water mark.

The western margin of the neck forms a regular curved trace in contact with Passage Group strata which include marine bands, red mudstones and interbedded tuffs (p. 115) and which are locally folded and turned down against the neck. The southern margin is covered by the sea, but probably does not extend far beyond low water mark.

The south-eastern margin, slightly displaced by a radial fault, has an irregular trace. The Passage Group sediments are turned down against it in some places, but strike at right angles to abut against it in others; at one locality the neck and the bedded sediments are separated by a displaced lobate raft of red mudstone extending for about 80 m along the margin.

Ruddons Point neck. Only the central and western parts of the Ruddons Point neck can be seen, in exposures extending 800 m from east to west and 600 m from north to south (Fig. 21a); the remainder, probably covering at least as big an area, is covered by beach deposits and sea. The neck is formed of approximately equal proportions of pyroclastic rocks and basaltic intrusions. The former consist of greenish bedded tuffs and agglomerates in which basaltic blocks predominate. At Shooters Point [NO 45490016] these dip south-westwards at angles varying from nearly vertical at the neck margin to 25°, suggesting centroclinal disposition around a centre point beyond low water mark. The record of tuffaceous sediments, presumably at the top of the sequence (Balsillie 1927), might be interpreted, like the Viewforth sediments, as representing deposition inside the breached tuff-ring prior to subsidence. The bedding farther north, around Ruddons Point itself, is however gently flexed in two dome-like structures, which are unlike the centroclinal arrangements of other necks.

A transgressive sill-like body of basalt within the bedded tuffs and agglomerates of Shooters Point is also unusual by comparison with other necks. However, the 4·5-m dyke farther north and the several plugs and bosses—some with columnar jointing—to the north-east, are exceptional only in their content of basic and ultrabasic nodules. These are up to 30 cm in diameter and contain olivine, chrome-spinel, chrome-diopside, pyroxene, magnetite and sporadic pockets of feldspar and biotite (Balsillie 1927). Most of the intrusions have ill-defined margins or pass by gradation through basaltic breccias to tuffs. This phenomenon, which probably reflects the presence of water contained in the bedded tuffs and agglomerates at the time of intrusion, is particularly well seen near high water mark in the north-eastern part of Shell Bay (Plate II).

The neck margin is exposed only on the east, where it is deeply indented into the neck, following a curved trace corresponding approximately to the western arm of Shell Bay. As the structures are complex and the exposures intermittent the lines shown in Fig. 21a are necessarily generalised. The margin is best seen at Shooters Point where adjacent Upper Limestone Group sediments are turned down at angles up to vertical along a north-westerly strike against and parallel to the bedded pyroclasts of the neck. Similar contacts can be seen in the north-western part of Shell Bay, where the margin trends E.–W. and where there is evidence of both a crush-zone and an outer parallel fracture. In the western part of Shell Bay, where the trace of the margin is most deeply indented into the neck outcrop, the crush-zone is only partly exposed, separating the neck from a zone of sediments, 150 × 220 m in extent, which are brecciated and permeated by tuffisite and bleached basalt. Some of the sandstone masses within the breccia are over 50 m long and have been folded, fractured and rotated. Inside the neck, near the area of maximum indentation, is a ragged-edged raft of red mudstone, about 120 m long. It belongs lithologically to the Passage Group or Upper Coal Measures rather than to the surrounding Upper Limestone Group and must therefore be displaced substantially downwards.

The evidence in and around the Ruddons Point neck suggests that the present level of erosion is not far below the original surface of eruption and that there has been

---

PLATE II

INTRUSION IN RUDDONS POINT NECK, SHOWING TRANSITION FROM BASALT THROUGH BASALTIC BRECCIA TO TUFF (UNDER HAMMER)

(*Photograph by E. H. Francis;* MNS 1742)

Geology of East Fife (*Mem. Geol. Surv.*) PLATE II

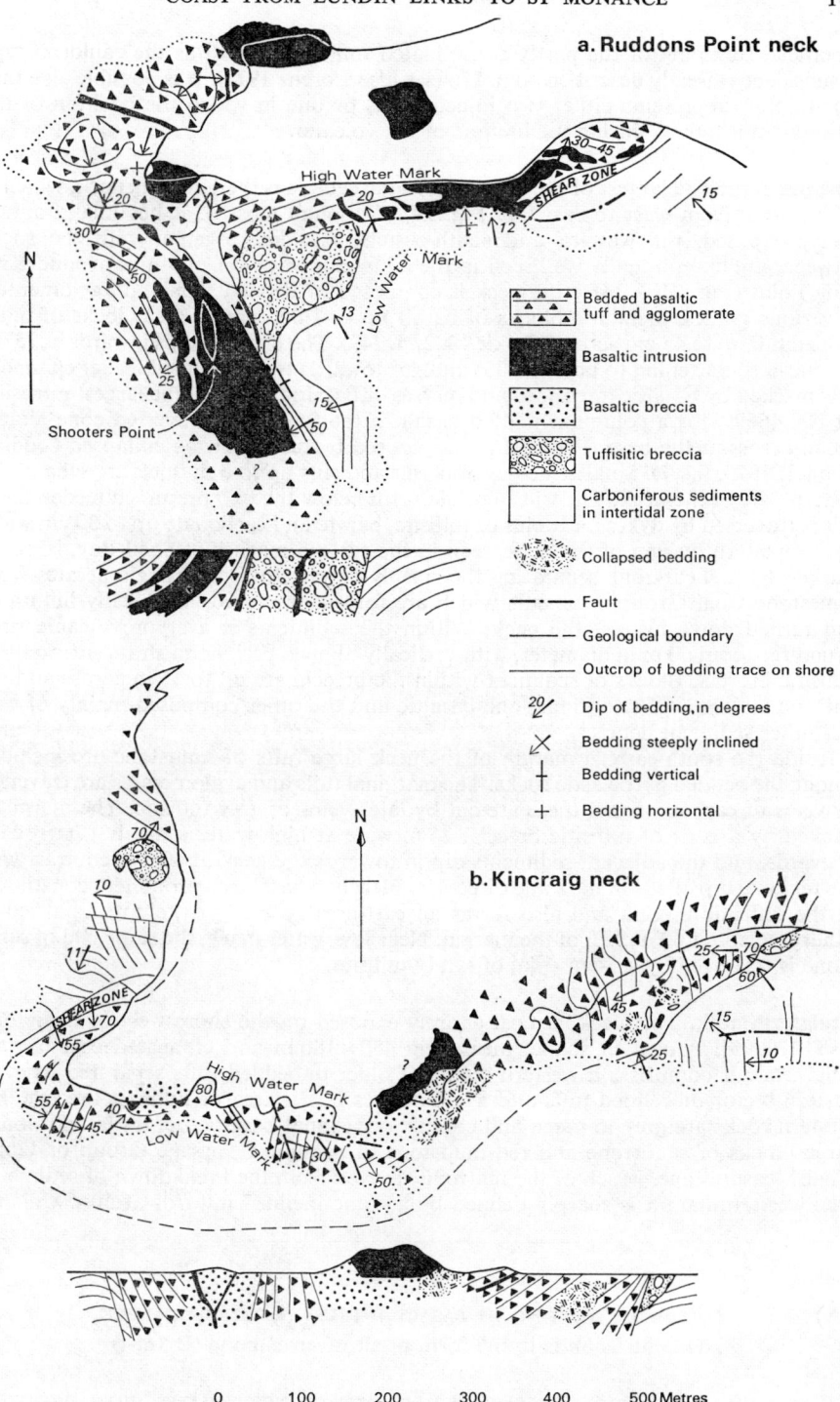

FIG. 21. (a) *Map and horizontal section of the Ruddons Point neck*
(b) *Map and horizontal section of the Kincraig neck*

superficial slumping of the partly consolidated tuff-ring as well as the cauldron type of subsidence already described on p. 176 (see also Lorenz 1973). It is possible also that the structure represents either two joined necks or one in which the position of the main conduit moved during the lifetime of the volcano.

Kincraig neck. The largest of the coastal volcanic centres is the Kincraig neck, which is 1·5 km long from west to east (see Fig. 19). Although most of it lies inland and is poorly exposed, the western and south-eastern margins, together with the rocks between, are exceptionally well seen in the cliffs and wave-cut platform around Kincraig Point (Fig. 21b). Here the neck is composed of bedded tuffs and agglomerates of various grades, including blocks of basalt up to 1·8 m across and blocks of older bedded tuff up to 45 cm across (Geikie 1902, p. 249). The bedding dips inwards at 75° at the margins, flattening to between 30° and 45° towards two presumed feeder channels, now marked by basaltic intrusions and masses of basaltic breccia. The largest intrusion [at NT 466998] is a columnar-jointed basalt in the form of an inverted cone almost 150 m across at the base (Plate IIIA) and floored by tuffs showing collapsed bedding (Francis 1970, fig. 7). Smaller bosses of basalt and sills up to 5 m thick are seen in the eastern part of the neck. The wave-cut platform below the two presumed feeder channels is traversed by dykes and veins of tuffisite, baryte and carbonate up to 30 cm wide.

The western margin of the neck, seen in the eastern reaches of Shell Bay, is partly marked by a shear-zone separating the steeply-dipping tuffs and agglomerates from Limestone Coal Group sediments which are folded, fractured and locally indurated and turned down against the neck. Within the sediments is a cryptovolcanic ring-structure, about 60 m in diameter, with vertically-aligned peripheral strata surrounding tuffisitic breccia. Blocks of sediment within the breccia are up to 15 m across and the tuffisitic matrix is of two kinds, one basaltic and the other composed mainly of coal and other sediment debris.

Inside the south-eastern margin of the neck large rafts of sandstone are included among the bedded pyroclastic rocks. The marginal tuffs and agglomerates are traversed by veins of carbonate and these are cut by later veins of grey tuffisite. The margin is flanked by a zone of tuffisitic breccia, 25 m wide at high water mark but narrowing seawards, and the adjacent sediments dip in towards the neck at 60°, in contrast with the regional dip of 15° in the same direction. Bleached basalt is a prominent constituent of the tuffisitic breccia and also forms an east-north-easterly dyke cutting a bed of sandstone about 60 m S.E. of the margin. Near low water mark, the same bed of sandstone is traversed by a 5-cm dyke of sandy tuffisite.

Craigforth neck. This neck, almost entirely exposed on the shore west of Craigforth [NT 479995], is oval in shape, measuring 450 × 180 m and elongated east to west (Fig. 22a). It comprises a western sector of older unbedded tuffs separated from an eastern sector of bedded tuffs and agglomerates by an area of tuffisitic breccia. The western rocks are grey to green and contain occasional basaltic bombs in addition to large blocks of sandstone and red mudstone of presumed Passage Group or Upper Coal Measures age. Much of the matrix is derived from the breakdown of sediments. The western margin is sharply defined by a plane inclined inwards at 50°, while the

---

PLATE III

(A)   COLUMNAR JOINTING IN BASALT INTRUSION, KINCRAIG NECK
The intrusion is in the form of an inverted cone (D 1684)

(B) IRREGULAR INTRUSION OF BLEACHED BASALT CUTTING SHALES WITH IRONSTONE BANDS ABOVE THE UPPER ARDROSS LIMESTONE, NEAR DAVIE'S ROCK, ST MONANCE
(D 1770)

(A)

Geology of East Fife (*Mem. Geol. Surv.*)

PLATE III

(B)

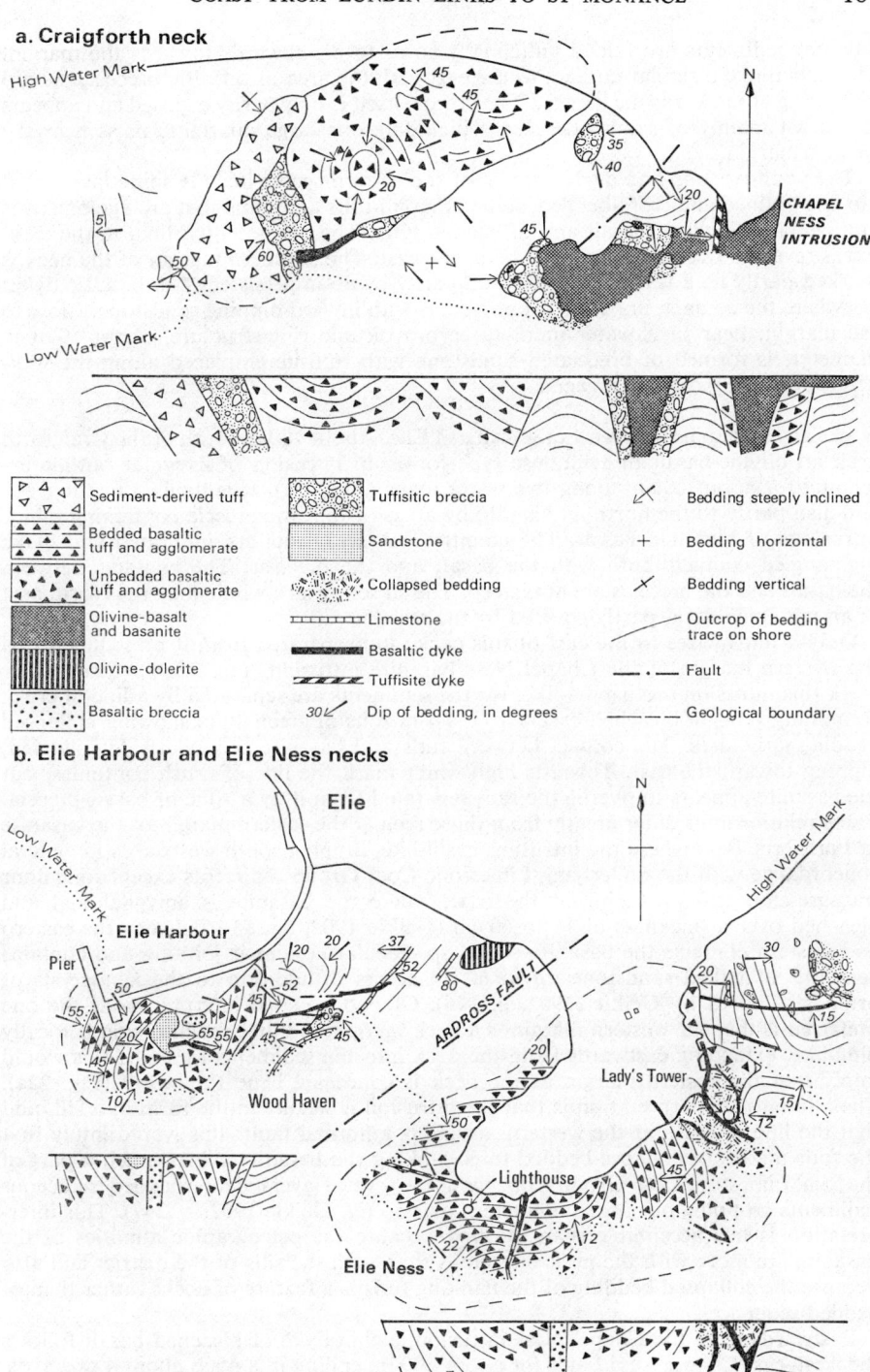

FIG. 22. (a) *Map and horizontal section of the Craigforth neck and structures between it and the Chapel Ness intrusion*
(b) *Map and horizontal sections of the Elie Harbour and Elie Ness necks*

adjacent sediments are folded anticlinally so as to dip inwards towards the margin. The tuffs make a similar contact to the east with the area of tuffisitic breccia, the dip here being 60° towards the breccia. The breccia itself is only partly exposed and appears to consist mainly of sandstone blocks including one large quartzitic mass near the western margin.

The younger tuffs and agglomerates of the eastern sector include some layers that are fine-grained and cross-bedded, some of lapilli-tuffs and some that are agglomeratic and contain blocks of basalt and of bedded tuffs. The dip is centroclinal in the west, but is reversed over an anticlinal axis in the east. The southern margin of the neck is marked partly by a tuffisitic breccia and partly by an impersistent 6-m basaltic dyke: elsewhere the contact, imperfectly exposed, is with inward-dipping sandstone. Close to the margin, near high water mark, a cryptovolcanic ring-structure, about 50 m in diameter, is formed of brecciated sandstone with tuffisite emplaced along the well-defined inwardly dipping margin.

Necks and intrusions between Craigforth and Elie. About 100 m south of the Craigforth neck an olivine-basalt of Hillhouse type forms an intrusion of irregular outline extending for about 200 m along low water mark (Fig. 22a). It is flanked to the west, and also partly to the north and south, by areas of tuffisitic breccia containing minor intrusions of bleached basalt. The country-rock sediments are indurated and make high-angled contacts both with the basalt and the breccias. The contacts between the basalt and the breccias are also steep. The structure as a whole may thus form part of an unnamed neck, partly covered by the sea.

Only a few metres to the east of this neck, and separated from it by sediments, is the western margin of the Chapel Ness basanite intrusion (Fig. 22a). At low water mark this intrusion too is plug-like, for the sediments are separated by a line of crush containing veins of bleached basalt from a 4-m zone of green tuffs showing collapsed bedding structures. The contact between tuffs and basanite is similarly high-angled, dipping toward the east. Towards high water mark the line of crush continues, but the basanite appears to overlie the tuff, separated from it by a zone of basalt breccia. These relationships differ greatly from those seen at the eastern margin of the basanite in Earlsferry Bay, where the intrusion is sill-like, dipping south-westwards in general concordance with the underlying Limestone Coal Group sediments except for minor transgression and puckering at the base, where the basanite is amygdaloidal and bleached over a thickness of 15 to 60 cm (Geikie 1902, fig. 53). Between the eastern and western contacts the basanite shows spectacular columnar jointing and contains xenoliths of baked sandstone up to 60 cm across in addition to the xenocrysts of orthoclase noted by Geikie (1902, p. 246). Of two possible interpretations the one preferred is that the western margin is a neck margin and that the intrusion is locally plug-like, extending eastwards from the neck into the sediments as a sill: this would imply that the eastern margin of the neck is concealed beneath the sill (Fig. 22a). The alternative interpretation is that the intrusion is wholly in the form of a sill, and that the line of crush at the western margin is a normal fault: this would imply that the four metres of collapse-bedded tuffs flanking the basanite are as much a part of the local Limestone Coal Group sequence as the several layers of interbedded tuffaceous sediments cropping out near the western margin (cf. Geikie 1902, p. 247). This interpretation is less acceptable than the first because the petrographic affinities of the basanite are more with the neck intrusions than with the sills of the district and also because the collapsed bedding of the flanking tuffs is a feature of necks rather than of bedded sequences.

Eastwards from the Chapel Ness intrusion, a thin dyke of bleached basalt follows the west–east Chapel Ness Fault for about 200 m, ending in a mass about 4 m across. A short distance along the line of the dyke and the fault, and near low water mark, is a ring-structure [NT 484996] about 65 m in diameter. It is filled partly by basanitic and bleached basalt intrusions of various shapes, and partly by green tuff, grey sediment-derived tuff and tuffisitic breccia. Cumming (1928, pp. 136–137) described it as a 'vent', but it is more likely to be a cryptovolcanic structure. Elie Bay also contains at least three more ring-structures interpreted as cryptovolcanic in origin, for they consist mainly of disturbed local sediments with inward dipping margins. The largest [NT 488998] is 60 m in diameter. Olivine-basalt of Hillhouse type exposed in an old quarry 250 m west of Elie Golf Clubhouse [NO 486002] is believed to belong to an intrusion lying in the southern part of a poorly-exposed neck otherwise composed of tuff and possibly about 250 m in diameter.

Elie Harbour neck. The Elie Harbour neck (Fig. 22b) occupies the promontory between Elie Bay and Wood Haven, and forms an oval outcrop measuring 135×225 m and elongated east to west. It consists mainly of green bedded tuffs and agglomerates traversed by a few thin sinuous impersistent basaltic dykes. The tuffs are mainly fine- to medium-grained on the eastern side, and lapilli-grade on the western side, with agglomerate containing large blocks of sediment and bombs of scoriaceous basalt up to a metre in diameter. Geikie (1902, p. 244) also noted fragments of wood and of anthracite. In the western part of the neck the bedding is centroclinally disposed, with dips decreasing inwards from 65° to 10°. This area is separated by a zone of vertical dips and a boss-like mass of basaltic breccia or agglomerate from an eastern area of outward, easterly dips. A large raft of tuffisitised sediment (mainly garnetiferous sandstone) lies near the centre of the neck. This caused Geikie (1902, p. 245, fig. 52) to interpret the neck as 'two vents which are partly united on the south side', but Cumming (1928, pp. 134–136, plate xxv) correctly saw it as a single neck.

The northern, western and southern margins are not seen, though they probably lie only a short distance beyond the present limits of exposure. The eastern margin is a ring-fault clearly marked by a trough-like feature along the shore. Inside it the pyroclastic rocks are heavily veined by carbonate minerals. Outside it is a narrow zone 3 to 7 m wide in which blocks of indurated sediments, including sandstone and limestone, are aligned with the nearly vertical margin and locally welded against it. This zone is bounded by a further ring-fracture, parallel to the margin, and the sediments adjacent to it are overturned so that they dip in towards the neck.

The narrow zone of highly-inclined sediments contains a thin dyke of bleached basalt which continues eastwards as one of two such intrusions following the planes of minor E.–W. faults. Three of the faults appear to be cut off near low water mark by an irregular mass of breccia consisting of sediments with a matrix of tuff and bleached basalt. The field relationships suggest a genetic connection between the bleached basalt of the breccia and that of the dykes. Cumming (1928, pp. 133–134, plate xxv) described the breccia as occupying an explosion fissure, but Francis and Hopgood (1970, p. 169) compared it with the structure along the Chapel Ness Fault (p. 188) and doubted if it had much, if any, subaerial expression. Both are probably cryptovolcanic ring-structures and are important as showing that, locally at least, some of the volcanism is later than some of the fracturing.

At high water mark in Wood Haven is an outcrop of dull greenish olivine-dolerite [NT 496997] which is bleached in a wide zone along its western contact. It is traversed, particularly in the eastern part of the exposure, by veins, up to 15 cm wide, of carbonate and grey tuffisite consisting mainly of sediment debris which is locally flow-aligned. Some of the veins close upwards. Except in the west, the contacts are covered by beach sand, but the general shape of the outcrop and the tuffisitic veining suggest that the dolerite is a boss or plug rather than a sill, though it is petrographically unlike other neck intrusions in the district and has been grouped with the olivine-dolerite sills (p. 139).

Elie Ness neck. The Elie Ness neck (Fig. 22b) is exposed over an area measuring 300×450 m and elongated N.E. to S.W., and is formed of green basaltic tuffs and agglomerates containing 'breadcrust' bombs of basalt and blocks of various sediments including red mudstone. The pyroclastic rocks are well bedded and dip centroclinally at angles decreasing inwards from 50° to 10° towards a presumed centre west of the lighthouse. The rocks have long been known (Heddle 1901; Geikie 1902; Balsillie 1927; Cumming 1928; Colvine 1968) for their content of basic nodules (amphibole and pyroxene) and for xenocrysts of garnet ('Elie ruby'), zircon and alkali-feldspars: some of the best localities for collecting are around the lighthouse. Cross-bedding, wedge-bedding and bomb-impact structures are well seen (Plates IVA and VA) in some of the more finely bedded layers east of the lighthouse. In a wide zone near the eastern margin of the neck the bedding has buckled, flowed or passed into massive rock of heterogeneous texture (Plate IVB), indicating breakdown of original bedding at a time when the rocks were still poorly cemented.

Near the presumed centre of the neck, as indicated by the inward dips, is an oval plug of breccia measuring 15 × 25 m and consisting of angular blocks of crystalline monchiquite up to 1·25 m across in a sparse matrix of fine green tuff and carbonate minerals. It is traversed by veins of tuffisite and carbonate up to 30 cm wide, which exhibit flow-foliation and extend eastwards beyond the breccia. The northern margin of the breccia dips inwards at 45°, separated from the bedded tuffs and agglomerates by a wedge of indurated sandstone (Geikie 1902, fig. 50), but to the south the breccia seems to rest irregularly on the top of the bedded rocks. It probably marks the site of a late-stage conduit. A further mass of basaltic breccia or agglomerate stands up above the wave-cut platform at Sauchar Point but it is poorly defined and is less confidently identified as marking the site of a former conduit.

The neck is cut by several minor intrusions. One lying about 50 m east of the lighthouse is a dyke which is 5 m wide at high water mark, tapering out southwards, and which consists partly of fine grey flow-foliated, locally sheared tuffisite simulating vertically bedded shale or siltstone, and partly (at high water mark) of basalt. Farther east, also at high water mark, one of the thinner tuffisite dykes trends E.–W. and is notable for including three separate generations of intrusion. In the eastern part of the neck, where the bedding structure is collapsed, north-westerly basanitic dykes are emplaced *en échelon* and form prominent linear stacks. One figured by Geikie (1902, fig. 51) wedges out downwards. Another, starting as a dyke near the neck margin at Lady's Tower, becomes irregular in outline at high water mark and makes ill-defined contacts with the tuffs.

The margin of the neck is seen only in the east where it follows a sinuous trace. The adjacent sediments, which are locally baked and partly brecciated and permeated by tuffisite, are turned down against the margin in some places, but are horizontal up to the sharply defined margin in others. The northern, western and southern margins are covered by sea and sand, though from the location of the presumed centre—the basaltic breccia—at the western limit of outcrop, the western margin may lie some distance offshore. Cumming (1928, p. 352) noted that the north-western edge of the exposed neck is approximately straight and thought that it might correspond to the trace of the Ardross Fault. Mapping by aerial photographs, however, suggests that the fault passes through Wood Haven farther to the west, beneath beach sand (Francis and Hopgood 1970, p. 169).

PLATE IV

(A) BASALTIC BOMB IN THE ELIE NESS NECK, SHOWING IMPACT EFFECTS IN THE UNDERLYING TUFFS

(*Photograph by E. H. Francis;* MNS 1635)

(B) COLLAPSED BEDDING IN TUFFS, SHOWING BUCKLING, MINOR FAULTING AND (RIGHT) TRANSITION INTO TUFFS OF HETEROGENEOUS TEXTURE, EASTERN PART OF ELIE NESS NECK

(*Photograph by E. H. Francis;* MNS 1636)

(A)

Geology of East Fife (*Mem. Geol. Surv.*) PLATE IV

(B)

(A)

Geology of East Fife (*Mem. Geol. Surv.*)   Plate V

(B)

The sediments between the Elie Ness and Wadeslea necks form an anticline which has a N.–S. axial trace, so that dips are inclined inward towards each neck margin (Figs. 22b, 23). These country rocks, probably lying stratigraphically below the Ardross limestones, are pierced by a group of small breccia structures, some of which have tuff matrices. They were described in detail by Cumming (1936, pp. 355–361) as 'intrusion breccias' and interpreted as post-volcanic emplacements caused by seismicity contemporaneous with the Ardross Fault. As noted by Francis and Hopgood (1970, p. 170), however, breccias of this kind have been recorded from other parts of the Forth volcanic province where seismicity engendered by faulting would be more difficult to invoke. Moreover, their affinity to larger cryptovolcanic structures described elsewhere in this account is emphasised by their close association with dykes of bleached basalt extending from the neighbouring necks. Most are roughly oval in outline, but the largest, on the crest of the anticline, also has a sill-like apophysis of grey sediment-derived tuffisite including coal debris and ragged-edged fragments of bleached basalt. A further small oval of grey tuffisite, without basalt, breaks the anticlinal crest farther south. Other structures, ranging in size down to 4×2 m, consist only of sediment breccias including blocks up to three metres across, with little or no basaltic material in the matrices. Near high water mark is a north-westerly dyke of tuffisite, eight centimetres wide, containing sediment debris and yellowish fragments of bleached basalt.

North-eastwards from this locality, for a distance of nearly two kilometres, the necks and sediments exposed along the shore are traversed by the Ardross Fault (Figs. 22b, 23a, 24), recognised first by Cumming (1936) who postulated a dextral strike-slip shifting parts of the necks which it traversed by 310 m: he thereby correlated the Elie Ness neck with the Ardross neck and the Wadeslea neck with the Coalyard Hill neck. Francis and Hopgood (1970) were unable to accept this correlation and showed that though the fault is partly a dextral strike-slip structure it has a strong vertical component. They also suggested, from the associated structures in the sediments and from alignments in the necks and related volcanic rocks, that this stretch of coastline coincides with a north-easterly fracture-zone which had a history of alternating volcanic and tectonic episodes, the formation of the Ardross Fault being the last.

Wadeslea neck. The area designated 'Wadeslea neck' by Cumming (1936) consists of two contrasted outcrops separated by an area of incompletely exposed sedimentary rocks (Fig. 23). The larger, southern outcrop is 400 m in diameter and is formed of basaltic tuffs and agglomerates which appear to be structureless over most of the area and which give rise to a featureless seaweed-covered platform. Near the northern margin they contain large rafts of sediment and near low water mark, in the same sector, they include blocks of crystalline basalt, up to a metre across, which are so abundant that the rock locally approaches the composition of basalt breccia. A dyke of sandy tuffisite, up to 30 cm wide, cuts the tuffs a short distance to the west. Towards the southern margin the pyroclastic rocks are centroclinally bedded at angles decreasing

---

PLATE V

(A) CROSS-BEDDING (LEFT CENTRE) AND WEDGE-BEDDING (LOWER CENTRE) IN TUFFS, ELIE NESS NECK
(*Photograph by E. H. Francis;* MNS 1637)

(B) CENTROCLINALLY BEDDED TUFFS IN THE KINKELL NESS NECK
The bedding traces curve round a central mass of vertically bedded tuffs which forms the stack on the left. Sandstone beds in the background dip obliquely to the left and are clearly truncated by the neck. The stack on the right is the Rock and Spindle (D 1672)

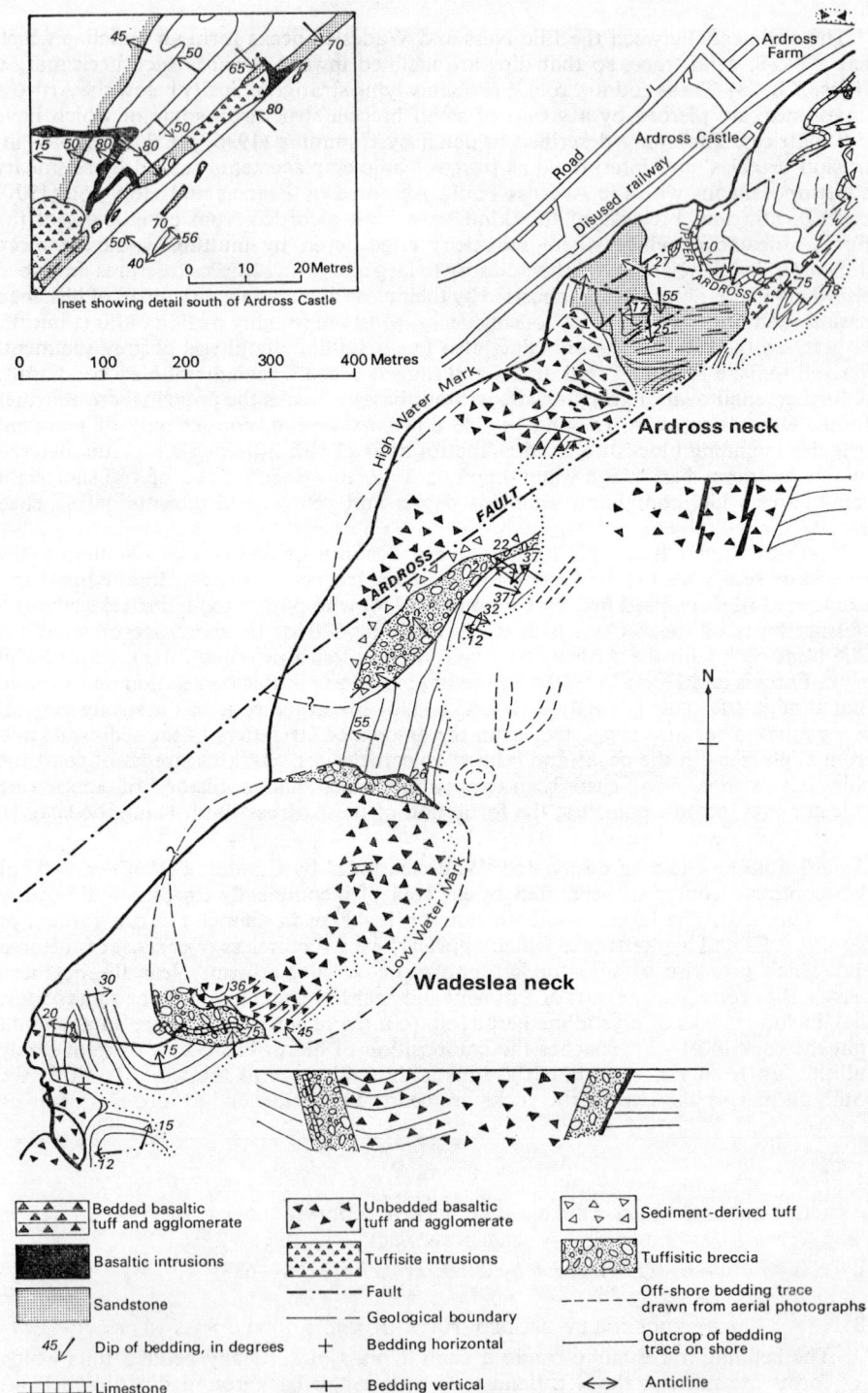

FIG. 23. *Map of shore between Elie Ness and Ardross Farm, showing the Ardross Fault, and plans with horizontal sections of the Wadeslea and Ardross necks*

from nearly vertical at low water mark to 35° approaching a centre now obscured by beach sand. They consist mainly of lapilli-tuffs with bombs of basalt and blocks of sediment, most of which are 6 to 15 cm across.

The neck is flanked to north and south by zones of tuffisitic sediment breccias. The southern zone passes gradually into the neck, but is cut off from the sediments to the south by a ring-fracture (Fig. 23; see also Cumming 1936, pp. 357–359). In places the material forming the breccia lies almost *in situ* (including large masses of what is probably the St Monance White Limestone), though tilted almost vertically, locally disorientated and penetrated by tuff. The breccia on the northern margin of the neck is cut off from the agglomerates by a clean break and is further separated from the sediments farther north by a fracture along which a dyke of bleached basalt has been emplaced.

At the time of the resurvey both the tuff–agglomerate of the southern outcrop and the northern flanking breccia were obscured by beach debris towards high water mark. It may be inferred from mapping by Cumming (1936, fig. 1 and plate xxi) that the northern breccia swings round through 90° of arc in continuity with a further zone of breccia which, together with a mass of sediment-derived tuff, comprises the smaller, northern outcrop of the Wadeslea neck. The breccia of this northern outcrop consists mainly of sandstones, collectively dipping north-westwards in continuity with undisturbed sediments near low water mark, but dislocated, slightly disorientated, and penetrated by tuffisite. The undisturbed sediments lie stratigraphically below the Ardross limestones and are aligned with the strike of similar beds to the south of the Wadeslea neck. Unlike the basaltic pyroclastic rocks of the southern outcrop, the tuffs lying to landward of the breccia are dark-grey in colour and consist mainly of comminuted sediments with ragged-edged, altered basaltic fragments which become increasingly abundant away from contact with the sediments. These tuffs are cut off by a north-easterly trench-like feature which Cumming (1936, p. 344) took to mark the trace of the Ardross Fault, beyond which is the mass of basaltic tuffs and agglomerates of the Ardross neck. The nature of the contact is obscured in the floor of the trench, but the feature itself, together with the contrast in type of tuff between the two sides, supports Cumming's interpretation.

Ardross neck. The more southerly of the two exposed parts of the Ardross neck (Fig. 23) is formed of coarse green basaltic tuffs and lapilli-tuffs with sporadic bombs of basalt up to 15 cm in diameter, blocks of bedded tuff up to a metre long and with local carbonate veining. The texture is mostly homogeneous, but in places ill-defined areas of collapsed bedding can be seen. These rocks are separated from the northern part of the neck by a 100-m embayment covered by sand, shingle and boulders, beneath which Cumming (1936, p. 344) established the continuation of the neck by digging. The northern area consists of material similar to that of the southern area, but is traversed by a rudely rectilinear pattern of joints. Along the dominant joint-trend, which is west-north-westerly, basaltic dykes up to two metres wide have been intruded *en échelon* and stand up as walls above the beach. Veins of carbonate and fine-grained tuffisite have been emplaced on joints parallel to the dykes and also along the subordinate group of joints at right angles. Near high water mark the basaltic rocks are penetrated by sediment-derived tuffisites and contain sediment rafts, and this suggestion of proximity to the north-western neck margin is confirmed by a small exposure of indurated sandstone just above high water mark. Near the north-eastern margin blocks of sandstone and limestone up to three metres long are included in the carbonate-veined tuffs. In places, the latter form an apparently vertical junction with tuffisitised sediments, though separated from them by a narrow band of hard sediment-derived tuffisite; in others the neck material is lobed irregularly into the sediments.

Two small bodies of tuffisite, 200 to 250 m south of Ardross Castle, are of significance in showing the relation between volcanism and tectonic structure (Fig. 23, inset). The smaller is elongated parallel both to the Ardross Fault and to the strike of the highly inclined sediments. The larger, several metres to the south-west, is also aligned with the fault though it cuts across the strike. The elongation of these masses is mirrored by the arrangement of their constituent particles, particularly in the larger body. These particles consist of comminuted sediments with frayed blebs of bleached basalt and large angular sedimentary blocks, all showing north-easterly flow-banding parallel to the margins and to the fault. The seaward margin of the larger mass coincides with the fault, but is not noticeably sheared by it; neither, on the other hand, does the tuff cross the fault. The evidence seems to indicate that emplacement of these intrusions was controlled by a north-easterly fracture line, though it cannot be determined whether the fracture was the final expression of the Ardross Fault as now exposed, or some earlier form. Certainly there was volcanism at an earlier time as is now shown by sheared tuffs in the fault zone a short distance to seaward.

Coalyard Hill neck. This neck consists of two elements, of different ages, which are so contrasted as to warrant their separate description as outer and inner necks (Fig. 24a). The outer neck, which is the older, is only about 100 m wide but it extends north-eastwards for a distance of about 700 m along the fault. It is exposed along a low, generally featureless wave-cut platform and is formed mainly of grey or brown tuff derived mainly from sediments, though some fine-grained basaltic material in the southern sector locally shows evidence of collapsed bedding and a knob of basaltic agglomerate measuring 7 × 15 m occurs near the centre of the northern sector. It is traversed by veins of grey tuffisite and by two thin dykes of bleached basalt in the south and small knobs of similar material are intruded near the north-western margin of both northern and southern sectors. Rafts of sandstone are also found near the margins; the largest, in the north-east, is permeated by flow-aligned tuffisite (Plate VIA).

The south-western margin of the older neck makes a clean break angled at a dip of 45° or more with the sediments and forms a prominent erosional scarp feature about a metre in height; large blocks of sediment attached to the marginal tuff are tilted towards the neck and suggest some local small-scale neck subsidence. Between 100 and 150 m east of Ardross Castle, the flanking sediments are cut by two dykes of tuffisite, one of which is in line with a fault farther west and was described as a sandstone by Wood (1887, p. 496) and Cumming (1936, p. 361). About 50 m east of the dyke is a small ring-structure filled with disorientated blocks of siltstone; it is cut by a dyke of bleached basalt extending along the line of a fault from a knob of similar material just inside the neck.

Around the north-eastern sector, by contrast, the margin is embayed into the surrounding sediments and the latter are intimately penetrated by tuffisites: both the tuffisites and the embaying tuffs of the neck show various lithologies. The penetration is particularly well seen at high water mark midway between Ardross and Newark Castle where, also, a dyke of bleached basalt crosses from neck into sediments without a break at the margin. From this locality southwards towards the Ardross Fault the eastern margin of the outer neck is followed by an impersistent dyke of hard tuffisite. Near the fault the linear trace of the margin is broken by some of the large rafts of

---

PLATE VI

(A) FLOW-BANDING IN TUFFISITE, NORTH-EASTERN PART OF COALYARD HILL NECK
   The tuffisite (below) is intruded into a large raft of sandstone (above). The pale fragments elongated parallel to the edge of the sandstone are of bleached basalt (D 1681)

(B)   TUFFISITE AT WESTERN MARGIN OF THE ST MONANCE NECK
   The tuffisite (dark) penetrates upwards into the sandstone (pale) to form a tuffisitic breccia. Edge of monchiquite dyke at bottom right (D 1768)

(A)

Geology of East Fife (*Mem. Geol. Surv*)   PLATE VI

(B)

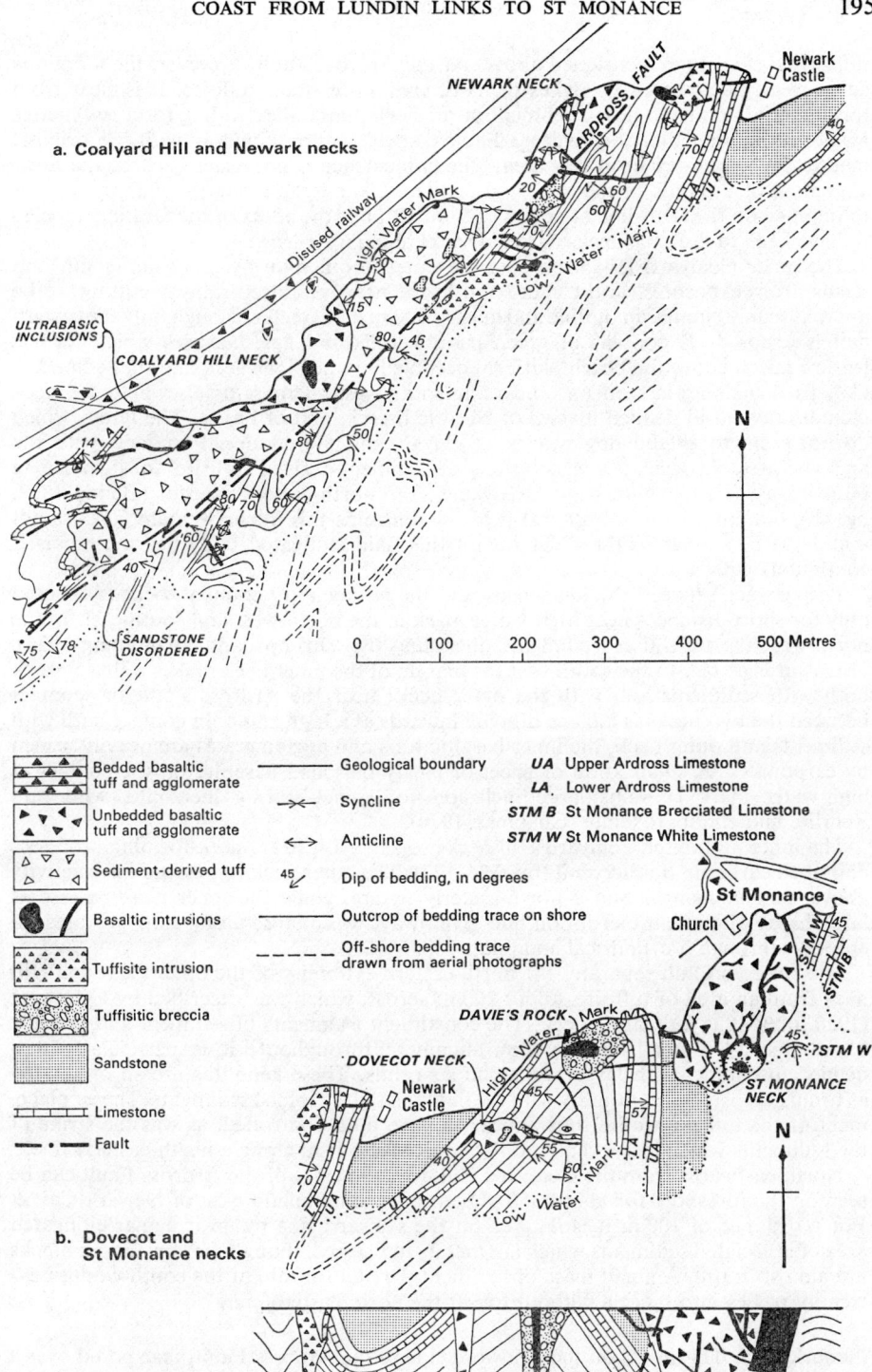

FIG. 24. (a) *Map of shore between Ardross Farm and Newark Castle, showing the Ardross Fault, Coalyard Hill and Newark necks and associated structures*
(b) *Map and horizontal section of shore between Newark Castle and St Monance, showing the Dovecot, Davie's Rock and St Monance cryptovolcanic structures and necks*

tuffisitised sandstone mentioned above. At the Ardross Fault, however, the margin is again clearly delineated by dykes of hard, erosion-resistant tuffisite. It is clear from this material and from flow-alignment of the neck material extending for a few metres inside the margin that the Ardross Fault coincides approximately with the original margin for about 100 m in this area. The coincidence is not exact because the flow-alignment of the tuffs in the neck and of the tuffisite dyke along the margin appears to impinge on the fault-line at angles of 5° to 10°, but the effect of the faulting appears to have been to drag the neck-marginal rocks along it *en échelon*.

The inner Coalyard Hill neck cuts the outer about midway along its length and seems, from exposures on the shore and in the nearby disused railway cutting, to be approximately circular in outline and at least 300 m in diameter, though only the eastern half is exposed. It consists of green basaltic tuffs and agglomerates which form a feature raised above the beach platform occupied by the older neck and the sediments. Collapsed bedding is seen only near the south-western margin: elsewhere the neck includes several ill-defined masses of basaltic breccia and of basalt. The latter, which do not seem to extend downwards as plugs, exhibit gradational contacts with the agglomerates. At least two generations of sediment-derived tuffisite are intruded in intimate association with a basalt dyke at the north-eastern margin (Geikie 1902, fig. 48), but from its lithology and field relationships this intrusive material does not appear to be similar to the older tuff of the main elongated body nearby, nor is it continuous with it.

The contact between the inner neck and the sedimentary country rocks is exposed only for short distances near high water mark to the north-west and south-east. In the north-west the tuffisitised sediments, including the 'shrimp band' (Cumming 1936), dip inwards at 18°. In the south-east the margin of the inner neck makes a sharp break both with sediments and with the outer neck: near the Ardross Fault the contact between the two necks is a plane dipping inwards at a high angle. In contact both with sediments and outer neck, the inner basaltic tuffs and agglomerates are heavily veined by carbonates. A small knob or sheet of partly bleached basanite on the contact at high water mark contains large inclusions of spinel-bearing lherzolite, with rare wehrlite and clinopyroxenite (Chapman 1974).

The inner and outer Coalyard Hill necks together illustrate neck-forming processes described earlier in this account (pp. 174–177). The outer neck represents early activity above magma rising along a north-easterly fissure, while the inner neck represents late phases when fissure eruption had given way, by collapse along ring-fractures, to eruption through a cylindrical conduit.

The Ardross Fault separates the north-eastern extremity of the outer Coalyard Hill neck from an area of tuffisite, about 200 m across, which was described by Cumming (1936, p. 342) as a separate neck. The constituent fragments of sediment and juvenile basalt display a north-easterly flow-alignment throughout; it is particularly well defined in contact with large sandstone xenoliths. These xenoliths are so orientated as to suggest original continuity with the flanking tightly folded sediments. The emplacement of the tuffisite would thus appear to have been controlled, as was the strike of the sediments which it has replaced, by earlier tectonism along a north-easterly line.

North-eastwards from the Coalyard Hill neck the trace of the Ardross Fault can be seen on the foreshore for about 350 m before it passes inland west of Newark Castle. For a distance of 100 m it is flanked on the seaward side by an irregular elongated area of brecciated sediments which are mainly tuffisitised, though the constituent blocks are almost *in situ*. A small mass of sediment-derived tuffisite at the south-western extremity passes into breccia without tuff at the north-eastern end.

Newark neck. This neck, on the landward side of the Ardross Fault, is exposed over a triangular area only 100 m across (Fig. 24a). Against the fault it is formed of material derived mainly from the breakdown of sediments, including some large rafts set in a matrix which contains fragments of juvenile basalt. This lithology suggests that the

Ardross Fault here approximates to part of the original neck margin. The sediment-derived tuffs pass northwards, without a clearly-defined break, into bedded basaltic tuffs and agglomerates. Near the margin, exposed only to the west, the bedding dips steeply inwards. This western margin is flanked by a zone of upturned, sheared sediments up to 1·3 m wide; traced seawards the margin is displaced by a north-westerly fault which also cuts the Ardross Fault. Veins of tuffisite cross the marginal bedded tuffs and are also seen, partly sheared, within the marginal zone of upturned sediments.

On the seaward side of the Ardross Fault hereabouts the folded sediments contain minor intrusions of tuff on broadly north-easterly lines parallel both to the fold axial trend and to the main fault. East–west basaltic dykes break across the folds in the sediments, but near the fault they veer along north-easterly-striking bedding planes parallel to the fault.

To the west of Newark Castle the sediments are traversed by two further masses of tuffisite. Both are mainly sediment-derived with small proportions of bleached basaltic fragments and both are associated with minor intrusions of bleached basalt. The larger, western mass of tuffisite is irregular in outline, with apophyses penetrating the folded sediments. The eastern mass takes the form of a sill which is 7·5 m thick at high water mark, overlying the Lower Ardross Limestone, but which wedges out seawards over a distance of 45 m and in doing so transgresses downwards to lie beneath the limestone.

Dovecot neck. The Dovecot neck, so called by Geikie (1902, pp. 236–237), is the westernmost of three volcanic masses which pierce the folded Ardross limestones and associated sediments along the shore between Newark Castle and St Monance (Fig. 24b). It consists of variable proportions of comminuted sediment and of juvenile bleached basalt. It also incorporates a raft of sandstone which measures 24×7 m. The sandstone is penetrated from below by flow-aligned tuffisite in a fashion which suggests that it is part of a stoped roof. The margins of the tuff are intimately welded to, and locally penetrate, the Ardross limestones and shales. The latter show no sign of having been turned down by post-volcanic subsidence, though there is local small-scale thrusting which may be related to the emplacement process. The evidence suggests that the neck had little or no surface expression as a volcano, but should rather be regarded as a large cryptovolcanic ring-structure. Thin east-south-easterly dykes of tuffisite and bleached basalt extend eastwards from the structure across the axis of an anticline which is also breached by a further ring-structure, 6×18 m across and filled with breccia.

Towards high water mark the anticlinal axis is further breached by an ill-defined area of breccia which includes Davie's Rock—a plug-like mass of nepheline-basanite measuring 15×45 m across and marginally bleached. A narrow zone of tuff can be seen in contact with the landward margin of this basanite, but the relationship of both rocks to the surrounding breccia, and much of the breccia and its contacts with sediment, were obscured by beach boulders at the time of the resurvey. Near its northern extremity it contains blocks of the Lower Ardross Limestone lying vertically; they are not in continuity, though they retain an approximate alignment with the more normally dipping outcrop of the bed on the western flank of the anticline. In the cliff near the edge of the breccia hereabouts the Upper Ardross Limestone is seen to be spectacularly crumpled and broken; together with the adjacent shales it is invaded by irregular tongues of bleached basalt (Plate IIIB) which are locally disintegrated to form intrusion breccias of similar lithology to some of the cryptovolcanic ring-structures of Earlsferry and Wood Haven (pp. 188 and 189). Because of this similarity, and because the margins are not typical of necks, the breccia is interpreted as representing exposure at a deep level within a large cryptovolcanic structure.

St Monance neck. About 100 m east of Davie's Rock the sediments on the eastern limb of the anticline mentioned above are interrupted by the St Monance neck (Fig. 24b), which is more than 200 m in diameter and is formed of green basaltic lapilli-tuffs and agglomerates in which the blocks and bombs up to 30 cm in diameter consist mainly of basalt. These rocks are structureless except at high water mark, where collapsed bedding structures can be seen passing without break into unbedded pyroclastic rock. Monchiquitic dykes, some standing above the beach, others forming depressions, ramify northwards and westwards, from an apparent centre at a plug-like mass in the southern part of the neck. Cross-cutting relationships show them to be of different ages and some thin out downwards (Geikie 1902, figs. 44, 45). One of the dykes crosses the western margin into the sedimentary wall-rocks and appears to be following the plane of an east–west fracture which has shifted the margin by about 15 m. Inside the neck the monchiquite is dark-grey and crystalline, but in the sediments it has been bleached. Thin dykes and sills of sandy tuffisite also traverse the neck.

Near the western margin rafts of sandstone are incorporated in the neck material, and rafts of red mudstone are similarly incorporated near the eastern margin. The margin of the neck is best exposed above high water mark on the western side, where the ring-fault crush is seen, together with intimate tuffisitisation of the sediments (Plate VIB). The latter are steeply turned down against the margin here, and also at the eastern margin, while to the south, below low water mark, a similar effect is indicated by the deflection of sediment bedding traces as seen in aerial photographs. Tuff and agglomerate exposed in the St Monance Burn, 50 to 100 m north of the church [NO 52270144], probably lie in the neck. E.H.F.

Coast from St Monance to Kingsbarns

On the coast between St Monance and Kingsbarns there are several cryptovolcanic ring-structures in which the country rocks have been folded, faulted, brecciated and largely or completely disorientated. Some of these structures consist only of disordered blocks of the country rock, but others contain a few small intrusions of bleached basalt, and the one at Balcomie Sands [NO 632102] is cut by a dyke-like body of basalt and by tuffisite derived from sediment. The structures range from only a few metres to at least 200 m in diameter. What appears to be the largest—but one which may possibly be tectonic rather than cryptovolcanic in origin—is situated just east of the Fluke Dub [NO 627107] and is fully 400 m by 270 m in size. One of the smaller ones is prominently exposed [NO 551024] just outside the east wall of Pittenweem Harbour. Another [NO 539022] obscures the outcrop of the Pathhead Upper Marine Band on the shore, and possibly explains why Kirkby (*in* Geikie 1902, p. 77) omitted this band from his table, noting only that the strata are 'disturbed'. Several small cryptovolcanic structures at Pittenweem Harbour [NO 54820235], described as 'vents' by Balsillie (1920b, pp. 83–85), contain tuffisitic breccia mixed with a little bleached basalt. I.H.F.

Kinkell Area

A group of necks, some with basaltic or basanitic intrusions, is exposed both inland and on the coast between St Andrews, Easter Balrymonth [NO 537141] and Babbet Ness [NO 592142] (Geikie 1902, pp. 207–221). The Kinkell Farm neck, exposed in the fields some 400 to 900 m north-west of Kinkell Farm [NO 540147], is the largest. It consists mainly of unbedded basaltic tuff but at its south-western side contains a mass of basanite. A member of the suite of Permo-Carboniferous quartz-dolerite dykes was reported (p. 168) to cut the tuff in the neck (Geikie 1902, pp. 198–199, 220, fig. 26), but the exposure is now too degraded to demonstrate this relationship. Smaller necks full of unbedded tuff are exposed at Easter Balrymonth Hill [NO 53211442], between Kinkell and Kingask [NO 541145], and in a cliff [NO 51951580] near Kinkell Braes. Basaltic tuff, with some sediment debris, which was encountered in a borehole

[NO 54611408] at Pitmullen, presumably also lies in a small neck. Several small intrusions of tuffisite have been detected on the shore (Balsillie 1920a, 1920b), some of them lying in the Maiden Rock Fault (p. 229) and associated fractures between the

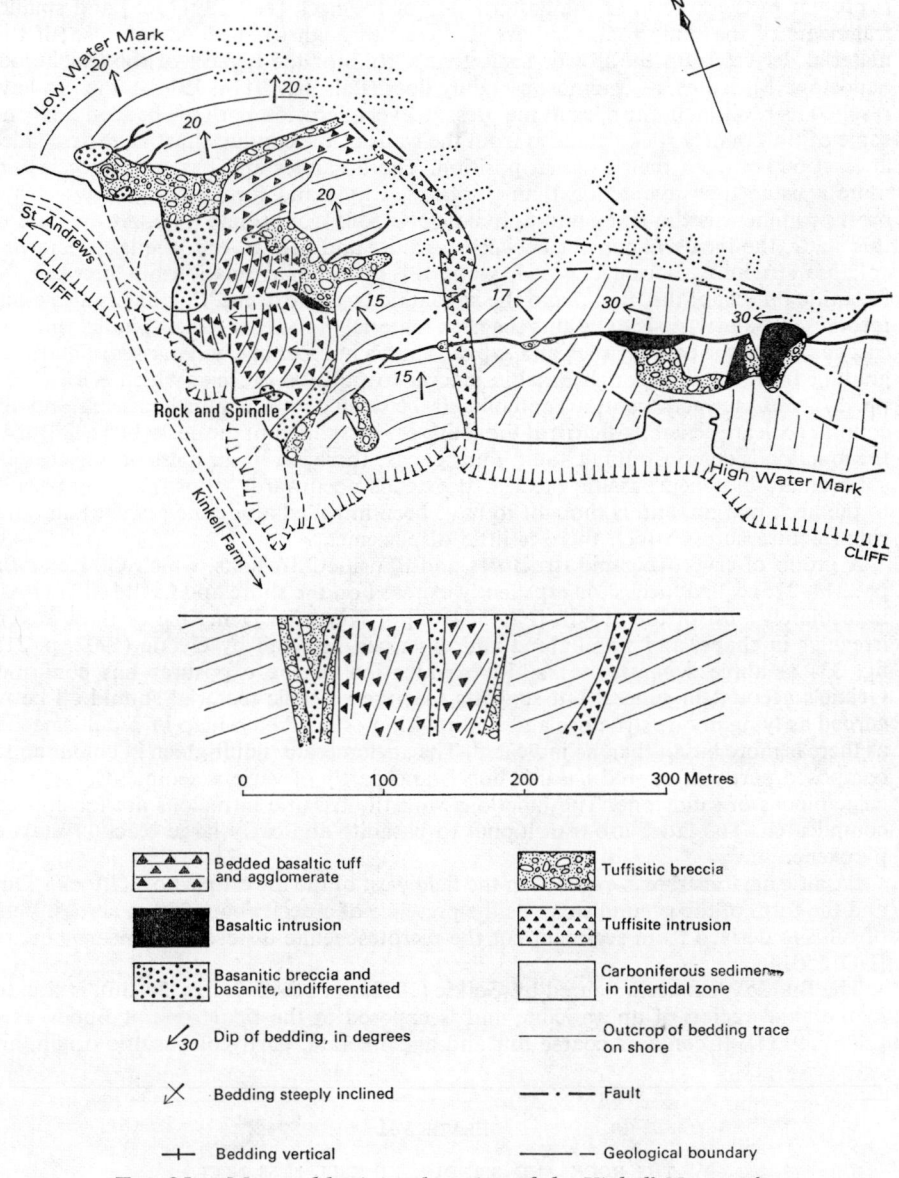

FIG. 25. *Map and horizontal section of the Kinkell Ness neck*

Maiden Rock [NO 52641579] and the southern end of the East Sands [NO 520160] at St Andrews.

The Kinkell Ness neck (Balsillie 1919; Geikie 1902, pp. 208–210) is well exposed on the shore just south of Kinkell Ness [NO 538158] (Fig. 25). It consists of tuff and agglomerate showing centroclinal bedding (Plate VB), with a central mass of similar material lying almost vertically. These rocks have been invaded by masses of xenolithic basanite (locally passing into basanitic breccia) and, near the neck margins, by tuffisitic breccia. A large block of crinoidal limestone (Balsillie 1919, p. 501; Kirk 1925, p. 374) is present in the breccia at the eastern side of the neck [NO 53911565] and smaller fragments of the same rock type are scattered through the bedded tuff also; if this material derives from the Lower Limestone Group or the highest of the Calciferous Sandstone Measures, as it almost certainly does (Balsillie 1911a, 1911b), then to have reached its position in the neck at the present level of erosion both the bedded tuffs and some of the country rock detached from the walls of the conduit must have descended at least 600 m from their original position. The basanites and basanitic breccias are more resistant to wave action than the bedded tuffs and form stacks, of which the most prominent is the Rock and Spindle[1] [NO 53851558] (Plate VII). At the base of this stack, the intrusive contact between basanite and bedded tuff, the latter showing collapse structure, is well exposed, while the gradational relationship between the basanitic breccia of the 'Rock' and the xenolithic basanite of the 'Spindle' is also easily recognised. A later basic dyke cuts the neck material at the Rock and Spindle, runs out eastwards into the country rock and expands into a number of small masses of material, grading from basalt through basaltic breccia to tuffisitic breccia, which Kirk (1925, pp. 375–382) termed 'vents'. The contacts between the Kinkell Ness neck and the country rock are generally sharp; at the south-eastern corner of the neck [NO 53861554], the junction is demonstrably a fault. Farther east there is a linear mass of tuffisite with subordinate bleached basaltic breccia. It extends northwards, approximately parallel to the neck margin, and is thought to have been intruded along the plane of an outer ring-fracture across which there is little displacement.   J.I.C.

A group of cryptovolcanic structures and tuffisitic intrusions, which Geikie (1902, pp. 214–216) called 'necks', is excellently exposed on the shore and in the cliffs at and immediately west of the mouth [NO 554101] of Kittock's Den. One of these is very irregular in shape, and was figured and described in detail by Geikie (1902, p. 215, fig. 33) as three separate necks. Re-examination of the exposures has confirmed Geikie's account in general but suggests that the volcanic material should all be regarded as lying in one structure and that revision of his sketch-map in detail is needed as there is more basalt than he indicated. The agglomerate is dull-green in colour and is composed partly of altered basaltic debris and partly of various sediments, including large blocks of sandstone. The junctions with the basaltic intrusions are locally very complicated. The latter are transitional to basanite and carry large black crystals of pyroxene.

Basaltic agglomerate is exposed in the field west of the lowest part of Kittock's Den, and the form of the ground suggests the presence of a neck about 200 m across. Veins of tuffisite derived from sediment cut the picroteschenite exposed in a nearby quarry [NO 550146].

The Buddo Ness neck, figured by Geikie (1902, pp. 216–217, figs. 34–36), is situated in the axial region of an anticline and is exposed in the tidal zone at Buddo Ness [NO 559154]. It contains coarse tuff and agglomerate, partly of basaltic origin, and

---

PLATE VII

THE ROCK AND SPINDLE, KINKELL NESS NECK

The basanitic breccia of the 'Rock' and the columnar-jointed basanite of the wheel-like 'Spindle' are part of an intrusion which cuts the bedded tuffs of the neck (D 1762)

---

[1] So called because of a supposed resemblance to a distaff ('rock') and spinning wheel ('spindle').

PLATE VII

partly derived from local sediments, including large blocks. The sandstones round the margins have been much veined, jointed and brecciated. Geikie figured the local downfolding of some of these beds. The structure, which is probably cryptovolcanic, is about 150 m long from north to south and reaches 40 m in breadth, but near the middle it is constricted to as little as five metres at one locality. The northern part is occupied by an olivine-basalt intrusion.

Geikie (1902, pp. 217–219, fig. 37) described one of the necks in this area under the heading 'Laws Castle'. This name no longer appears on Ordnance Survey maps and the name 'Salt Lake neck' is therefore used here. This structure, which is probably cryptovolcanic, lies on the shore 400 m east of the mouth of the Kenly Water. It is a very elongate body, at least 200 m long in a north-north-easterly direction but less than 10 m wide except near high water mark [NO 58511439] where it expands to about 30 m. The neck is composed mainly of basic agglomerate containing some sedimentary rock debris, and there are several small intrusions of basanitic olivine-basalt with large crystals of black augite.   I.H.F.

Cairnsmill Area

A large body of bedded basaltic tuff and agglomerate at Wester Balrymonth [NO 501143] is believed to lie in a neck but may possibly be interbedded with sedimentary rocks at a horizon high in the Calciferous Sandstone Measures. Up to five metres of tuff are exposed in two quarries, one [NO 50101462] near the summit of Wester Balrymonth Hill and the other [NO 50781441] near Scooniehill (Geikie 1902, pp. 221–222). Unbedded tuff and agglomerate, well exposed in a stream [NO 498141] south-west of Wester Balrymonth, probably lie in the same neck (Geikie 1902, pp. 222–223).

Unbedded basaltic tuff appears to be the major constituent of several smaller necks in the surrounding area: most of these have been described in detail by Geikie (1902, pp. 221–225). Well-exposed examples are seen in stream sections immediately to the east and north of Bogward [NO 492156], 200 m north-west of Cairnsmill [NO 497149] and 150 m north of Priorletham [NO 497127]; less well-exposed examples have been mapped at Feddinch [NO 485135], 350 m north-east of North Lambieletham [NO 504132], 300 m south of Allanhill [NO 520141], and around South Lambieletham [NO 504126]. Four small poorly-exposed necks near Strathtyrum House [NO 490172] contain mixtures of basanite, sandstone and bleached basalt. A small neck north-west of New Grange is known only from a borehole [NO 51171502] (Balsillie 1920b, pp. 82–83). Bedded basaltic tuff and agglomerate in a small neck are well exposed in stream banks at The Den [NO 48551410], about 700 m south of Lumbo, and bedding was detected in tuff samples recovered from a borehole [NO 48411345] in the Feddinch neck.   J.I.C.

Coarse basaltic tuff and fine agglomerate at least four metres thick, with occasional blocks including one of dolerite over 0·3 m in size, can be seen in an old quarry [NO 505123] on the north bank of the Cameron Burn. The pyroclastic rocks show traces of near-vertical banding. Together with the scattered exposures of similar material around South Lambieletham they are assumed to lie within the Lambieletham neck, which must on this basis have a north–south diameter of almost 700 m and a west–east diameter of about 600 m.   I.H.F.

Ladeddie Area

Within a radius of about three kilometres around Backfield of Ladeddie [NO 440138] numerous small necks of basaltic tuff, some of them described in detail by Geikie (1902, pp. 224–225), occur in an area heavily invaded by quartz-dolerite and olivine-dolerite sills. The age-relationships between the necks and these intrusions cannot be proved, however, except in the case of the Ladeddie neck, where an associated intrusion of olivine-basalt of Hillhouse type appears to cut the olivine-dolerite of the Drumcarro intrusion. Small necks containing unbedded, mainly basaltic, tuff and agglomerate are exposed by the roadside [NO 44761528] 400 m east of Nether Magask, near a quartz-dolerite intrusion in a ploughed field [NO 43051372] 300 m north of Newbigging of Blebo, on a low knoll [NO 45711407] 350 m north-east of Denork, and in the banks of a small stream 500 m south-west of Newbigging of Blebo [NO 430135] (Craig 1912, pp. 83–87). Bodies of similar rock have been encountered in boreholes a short distance to the east and south-east of Drumcarro Farm [NO 453129]. Bedded basaltic tuff, probably in small necks, is exposed in a quarry [NO 43511582] at Clatto Hill and on a small knoll [NO 45391349] just north of Drumcarrow Craig. Two small necks on the hillside 150 to 500 m south-west of Morton of Blebo [NO 432148] consist of mixed basaltic and sediment-derived tuff; in a quarry [NO 43001442] in the more southerly neck (Craig 1912, pp. 87–89) over five metres of grey, unbedded, mainly sediment-derived tuff and agglomerate rest on baked calcareous shale and limestone, probably a large mass detached from the country rock. Dark-grey to black sediment-derived tuff, probably in small necks, was encountered to depths of 18 m below surface in Denhead Nos. 1 and 2 boreholes [NO 46881396 and NO 46941391] put down by the Geological Survey in 1966, and tuff of the same nature is exposed in a ditch [NO 44061335] on the hillside about 450 m south of Backfield of Ladeddie.

The Ladeddie neck is the largest in this area and contains unbedded basaltic tuff and agglomerate, with several intrusive masses of olivine-basalt of Hillhouse type. One of these intrusions, exposed [NO 44571322] near the top of Ladeddie Hill, appears to cut the olivine-dolerite of the adjacent Drumcarro intrusion (p. 142). An isolated knob [NO 46051379] of basanite exposed in a field west of Elderburn Farm may lie in a neck, but no tuff is exposed in the vicinity.   J.I.C.

Exposures of grey, mainly sediment-derived, tuff in the Kinninmonth Burn and a small left-bank tributary some 250 m south-west of Kinninmonth [NO 424124] are taken to indicate the presence of a small neck. An unexposed neck was discovered in 1953 when Callange No. 5 Borehole [NO 42041196] penetrated 18 m of tuff, largely derived from sedimentary rocks, beneath 23·5 m of superficial deposits. In the Wilkieston Burn east of Nether Baldinnie [NO 436121] exposures of coarse basaltic tuff, with blocks up to seven centimetres across and showing traces of steeply-inclined bedding, indicate the presence of a neck at least 200 m in diameter. Coarse basaltic tuff exposed in a streamlet about 200 m north-north-west of Northbank [NO 447115] is presumed to mark the site of a small neck and exposures [NO 45341241] of coarse basaltic tuff in the Kinninmonth Burn suggest the presence of a small neck there also.

Largoward Area

Scattered exposures of coarse tuffs and lapilli-tuffs in the fields about 500 m south-west of Bruntshiels [NO 435102] suggest that the Bruntshiels neck is probably some 500 m by 350 m in extent. The outcrop includes an elongate body of quartz-dolerite but the age-relationship between the latter and the neck material is not known. The reported occurrence of 'ash' in bores at Cairnhill [NO 468103] may indicate the presence of an unexposed neck in that area. A volcanic neck forms Falfield Bank [NO 455088], but the only exposures are of grey medium- to coarse-grained tuffs, 200 m west-south-west of the farm of that name. Boreholes put down by the Opencast Executive of the National Coal Board have revealed the presence, but not the shape or size, of several unexposed necks around the Bungs of Cassingray [NO 476076]. The old quarry 250 m north-east of North Cassingray [NO 483083], in which 'green mudstone-like tuff and agglomerate' were formerly exposed (Geikie 1902, p. 226), is now filled in. The dimensions of the neck to which they belong are not known.

The Pepper Knowe neck probably measures about 500 m by 300 m. There are ex-

posures in old quarries on the northern slopes and around the summit [NO 48600766] of the hill, showing purple and green agglomerate and lapilli-tuff, with masses of basalt which may be large blocks or small intrusions (Geikie 1902, p. 226). About one and a half kilometres to the south-west, the presence of another neck is suggested by exposures of basalt and of tuff composed largely of sedimentary material, on the north side of the disused railway 400 m south of West Cassingray [NO 475072]; a bore put down about 300 m to the south-east of these exposures is said to have encountered nearly 150 m of 'ash', possibly belonging to the same neck. In a railway cutting 300 m south-south-west of East Cassingray [NO 491070] medium-grained to lapilli-grade tuffs with rare small bombs up to 15 cm across are thought to form part of a neck about 300 m in diameter.

Kinaldy Area

Green tuffs formerly exposed 130 m north of Lathockar [NO 493110] probably lie within a neck of unknown dimensions. Two small necks have been detected in a stream, some 400 m and 600 m respectively west of Kinaldy Farm [NO 507107]. The eastern one is the better exposed and is composed of basaltic tuff; the western one appears to contain a good deal of sedimentary material. The Kinaldy neck is exposed in the Kinaldy Burn south of Kinaldy House [NO 51221043]. It is almost 400 m across and includes a central plug of monchiquite about 50 m in diameter. This body has well-developed columnar jointing, a vertical, sharp, eastern contact with the agglomerate and a complex western one involving a zone of brecciated monchiquite. Coarse tuff and fine agglomerate, showing traces of bedding and containing occasional large blocks, are well exposed in the cliffs on both sides of the stream. The south-western part of the neck is cut through one of the teschenitic olivine-dolerite sills, and bodies of tuffisite derived from sediments can be seen to penetrate the dolerite. Another neck about 200 m in diameter is exposed on the south bank of the stream about 600 m south-east of Kinaldy House; it is composed of green basaltic agglomerate locally containing many sandstone fragments.

The Dunino Burn traverses the outcrop of a neck for a distance of about 250 m at Lingo Burnside [NO 511087]. The exposures consist partly of a xenolithic monchiquitic rock and partly of grey coarse to lapilli-grade tuff which locally consists of sedimentary material but also includes large masses of dolerite. The latter were probably derived from the adjacent teschenitic olivine-dolerite sill which, although the field evidence is not conclusive, this neck appears to penetrate.

The Dunino neck is exposed in a steep bank on the east side of the Dunino Burn, 400 m north-west of Dunino Church [NO 54101093]. It consists of dull-green basaltic agglomerate, with steeply-inclined bedding in places. The neck contains an irregular basic dyke six metres wide.

Largo Law Area

The volcanic neck of Largo Law (Plate VIII) stands 292 m above O.D. and forms a conspicuous hill, with its summit cleft by a west–east depression. The break in slope

---

Plate VIII

Largo Law, a complex volcanic centre

The hill is built mainly of tuff and agglomerate with intrusions of basalt, basanite and monchiquite. The break in slope in the middle distance marks the approximate margin of the neck, the lower ground being formed of Carboniferous sediments and bedded tuffs and agglomerates (D 1765)

which is present round most of the hill, at about the 500-ft (152·4-m) contour, has been taken to mark the approximate margin of the neck, which on this basis is nearly circular in shape with a diameter of between 1·2 and 1·8 km. No contact is exposed, but the presence of sedimentary rock debris and signs of trial pits close to exposures of tuff mark the position of the margin quite closely on the north-west side. The bedded tuffs and agglomerates which, with a few intercalations of lava, lie adjacent to the neck on its southern and eastern sides, are described in Chapter 7.

Much of the hill is grass-covered and in the eastern part, especially, exposures are few, but the neck appears to be composed largely of green, yellow or purple coarse tuff and fine agglomerate, with basaltic bombs up to 0·3 m or more in diameter. Intrusions within the outcrop area of the neck are numerous and many are conspicuously exposed, as at Craig Rock [NO 434049]. Geikie (1902, p. 260) noted that those near the summit tend to be sheet-like whereas those on the lower slopes are usually in the form of dykes. The sheets include a body of olivine-basalt of Hillhouse type, which crowns the northern summit [NO 42710498] and rings the southern one. Prominent among the other intrusions is a very irregular dyke of monchiquite on the southern slopes. Craig Rock is a prominent feature some 270 m by 145 m. It is made of basanite, and the Largo Law neck therefore contains at least three allied but distinct petrographic types among its intrusions.

The Boghall and Gilston burns provide abundant exposures in the Balhousie neck. Grey and green tuffs, with occasional bombs up to a metre across, are visible in both streams for about 600 m upstream from where they join [NO 42000629] to form the Keil Burn, and for 200 m downstream from the confluence. Grey and yellow coarse tuffs and lapilli-tuffs are also exposed in an unnamed streamlet which flows westwards to join the Keil Burn at locality NO 42030619. The contacts of the neck with the country rocks, where visible, are almost vertical. The neck contains a central intrusion of basanitic olivine-basalt about 100 m in diameter which is prominently exposed in disused quarries [NO 423063] 250 m west of Balhousie. Exposures of grey tuff, mainly derived from sedimentary rocks, in a streamlet 250 m S.S.E. of Balhousie and in the Boghall Burn at Gils Law [NO 427069] suggest the presence of small necks separated from the Balhousie neck by thin strips of sediment.

Grey and green lapilli-tuffs, with bombs up to 10 cm across, are exposed around Balmain [NO 417059]. The extent of the Balmain neck is limited to the south and east by exposures of sediments in the nearby Keil Burn, but its extent to the north and west is less certain. A report of 'ash' in an old bore doubtfully sited in a field north-west of the road from Balcormo suggests that this neck may have a diameter of a few hundred metres. It does not extend as far as the Balcormo Wood Borehole [NO 41320575], but the unbedded agglomerate at the top of that bore must have been derived from a nearby neck, most probably that at Balmain. The occurrence of grey lapilli-tuffs in a ditch 200 m east of Branxton [NO 408065] suggests the presence of a neck there.

Unbedded grey tuffs, with basalt bombs up to 15 cm across, are exposed in a streamlet 200 m north-east of Auchindownie [NO 420056]. Several bores sited between the farmhouse and the Keil Burn also recorded tuffs. This suggests the presence of an elongated neck, about 550 m by 200 m in size, but it may be that two separate smaller necks are present.

The railway cutting 350 m W.S.W. of Wester Lathallan [NO 447063] provides a section over 200 m long through apparently unbedded grey and green coarse tuffs and lapilli-tuffs with basalt bombs up to 0·3 m in size, presumably part of a neck. A boss of monchiquite six metres wide is also exposed [NO 44410618]. A small intrusion of basanite can be seen in an old quarry just north of Easter Newburn [NO 446049]. Basanite full of xenocrysts is exposed in an old quarry [NO 43960339] 200 m north-west of Drumeldrie, where similar material was formerly seen in another old quarry [NO 441032]. The full extent of this intrusion is not known, but it probably exceeds 300 m in diameter.

The Rires neck is taken to be about two kilometres in diameter and forms hilly ground which locally rises to more than 180 m above O.D. Exposures of pyroclastic rocks are scattered and generally poor, the best being on the slopes west and north of Charleton House [NO 460039], where bedded tuffs, ranging in grain-size from fine to coarse and lapilli-grade, are locally visible. Some of the intrusions, on the other hand, are excellently exposed, for example the olivine-basalt of Hillhouse type which is at least 200 m in diameter and is prominently displayed in a columnar form on Rires Craigs [NO 456043]. Another, smaller, intrusion of the same material is visible in a disused quarry [NO 448045] near Blinkbonny. On Flagstaff Hill [NO 451046] there are several intrusions of basanitic olivine-basalt. A dyke-like body of basanite, at least 130 m long, lies near the southern margin of the neck [NO 456039].

The Geological Survey Charleton Borehole (1964) [NO 46370319] encountered volcanic agglomerate (Forsyth and Chisholm 1968, p. 74), with fragments up to 20 cm across, mainly of basalt with some of dolerite, limestone, sandstone and shale. The material is bedded, with dips of 40° to 90°, and there can be no doubt that it lies in a hitherto unknown neck.

Balniel Area

Three necks are known in Balniel Den. The best exposed lies about 400 m south-east of Balniel [NO 474056]. It is about 300 m across and consists of unbedded green medium-grained tuffs and lapilli-tuffs with rare basalt bombs up to 0·6 m in size. The presence of another, smaller, neck is indicated by scattered exposures of tuff about 400 m west of Balniel. The presence of the third one is inferred partly from the brecciation of the olivine-basalt sill exposed in the stream 900 m west of Balniel, and partly from the presence of over 45 m of 'ash' near the top of the Lathallan Borehole [NO 46440579] and several thinner 'ash' bands lower down the bore. It is possible that the 'ash' is interstratified tuff, but there are few undoubted records of volcanism in east Fife at the period concerned (late Lower Limestone Group and early Limestone Coal Group times) and none that indicate activity on the scale that the Lathallan Borehole would suggest. It seems more likely, therefore, that the bore began in the neck responsible for the brecciation of the basalt noted above, but gradually diverged from it, the lower 'ash' bands being tuffisites.

The prominent hill of Kilbrackmont Knock [NO 484054] is believed to be entirely composed of tuff and agglomerate lying in a neck about 700 m across, but the exposures are poor. A small neck, full of coarse and lapilli-grade basaltic tuff with dolerite blocks, is situated some 450 m north of Gibliston [NO 494055]. Two small dykes of bleached basalt are visible in the stream nearby. In Balcarres Den there is a 12-m bluff of agglomerate [NO 48550404] which, together with one or two small exposures of the same material and a marked disturbance of the adjacent sediments, indicates the presence of a neck, possibly about 400 m in diameter. In the den nearby [NO 48450392] there is an oval-shaped intrusion of bleached basalt, 60 m by 30 m in size.

Kellie Law Area

It is now thought that Kellie Law and Carnbee Law both lie within a single large neck, for which the name Kellie Law neck is used here. It has an irregular, roughly oval, shape with a maximum diameter, from south-west to north-east, of over a kilometre and a minimum diameter of about 0·4 km. Following Geikie (1902, p. 227) it is assumed that the neck margin approximately coincides with the break in slope round the two hills. The best exposures are on Carnbee Law [NO 522068], where coarse tuffs and lapilli-tuffs are seen to include basalt bombs up to 1·2 m across. The exposures are poorer on Kellie Law, but the material appears to be similar. Several basic intrusions cut the agglomerate, including one of monchiquite which has been analysed

(Table 10) and others of basanite and basanitic olivine-basalt. Some are dyke-like, but the size and shape of others, including the one mapped in the summit region, are uncertain.

Five necks are known in the Dreel Burn. Two small ones lie respectively 250 m west and 350 m S.S.W. of Over Kellie [NO 510069] and there is another small one at Dean Bridge [NO 50840545]. The exposures of 'coarse basalt-breccia' enclosing 'at least one huge block of sandstone' which Geikie (1902, p. 230) recorded at the western margin of the Gillings Hill neck have been obscured, and the only part of the neck now visible is the central plug of columnar olivine-basalt of Hillhouse type exposed in a quarry [NO 51100626] near the stream. The southernmost of the five necks exposed in the Dreel Burn lies some 400 m north of Kellie Mill [NO 505046]. It is at least 300 m in diameter and consists of green and grey unbedded lapilli-tuff with pieces of tuff, limestone and sandstone, mainly less than 0·3 m across but including a sandstone block measuring 2 m by 1·2 m. Exposures [NO 50520605] of coarse basaltic tuff suggest the presence of another neck about 400 m west of the Dreel Burn.

Grangemuir No. 3 Borehole [NO 53990361] encountered almost 18 m of 'whin ash' which is probably tuff and agglomerate in an otherwise unknown neck.

A neck at least 250 m across has been laid bare in the railway cutting a kilometre south-west of Lochty Goods Station [NO 52160797], where coarse basaltic tuffs with cattered blocks up to 0·3 m across, mostly of dolerite, are exposed. Near the north-east margin, the only one visible, the tuffs contain larger blocks of sandstone. Three other necks of unknown dimensions have been detected around Lochty [NO 526081] by bores for the Back and Fore coals.

Minor Intrusions outside the Volcanic Necks

Within the general area in which the volcanic necks occur there are a few scattered basic to ultrabasic minor intrusions which, although not associated with any known volcanic necks, have strong petrographic affinities with many of the neck intrusions. It seems probable therefore that they are related to the episode of volcanic activity during which the neck intrusions were emplaced. The minor intrusions take the form of dykes or steeply-inclined sheets, most of them no more than 1·5 to 3 m in thickness. The following examples may be cited; their petrography is discussed along with that of the neck intrusions in the following section.

A dyke of nepheline-monchiquite 1·5 m wide, with east-south-easterly trend, is exposed [NO 47010976] in Radernie Quarry, where it cuts Lower Limestone Group strata. In the quarry [NO 451083] on Dunicher Law, a monchiquite dyke about two metres thick, with offshoots, cuts an olivine-dolerite sill (Walker and Irving 1928, pp. 10–11). A west–east dyke of olivine-basalt, 1·5 m wide, crosses the Keil Burn [NO 41520528] 600 m south-west of Auchindownie, probably occupying the plane of a fault. The Dreel Burn between Gillings Hill and Kellie Mill is crossed by several basaltic or basanitic dykes and sheets up to three metres thick. Some are highly altered, but one, at locality NO 50890574, has been determined as a basanitic olivine-basalt. A basanite dyke at least 1·5 m thick, with south-easterly trend, is exposed [NO 54520470] in one of the tributaries of the Dreel Burn, 530 m W.S.W. of Clephanton. I.H.F.

Petrography of the Neck Intrusions

The neck intrusions of east Fife belong, with only a few exceptions, to a suite of rocks ranging from mafic olivine-basalt allied to the Hillhouse type, through analcime-basanite to monchiquite (analcime-basalt). Nepheline-basanites and nepheline-monchiquites also occur. In many instances the rocks are so chilled and some are so decomposed that specimens from individual intrusions cannot be assigned precisely within the suite. Rocks of this suite also form small intrusions not related to any known necks. These intrusions have been included in this account.

A number of the intrusions contain xenoliths of peridotite (lherzolite) and pyroxenite, and xenocrysts of anorthoclase, kaersutitic hornblende, biotite, orthopyroxene, spinel and apatite have been recorded. Xenoliths of sedimentary origin, including quartz grains, are common, particularly in the marginal phases of some of the intrusions.

A few neck intrusions of different type occur. These include the unusual xenolithic basalt intrusions of the Lundin Links neck, olivine-basalt at Balcomie Sands and Kittock's Den, and non-ophitic olivine-dolerite at Feddinch. The rock from the last locality is very similar to that of the neighbouring sills (pp. 157, 217–218).

Nomenclature and classification

The nomenclature of the feldspathoidal end-members of the suite presents certain difficulties. Tomkeieff (1952, p. 355), in his description of the nepheline-basanite intrusion of Southdean, pointed out that the term monchiquite has been used in Scotland, and particularly for rocks in the Midland Valley, in a sense not in accord with the original definition of Rosenbusch, namely that monchiquite is a hornblende-bearing rock. In this sense the monchiquitic rocks of east Fife would be more correctly described as analcime-basalts (olivine-analcimites) and nepheline-basalts (olivine-nephelinites) but it seems reasonable in view of the long-established usage in Scotland to refer to the feldspathoidal rocks as monchiquites. They have many of the characteristics of rocks called by that name elsewhere in the Midland Valley (Macgregor and MacGregor 1936, pp. 63, 65–66). Though the monchiquitic rocks are not characterised by the presence of hornblende this material has been recorded as xenocrysts in a number of the intrusions. The term limburgite was used to describe a number of the east Fife rocks by Seymour and Flett (*in* Geikie 1902, pp. 396–404) but since the rocks described by that name are usually very chilled, and since similar chilled rocks may be found at the margins of intrusions of basalt, basanite and monchiquite, the term limburgite has not been used in this account.

According to the chemical classification proposed by Macdonald and Katsura (1964, p. 89) and adopted, with slight amendments, by Green and Ringwood (1967, p. 106) and Coombs and Wilkinson (1969, p. 471), the distinction between alkali olivine-basalt and basanite (and basanitoid) is made on the presence of 5 per cent. or more of normative nepheline in the latter; basanitoids are further distinguished, by the absence of modal nepheline, from basanites. The percentages of normative nepheline in the analysed rocks from east Fife (Table 10) are: I, 3·82; II, 4·94; III, 8·51; IV, 3·95; V, 13·38; VI, 12·03. In accordance with the proposals outlined above, the glassy olivine-basalt of Hillhouse type (Anal. I) and the analcime-bearing olivine-basalt (Anal. II) would be classed as alkali olivine-basalts, the basanite of Craig Rock (Anal. III) as basanitoid and the nepheline-basanite of the Chapel Ness intrusion (Anal. V) as basanite. Macdonald and Katsura stated that the term 'nepheline-basalt' has its usual significance and they did not extend their chemical classification to this type. The St Monance intrusion (Anal. VI) which has been described as a nepheline-monchiquite (cf. nepheline-basalt and olivine-nephelinite) would therefore

TABLE 10

Analyses of neck intrusions and comparable rocks

	I	II	A	III	IV	B	C	V	VI	D	E
SiO_2	42·60	44·38	42·49	42·53	41·51	45·13	43·55	43·26	41·92	43·70	40·2
Al_2O_3	13·28	13·80	13·85	13·05	13·50	13·07	14·60	13·70	13·13	13·63	12·8
Fe_2O_3	2·93	3·05	2·59	3·71	6·24	2·55	5·34	3·06	3·13	3·68	4·0
FeO	8·49	7·99	9·32	7·68	6·59	7·97	7·46	8·08	7·99	7·87	10·4
MgO	11·14	9·65	11·21	10·30	8·36	12·65	4·77	9·65	10·20	9·44	11·9
CaO	11·10	9·30	9·76	10·67	9·71	10·18	7·94	10·21	10·57	10·20	10·4
Na_2O	2·75	3·08	2·39	3·27	2·77	2·32	5·01	4·18	3·95	2·95	2·7
K_2O	0·42	1·84	0·87	1·12	1·56	1·12	1·40	1·31	0·90	1·30	0·8
H_2O+	2·42	1·98	3·35	2·77	3·71	2·44	4·58	2·20	3·19	2·70	—
H_2O-	0·75	0·67	0·47	0·75	1·41	0·17	0·63	0·31	0·42	0·30	—
TiO_2	2·37	2·55	2·51	2·76	3·06	1·57	2·52	2·78	2·86	2·70	2·9
P_2O_5	0·50	0·69	0·61	0·81	0·75	0·49	0·88	0·73	0·96	1·15	—
MnO	0·21	0·17	0·29	0·19	0·37	0·38	0·55	0·18	0·20	0·45	—
CO_2	0·72	0·45	0·22	0·06	0·09	0·02	0·56	0·10	0·14	0·00	3·4
Loss on ignition	—	—	—	—	—	—	—	—	—	—	—
Allow for minor constituents	0·35	0·31	—	0·32	0·33	—	—	0·32	0·32	—	—
Other constituents	—	—	0·13†	—	—	0·07†	0·63†	—	0·15†	—	—
TOTALS	100·03	99·91	100·06	99·99	99·96	100·13	100·42	100·07	100·03	100·07	99·5
	p.p.m.	p.p.m.		p.p.m.	p.p.m.			p.p.m.	p.p.m.		
*Ba	410	430		440	780			380	500		
*Co	36	28		30	26			33	33		
*Cr	270	200		210	85			210	160		
*Cu	40	22		30	16			30	27		
*Ga	11	12		11	11			14	14		
Li	10	40		11	19			<10	25		
*Ni	250	190		170	110			160	170		
*Sr	660	540		610	520			610	620		
*V	260	190		200	180			250	230		
*Zr	290	300		290	290			260	280		
B	15	3		3	9			<5	16		
F	760	680		900	950			900	900		
S	250	300		150	300			300	—		

DETAILS OF ANALYSED ROCKS OF TABLE 10

I. Olivine-basalt, glassy (cf. Hillhouse type); plug. Rires Craigs, Fife. [NO 45600438]. S 50482. Lab. No. 1997. Anal. J. M. Nunan and W. H. Evans, spectrographic work by C. Park. *A. Rep. Inst. geol. Sci. for 1967*, 1968, p. 108.

II. Olivine-basalt (cf. Hillhouse type); dyke in neck. Flagstaff Hill, 520 m north of East Coates, Fife. [NO 45120440]. S 50489. Lab. No. 2017. Anal. J. M. Nunan and W. H. Evans, spectrographic work by C. Park. *A. Rep. Inst. geol. Sci. for 1968*, 1969, p. 107.

A. Olivine-basalt (Hillhouse type); sill. Hillhouse Quarry, near Linlithgow, West Lothian. Anal. E. G. Radley. Clough and others 1911, p. 134.

III. Basanite; intrusion in neck. Craig Rock, Largo Law, Fife. [NO 43340487]. S 50485. Lab. No. 2016. Anal. J. M. Nunan and W. H. Evans, spectrographic work by C. Park. *A. Rep. Inst. geol. Sci. for 1968*, 1969, p. 106.

IV. Analcime-basalt; intrusion in neck. Kellie Law, 400 m east of Gillingshill, Fife. [NO 51580627]. S 50490. Lab. No. 2018. Anal. J. M. Nunan and W. H. Evans, spectrographic work by C. Park. *A. Rep. Inst. geol. Sci. for 1968*, 1969, p. 107.

B. Olivine-basalt (monchiquitic); dyke cutting ash neck. Patna Hill, 1030 m W.9°S. of Downieston, Patna, Ayrshire. Anal. E. G. Radley. *Summ. Prog. geol. Surv. Gt Brit. for 1928*, Part I, 1929, p. 98.

C. Monchiquite; dyke. West side of Camas an Fhais, 800 m N.E. of Rudha Fionn-aird, Ardmucknish, Argyll. Anal. E. G. Radley. Kynaston and Hill 1908, p. 128.

V. Nepheline-basanite; intrusion. Shore 55 m south of Earlsneuk, Fife. [NT 48029937]. S 50484. Lab. No. 1999. Anal. J. M. Nunan and W. H. Evans, spectrographic work by C. Park. *A. Rep. Inst. geol. Sci. for 1967*, 1968, p. 108.

VI. Nepheline-monchiquite; dyke in neck. Shore 150 m S. by W. of St Monan's Church, Fife. [NO 52280127]. S 50488. Lab. No. 2002. Anal. J. M. Nunan and W. H. Evans, spectrographic work by C. Park. *A. Rep. Inst. geol. Sci. for 1967*, 1968, p. 108.

D. Nepheline-basanite; intrusion. Southdean, Roxburghshire. Anal. W. H. Herdsman. Tomkeieff 1952, p. 354.

E. Nepheline-basanite (monchiquitic); sill. Chester's quarry, Whitelaw Hill, East Lothian. Anal. J. H. Player. Hatch 1892, p. 117.

retain its name. On the other hand Green and Ringwood characterise olivine-nephelinite as lacking normative albite, of which the St Monance rock contains 11·20 per cent. In view of the oxidation of the analcime-basalt of Kellie Law (Anal. IV) as shown by the high Fe_2O_3 content, which in the calculation of the norm has the effect of lowering the content of normative nepheline (Coombs and Wilkinson 1969, p. 472), it does not seem appropriate to use the chemical classification for this rock.

A chemical classification of the type described may prove a useful supplement to, or substitute for, a mineralogical classification in comparing suites of rocks. It has, however, relatively small application in an area such as east Fife from which there are few analyses and in which many of the rocks are altered. For these reasons a mineralogical classification has been used throughout this account. A further limitation to the chemical method results from using variation in percentage of a single normative mineral, rather than variation in the ratio between two such minerals, as a criterion for classification. This causes difficulties in classifying rocks which lie close to the arbitrary borderline. For example, the percentages of normative nepheline given for east Fife rocks are based on a calculation in which CO_2 is computed as normative calcite. If this is not done the percentages of normative nepheline for Anals. I and II become 5·57 and 6·03 respectively and these higher values would result in the rocks being classed as basanitoids rather than alkali olivine-basalts. Furthermore, variations in the percentage of water in individual analyses may lead to resultant variation in the percentage of normative nepheline. A more valid comparison might be obtained by using the percentage of nepheline calculated from analyses from which water, etc. have been removed and the other constituents recalculated. It might even be more useful to use a dividing line based not on the percentage of one mineral but on the relative proportions of normative nepheline to normative albite (or to total normative feldspar).

Details

BASALTS OF HILLHOUSE TYPE, BASANITES AND MONCHIQUITES

The rocks of this suite have certain features in common. In general they are mafic and are characterised by the presence of many microphenocrysts, and locally phenocrysts, of augite and of olivine or pseudomorphs after that mineral. The relative proportions of olivine and augite are variable. The groundmass is rich in small prisms or grains of augite and iron ore accompanied by feldspar, analcime or nepheline and in some instances by glass.

The smaller phenocrysts or microphenocrysts of augite and the groundmass augite usually have a faint purple tinge, although in some slices the mineral is deep purple in colour. The larger phenocrysts commonly show zoning with an outer zone of similar colour to that of the small crystals and a pale, almost colourless, core. In some rocks (S 8919, 16185, 34049, 42200, 45992, 46002) a central core of green clinopyroxene occurs in the larger augite crystals. In one specimen (S 16173), however, an augite microphenocryst has a green core but the purple rim is discontinuous. The smaller augites are in general euhedral but the larger phenocrysts are variable in shape, some being euhedral and others subhedral while some show marked corrosion locally with channels leading from the periphery to patches of the matrix of the rock enclosed within the augite. An augite phenocryst in a specimen (S 42200) from an intrusion in the Kincraig neck shows euhedral form but the core is corroded and filled by the fine-grained matrix of the rock. Augite phenocrysts also occur with cores sieved by small inclusions of iron ore. Flett (in Geikie 1902, p. 398) commented on the characteristic corroded interiors of the augite phenocrysts in a specimen (S 8912) from the plug in the Gillings Hill neck. The phenocrysts in the intrusion in the Salt Lake neck (S 1878, 8913–4, 47655) are very distinctive. The large augites have a clear core of irregular shape, from which feathery lines of inclusions branch out in an intricate manner, and a clear outer rim. The outer zone, though euhedral against the matrix of the rock, consists of a series of stepped outgrowths all of which are in optical continuity and which give rise to a somewhat skeletal appearance.

The olivine is commonly fresh but in many rocks is replaced by a variety of alteration products including serpentine, bowlingite and carbonate. The larger olivines are sometimes intensely corroded (S 50489, 56192, 59703) but the smaller phenocrysts may be euhedral or anhedral and corroded within the same rock. In one specimen (S 45990) of glassy olivine-basalt of Hillhouse type a small euhedral crystal of olivine contains a core of glass similar to that in the matrix of the rock and presumably connected to it by corrosion channels not seen in the plane of the section.

Flow-structure is marked in some rocks. In specimens (S 49550, 50482, 56192) from the plug of glassy olivine-basalt allied to Hillhouse type that forms Rires Craigs (p. 205) the olivine microphenocrysts are very elongate and show marked flow alignment.

Nests or clusters of augite occur in many of the rocks, particularly the fine-grained varieties. Undoubtedly these mainly represent reaction rims to xenocrysts of quartz (p. 216). Clusters of augite of rather different texture have been recorded in some specimens (S 34036, 50484, 56275, 59764). The clusters are somewhat larger than the nests of augite associated with xenocrystic quartz, the converging prisms of augite are coarser, and range in colour from purplish near the edge to pale or colourless at the core. Some glass or analcime may occur interstitially. The margins of such clusters tend to be irregular and they are commonly surrounded by areas of glassy rock. It may be that they represent recrystallised reaction rims to quartz grains but no evidence has been found to confirm this. Bailey (*in* Clough and others 1910, p. 109) described similar radiating clusters of augite some of which he interpreted as spherulitic growths.

Included in this suite are a number of olivine-basalts allied to Hillhouse type but differing from specimens from the type locality in the presence of a patchy brown glassy mesostasis. Good examples of this variety occur in the plug that forms Rires Craigs (S 49550, 50482, 56192; Anal. I, Table 10; p. 205), the plug in the Gillings Hill neck (S 42199; p. 206), some of the intrusions in the Kincraig (S 34054) and Ruddons Point (S 45990) necks, and the small intrusion (S 46002; p. 188) west of the Chapel Ness intrusion. In this variety there is a pervading brown glassy mesostasis which is locally concentrated in patches. Delicate crystallites (trichites) of iron ore are enclosed in the glassy patches. Many of the minor intrusions associated with the necks may be classed as olivine-basalts of Hillhouse type and are more similar to rocks from the type locality. They commonly contain a little analcime and may also contain small patches of glass with trichites of iron ore. Examples of this variety may be found in the Flagstaff Hill intrusion (S 50489; Anal. II, Table 10) in the Rires neck, the intrusion (S 49511) which caps the north summit of Largo Law, and one of the intrusions (S 48042) in the Ladeddie neck. The sill of basalt of Craiglockhart type near West Coates (p. 158) is probably related to these rocks.

With increasing proportion of analcime and decrease in plagioclase the basalts pass into basanites and in turn into monchiquites or analcime-basalts. Because of the fine-grained nature of many of the rocks and because of analcimisation of original plagioclase it is commonly difficult to give precise names to individual specimens within this transitional series. In addition variation of rock type within individual intrusions is known to occur. Wallace (1916, p. 360) noted that some parts of the Chapel Ness intrusion might be described as nepheline-basanite, some as nepheline-basalt and others as monchiquite.

Examples of analcime-basanite may be found in the Craig Rock intrusion (S 50485; Anal. III, Table 10) of the Largo Law neck, an intrusion (S 34053) in the Ruddons Point neck and the intrusion (S 34037) in the quarry 200 m north-west of Drumeldrie. The basanite (S 34036, 49539, 50485) of the Craig Rock intrusion contains some small patches of glass with trichites of iron ore and bears a strong resemblance to some of the analcime-bearing olivine-basalts. The proportion of feldspar is very low in the analysed rock (S 50485) but this may be only apparent and be caused by analcimisation. A basanite (S 49542) from the small intrusion at Easter Newburn (p. 204) differs from other rocks of this group in that it is much more leucocratic than normal. Small flakes of biotite are scattered throughout the base.

Analcime-basalts or monchiquites with at most only sparse microlites of plagioclase occur in one of the minor intrusions (S 49534) on the south slopes of Largo Law (p. 204), the boss (S 19072–5, 56272) exposed in the railway cutting near Wester Lathallan

(p. 204), an intrusion (S 50490; Anal. IV, Table 10) in the Kellie Law neck (p. 205), the plug (S 34040) in the Kinaldy neck (p. 203) and in the dyke (S 56273) cutting the Dunicher Law sill. Small needles of brown hornblende were noted in the last-mentioned rock. The Kinaldy rock contains numerous small flakes of biotite.

Small needles of brown, probably kaersutitic, hornblende have been observed in the groundmass of a number of the basanites and monchiquites but are generally not conspicuous and are more common in the feldspathic patches and ocelli which occur in many of the rocks (see below).

Occurrence of nepheline. The occurrence of nepheline, or of pseudomorphs after that mineral, has been recorded in a number of the basanite and monchiquite intrusions of east Fife. The first record of nepheline in this area was made by Wallace (1916, pp. 360–361) in the Chapel Ness basanite intrusion (S 16182, 50484, 56275, 59764); she also recorded (1916, p. 356) the occurrence of pseudomorphs after that mineral in dykes in the St Monance neck (Wallace 1916, p. 356). Nepheline is plentiful and feldspar relatively sparse in the analysed rock (S 50484; Anal. V, Table 10) from the Chapel Ness intrusion and this specimen approaches a nepheline-basalt or nepheline-monchiquite in composition.

Fresh nepheline has been observed as large plates in the coarse varieties (S 37107A, 59762) of the nepheline-basanite intrusion at Davie's Rock (p. 197). In another specimen (S 16197) from this locality nepheline has not been confirmed but locally the groundmass has a fibrous structure suggestive of pseudomorphs after that mineral. Pseudomorphs after nepheline have also been observed in the nepheline-monchiquite dyke (S 42192, 49498, 56193) of Radernie (p. 206), in the St Monance neck intrusion (S 46012, 50488; Anal. VI, Table 10; p. 198) and in an intrusion of nepheline-basanite (S 16191–2) near Ardross Castle. Fresh nepheline has also been observed in some of these rocks. The type of material replacing the nepheline varies. In most examples it consists of massive pale-green chloritic material of low birefringence and slow elongation parallel to the cleavage, but in the St Monance neck intrusion the undoubted pseudomorphs after nepheline are composed of a fine-grained colourless aggregate of a mineral of slightly higher birefringence. Good examples of nepheline-monchiquite (S 8899–8900) have been recorded as blocks in the Elie Ness neck (p. 190). A number of other rocks occur in which the original presence of nepheline has been suspected but not confirmed. These include the basanite dyke (S 47650) near Clephanton and a basanite dyke (S 16185) in the Elie Ness neck. The nepheline-monchiquite of Radernie differs from other members of the group in its relative lack of olivine. The nepheline and its pseudomorphs vary in habit. In some rocks it forms quite small equant crystals (S 8900, 46012, 50488) and in others (S 37107A, 42192, 49498, 56193) large tablets. Though moulded on the pyroxene and olivine it is euhedral against the analcime base.

The intrusion at Davie's Rock shows some variation in composition. One specimen (S 16197) is a fine-grained basanite with numerous small microlites of plagioclase whereas coarse varieties (S 37107A, 59762) contain large plates of nepheline and relatively sparse plagioclase, hence approaching a nepheline-monchiquite in composition.

Feldspathic patches and veinlets. Leucocratic feldspathic ocelli occur in many of the neck intrusions. In general they are characterised by small laths or, in some instances (S 42190, 59762), feathery crystals of feldspar, with flakes of biotite, small prisms of brown hornblende and locally of augite and scattered grains of iron ore. Chlorite, carbonate or, locally, analcime occurs interstitially, and quite commonly there is a central amygdaloidal patch in many of the ocelli which is infilled by these minerals. The type of feldspar may vary. In some rocks small laths of plagioclase (S 1036, 47650, 59762) have been confirmed but in others (S 59763) the feldspar appears to be mainly or entirely alkali-feldspar. Though many of the feldspathic patches have a somewhat ovoid appearance some are very irregular in form and have a vein-like appearance. In addition, however, some of the very chilled rocks (S 1861, 46674, 47655, 49237) are intricately dissected by ramifying leucocratic veinlets composed mainly of alkalifeldspar but locally with biotite and quite commonly with interstitial chlorite. It seems unlikely that such veinlets have developed as segregations in the chilled edges of the intrusions but they may represent segregation material generated in and injected from the still fluid central part of the intrusion. The veinlets are however particularly common in specimens in which sedimentary xenoliths show thermal alteration and, in some cases, the development of feldspathic material (p. 215). While these may represent feldspathised xenoliths it is also possible that the veinlets represent mobilised sedimentary inclusions. Conclusive proof of this has not however been found. It may be noted that in some specimens (S 47660) a few rather angular feldspathic patches occur which are reminiscent of the alteration seen in some sedimentary xenoliths.

Igneous xenoliths and xenocrysts. The neck intrusions and agglomerates of east Fife are well known for the relatively common occurrence and the variety of ultrabasic igneous xenoliths and xenocrysts.

Among the xenoliths recorded is peridotite (lherzolite) composed of olivine, orthopyroxene, clinopyroxene and reddish-brown spinel. The freshest examples of this type have been found in a basaltic intrusion (S 17895) at the side of the Coalyard Hill neck and in an intrusion (S 45994) in the Ruddons Point neck. The first record of orthopyroxene from a block or inclusion, other than of quartz-dolerite, in the east Fife neck intrusions was made by Balsillie (1927, p. 489) who recorded its presence in a basic xenolith ('segregation') from one of the intrusions in the Ruddons Point neck. Rather altered peridotite fragments in which orthopyroxene has not been confirmed have been found in nepheline-basanite from near Ardross (S 16191). In one altered specimen (S 45995) from an intrusion in the Ruddons Point neck large plates of biotite occur. Peridotite blocks (S 16195-6) have also been recorded from the agglomerate of the Newark neck.

Small xenocrysts of colourless orthopyroxene have been recorded in some of the intrusive rocks, including a basalt dyke (S 1036) near Ardross Castle, the nepheline-basanite (S 16197) of Davie's Rock, the basanite dyke (S 47650) near Clephanton, the intrusion of analcimic olivine-basalt (S 50489) of Flagstaff Hill and a basanite intrusion (S 56276) in the Coalyard Hill neck. The orthopyroxene commonly occurs as small inclusions within the microphenocrysts of augite and is mantled by a thin corona of commonly chloritised, finely granular augite. Small xenocrysts of reddish-brown spinel with a dark margin also occur in rocks with orthopyroxene xenocrysts (S 16197, 47650B, 56276). Since small composite xenoliths composed of varying combinations of olivine, orthopyroxene, augite and spinel occur in some of these rocks (S 16197, 47650A, B, 50489) and since both orthopyroxene and spinel have been recorded in the peridotite xenoliths, it seems likely that they are derived by the fragmentation or dissociation of such xenoliths. Though the xenocrysts of spinel are normally brownish in colour a large rounded xenocryst of dark greenish-grey spinel has been observed in a specimen (S 8914) from an intrusion in the Salt Lake neck and in addition two small inclusions of greenish-grey spinel are enclosed in a large phenocryst of augite. Pale greenish-brown spinel has been recorded by Chapman (1974, p. 225) in lherzolite xenoliths from an intrusion at the margin of the Coalyard Hill neck.

Xenoliths of pyroxenite composed of large anhedral plates of pale clinopyroxene and much less common pseudomorphs after olivine also occur. A sample (S 49534) of this variety has been recorded from an intrusion of analcime-basalt or monchiquite in the Largo Law neck. It must be noted however that in a number of rocks (S 16183, 48042, 59763) aggregates of clinopyroxene occur which may represent glomeroporphyritic aggregates rather than xenoliths.

Several xenoliths are composed of an aggregate of deep brown, probably kaersutitic, hornblende and clinopyroxene. One sample (S 8857) from an intrusion in one of the Elie necks contains very large crystals of brown hornblende with smaller but coarse anhedral plates of clinopyroxene and patches of carbonate. In a xenolith (S 16189) from an intrusion in the Ardross neck the hornblende occurs as large poikilitic plates and the pyroxene as small plates. Similar coarse hornblende-pyroxene rocks (S 37112–3) occur as blocks in the agglomerate of the Ardross and Elie Ness necks, in one instance (S 37112) with biotite.

Some xenoliths or aggregates are marked by structures suggestive of sudden chilling and rapid crystallisation. One specimen (S 34027) from an intrusion in the Largo Law neck is composed of large plates of pale clinopyroxene speckled with inclusions of iron ore and contains intersertal patches of pale-brown glass which enclose small, largely serpentinised, olivines and rare laths of plagioclase. Skeletal growths of pyroxene project from the plates into the glassy areas. Locally there are patches composed of a network of rods of ore and a deep brown, highly refracting mineral. In some of these augite occurs in the intergrowth and locally is in optical continuity with an outer rim of augite. A xenolith (S 46002) from the small intrusion of glassy olivine-basalt of Hillhouse type west of the Chapel Ness intrusion (p. 188) contains plates of pale augite and rare plagioclase with some small serpentinised olivines and patches composed of a network of iron ore rods and skeletal purple augite with brown glass. In a specimen (S 59762) from the intrusion of Davie's Rock the skeletal plates of augite are up to about 4 mm in diameter. Two small xenoliths (S 42200, 49511) have been observed in which probably kaersutitic hornblende is associated with a network of rods of iron ore and deep purple augite.

In addition to the hornblende occurring with pyroxene as xenoliths, the presence of large crystals of hornblende has been recorded by previous workers in many of the neck intrusions and agglomerates of east Fife. In a specimen from an intrusion (S 1897) in the Kellie Law neck there is a small xenocryst of hornblende which is rimmed by, or altered peripherally to, chlorite and leucoxene with an outer rim of augite and iron ore. In sliced rocks (S 19072B, 19074) from the boss near Wester Lathallan (p. 204), xenocrysts of hornblende have been observed, mantled in one instance (S 19072B) by analcime. The analcime patch enclosing the hornblende is similar in size to the amygdales in the rock and presumably the hornblende crystal is enclosed in a steam hole. Xenocrysts of biotite have been recorded in neck intrusions and agglomerates.

Apatite occurs as small needles and prisms in the neck intrusions other than the chilled rocks. Locally, however, large crystals have been noted up to about 0·5 mm long, some clear and rounded (S 8918, 42199) and some (S 46004) with striae of dark inclusions. In one specimen (S 8916) small apatite crystals dark with inclusions are enclosed in a large turbid pseudomorph possibly after spinel, and in another (S 45992) an inclusion of apatite, slightly dusky with inclusions near the periphery, is enclosed in a chloritic patch in a phenocryst of augite. A prism of pleochroic brown apatite, with dark inclusions arranged in striae parallel to the crystallographic axis, occurs enclosed in a viriditic pseudomorph after a ferromagnesian mineral (S 49511). Balsillie (1927, p. 490) recorded crystals of apatite, some of considerable size, in some of the feldspathic xenoliths from agglomerates in east Fife. Large crystals of apatite originally considered to represent fish teeth were recorded by Balsillie (1919, pp. 503–504) from the intrusion of the Spindle, in the Kinkell Ness neck (p. 200). As shown by MacGregor (1939), however, these are typical of the large crystals of apatite which have been recorded from monchiquitic intrusions elsewhere in Scotland.

Zircon has been recorded by Balsillie (1927, p. 490) in some of the feldspathic or syenitic xenoliths from the neck agglomerates of east Fife and has also been recorded by Traill (quoted by Heddle 1901, p. 55) and confirmed by Cumming (1928, p. 138) from agglomerate in the Elie Ness neck. Balsillie (1927, p. 491) suggested that the individual xenocrysts of the zircon in the agglomerate may derive from disintegration of the feldspathic or syenitic xenoliths.

Potash-feldspar has been noted by previous workers (p. 207) in many of the neck intrusions and agglomerates of east Fife. It has been recorded variously as orthoclase, sanidine or soda-microcline but it appears to have now been accepted as anorthoclase.

An early analysis of the feldspar is given by Heddle (1901, p. 2) and more recent analytical and structural data for anorthoclase from the Kellie Law neck and one of the necks east of Brownhills have been published by Carmichael and MacKenzie (1964). The compositions of anorthoclase xenocrysts from some of the neck intrusions in the Elie area have been determined by Chapman (1974, p. 226). Balsillie (1927, pp. 489–490) recorded the presence of blocks of highly feldspathic syenitic rock in the agglomerates and neck intrusions of east Fife, including those of Elie Ness, and suggested that the xenocrysts of anorthoclase were derived by the disintegration of rocks of this type. In general the xenocrysts of anorthoclase show little or no reaction against the enclosing igneous rock but locally some variation occurs. A xenocryst of anorthoclase from an intrusion on Kellie Law (S 8917) is mantled by a very thin rim of feldspar of slightly higher refractive index. In a specimen of basanite (S 37109) from a dyke at Elie Ness two anorthoclase xenocrysts occur. One shows no apparent reaction but the other is partly mantled by a zone of small plagioclase laths and iron ore. Three small xenocrysts of anorthoclase have been observed in xenolithic fine-grained basalt or basanite (S 46676) from one of the small necks near Strathtyrum House (p. 201); these are margined by an outer rim of feathery devitrified glass containing tiny flakes of biotite, some prisms of augite and a little feldspar. The glassy zone is, however, similar to the material composing ocellar patches which occur elsewhere in the rock and it seems likely that the xenocrysts have been trapped in such areas.

Though probably the best-known locality in this area for the occurrence *in situ* of pyrope (Elie ruby) is the agglomerate of the Elie Ness neck (p. 190), Heddle (1901, p. 48) also recorded the presence of pyrope in two dykes cutting tuff about 1·5 km east of Elie Ness and in basalt at Ruddons Point. Recent chemical work on pyrope xenocrysts from the Elie Ness neck by Colvine (1968, pp. 284–285) suggests that in view of its chemical composition, and of the fact that it has not been reported as a constituent in any nodule or xenolith of this neck, the pyrope is not derived from the fragmentation of a pre-existing garnetiferous rock. He considers that it is more likely to have originated as a primary precipitating phase from a suitable magma at depth.

Sedimentary xenoliths. Small xenoliths of mudstone have been observed in sliced rocks from a number of the intrusions. In some instances these show signs of bloating, with the development of amygdaloidal patches of chlorite near the periphery of the xenoliths (S 8905, 48073) though in one instance (S 49237) the amygdaloidal chloritic patches are restricted to the core of the xenolith. One sedimentary xenolith (S 59703) contains tiny grains of green spinel set in a cryptocrystalline, lowly-birefringent base, and in another specimen (S 49238) tiny prisms of colourless pyroxene occur. In some there appears to be a development of fine-grained feldspar and there may be a genetic relationship between such xenoliths and some of the feldspathic patches and veinlets seen in the finer-grained intrusive rocks. In a specimen (S 46676) from an intrusion in one of the small necks near Strathtyrum a large xenolith of silty chloritic mudstone contains angular grains of quartz and small prisms and turbid dusty tabular crystals of low birefringence, possibly cordierite, set in a base of yellow-green 'gel-like' chloritic material. This passes outwards at one side of the xenolith into a zone in which the base is fibrous radiating chlorite. Towards the margin of the xenolith small crystals of alkali-feldspar, locally skeletal in form, are developed and small prisms of augite occur near the edge of the xenolith. Another small xenolith of chloritic mudstone in this rock contains tiny grains of augite and has a rim rich in granular augite.

Xenoliths of sandstone have been recorded in a few specimens. In a sliced rock (S 49237) from a small neck near Kingask the relations between the sandstone and the chilled intrusion are complex and the sandstone occurs in vein-like areas. Locally it has a calcareous cement and shows no apparent thermal alteration. In places however there is a development of much radiating fibrous material of low refringence and birefringence in the matrix of the sandstone, and small prisms of augite are developed around the quartz grains and in the matrix. Locally the patches of sandstone are separated from the chilled basaltic rock by a thin zone of small alkali-feldspar crystals with some prisms of augite and, locally, tiny flakes of biotite. Similar feldspathic material occurs as veinlets in the rock. Spherulitic, lowly-birefringent material is also developed in the matrix of a sandstone fragment from an intrusion (S 46677) in the Kinkell Ness neck. In a specimen (S 1878) from the Salt Lake neck a very irregular contact is seen between chilled basaltic rock and sandstone but thermal alteration was not detected. Outgrowths of tridymite mantle the quartz grains in a sandstone fragment from one of the intrusions (S 59703A) in the Largo Law neck and the residual core of quartz grains is mantled by deep yellow glass of low refractive index with an outer zone of prisms of augite.

Quartz xenocrysts occur in many of the neck intrusions. Flett (*in* Geikie 1902, p. 404) gave a detailed description of the quartz xenocrysts and their reaction rims in a sliced rock (S 9966) from the Dunino neck. In this rock the quartz grains are surrounded by a shell of pale glass containing small prisms of augite, this zone in turn being surrounded by a rim of granular augite. Elsewhere in the rock there are numerous nests or patches of tiny converging prisms of augite; these undoubtedly represent either completely absorbed quartz grains or the reaction rims of grains not in the plane of the section. This phenomenon has been seen in many of the finer-grained rocks (S 8913–4, 34050, 46676, 47655, 49511) of the neck intrusions. In some instances (S 47658) the intervening zone of pale glass between the pyroxene rim and the quartz core has not been detected. In a number of sliced rocks the augite nests have a core, not of glass, but of chlorite or saponite (S 8914, 19611, 46179, 46681) or, less commonly, of turbid isotropic material, possibly analcime (S 37109, 34040). Quartz xenocrysts as such are absent in many specimens and only the small clusters or nests of augite are seen. In one specimen (S 46679) an augite nest is truncated against an adjacent irregular amygdaloidal patch of chlorite.

OTHER BASALTIC AND DOLERITIC INTRUSIONS ASSOCIATED WITH THE NECKS

Lundin Links. Detailed petrographical descriptions of sliced rocks from the minor intrusions associated with the Lundin Links neck (p. 180) have been published by Flett and Seymour (*in* Geikie 1902, pp. 400–402). The intrusions are composed of fine-grained basalt rich in xenocrystic grains of quartz. They contain lathy microphenocrysts, up to about 1 mm long, of zoned labradorite (An_{68} at core) with less common subhedral prisms of pale clinopyroxene, the latter commonly as aggregates locally with feldspar. In some specimens, olivine, or pseudomorphs after that mineral, occur as small microphenocrysts. In a number of specimens the groundmass, though fine in grain, is quite crystalline and is composed of microlites of plagioclase, tiny granules of pyroxene and iron ore. In other specimens the matrix is very fine-grained and chilled. The distribution of olivine is very variable and apart from two specimens (S 45981, 56270) it is present only as pseudomorphs in carbonate or bowlingitic material. Though never abundant it generally seems more common in the very fine-grained chilled rocks (S 8923–4, 9962–3, 45981, 56270) and is rather sparse (S 1361, 45982, 56271) or has not been observed (S 1362) in the slightly coarser rocks. The pyroxene of the small microphenocrysts is pale in colour and in view of its relatively small optic axial angle ($+2E$ ca. 77°) may be pigeonitic. Some specimens (S 1312–3, 9964) show considerable alteration, with replacement of plagioclase by kaolinite and carbonate, and replacement of the pyroxene and the matrix by carbonate.

As described by Flett (*in* Geikie 1902, pp. 401–402) the abundant quartz xenocrysts are mantled by a zone of fine-grained augite. Such coronas are most marked and widest in the coarser varieties of basalt. Locally in thin section (S 45982) small aggregates of commonly radiating prisms of augite mark the original presence of xenocrystic quartz grains now either absorbed or not in the plane of the section. Locally (S 56270) a little interstitial glass can be seen in the augite corona or between the corona and the grain of quartz. Less common small xenocrysts of grains of alkali-feldspar occur and these are sieved by an intricate network of glass. Flett ascribed the network of glass in the feldspar xenocrysts to infiltration of the basaltic magma along cracks in the feldspar rather than to incipient fusion of the feldspar but the evidence is not clear. It seems probable that the xenocrysts of quartz and feldspar were derived from the local sediments and tuffs.

The intrusions associated with the Lundin Links neck are unique among the neck intrusions of east Fife. They do not resemble the normal suite of neck intrusions. These may indeed, like the Lundin Links intrusions, be xenocrystic, enclosing quartz grains with augite reaction coronas, but olivine (or its pseudomorphs) is invariably present, microporphyritic plagioclase is absent even in the basaltic members of the suite and the smaller grains of pyroxene normally have a purplish hue, particularly at the margins of the crystals.

In many respects the basalts from Lundin Links have affinities with and resemble the basalt of Binny Craig type from West Lothian, described by Watts (*in* Geikie 1897, p. 419) and Lunn (1932). Though finer in grain than most of the sliced rocks from Binny Craig in the IGS collections, some specimens (S 45982, 56271) from Lundin Links are very similar in texture and appearance to a fine-grained specimen (S 40159) from the Binny Craig intrusion, particularly in that all are characterised by small microphenocrysts of plagioclase and pale clinopyroxene and of quartz xenocrysts mantled by augite. By contrast, however, olivine has not been recorded from the Binny Craig intrusion and Lunn records the presence of orthorhombic pyroxene or its pseudomorphs in that rock. Orthorhombic pyroxene has not been detected in any of the sliced rocks from the Lundin Links intrusions. The Binny Craig intrusion was regarded by Lunn and by Macgregor and MacGregor (1936) as a rapidly-chilled intrusion of the quartz-dolerite magma.

It is of interest to note that the tuffs (S 8898, 9959–9961) from the Lundin Links neck contain numerous basaltic fragments, many of which, as Flett recorded (*in* Geikie 1902, pp. 395–396), are similar to the minor intrusions associated with the neck; also, the quartz and feldspar grains in the tuffs are similar to those which occur as xenocrysts in the intrusions. Of the other basaltic fragments in the tuff, most are chilled, altered, locally highly scoriaceous, basaltic glass and might be products of the same magma as the intrusions. The intrusions within and near the neck are therefore closely related to the explosive activity of the vent.

Balcomie Sands. A specimen (S 1883) from the chilled edge of the intrusion in the area of brecciated rocks at Balcomie Sands (p. 198) is a fine-grained olivine-basalt with a most unusual texture. It contains numerous pseudomorphs, in pale viriditic material and carbonate, after olivine as microphenocrysts in a fine-grained matrix of tiny prisms of augite which commonly form stellate radiating clusters, small microlites of plagioclase and small patches of glass and chlorite. In thin section the grain-size of the rock decreases towards the contact of the intrusion where there is a zone of pale brown tachylite.

Neck west of Kittock's Den. The plug in the neck west of the mouth of Kittock's Den (p. 200) is composed of olivine-basalt of Dalmeny type. One sliced rock (S 47661) from the plug is very fresh and contains numerous pseudomorphs after microporphyritic olivine, laths of zoned labradorite (An_{68} at core), prisms of pale purplish augite with deep purple rims, small plates of iron ore and an intersertal glassy mesostasis containing crystallites of iron ore and locally of augite. Another sliced rock (S 1894) from the intrusion is of similar type but is highly altered and lacks the glassy mesostasis.

Feddinch neck. The tuffs and agglomerates of the Feddinch neck (p. 201) are cut by a dyke-like body of olivine-dolerite. The dolerite (S 48038) contains numerous crystals, including some phenocrysts, of olivine, stout subhedral to euhedral crystals of augite, laths, up to about 1·5 mm long, of zoned labradorite, and plates and rods of iron ore. Analcime fills small intersertal patches and to a slight extent replaces the feldspar. The rock is very similar to the non-ophitic olivine-dolerites (S 48040, 50486) from the nearby Drumcarro intrusion (p. 141). An early specimen (S 8911) presumed to come from the same intrusion is different in character and contains numerous small microphenocrysts of fresh olivine and, less commonly, of augite, with small crystals of olivine and prisms of augite in a dense chilled base. Minute granules of augite and scattered tiny microlites of plagioclase occur in the base. This rock is much more like some of the chilled rocks of the main suite of neck intrusions and was described by Flett (*in* Geikie 1902, p. 398) as olivine-basalt approaching limburgite in type.   R.W.E.

REFERENCES

BALSILLIE, D. 1911a. Note on the limestone fragments in the agglomerate of the 'Rock and Spindle' volcanic vent, St Andrews, Fife. *Geol. Mag.*, **8**, Dec. 5, 201–202.
—— 1911b. The limestone fragments in the agglomerate of the 'Rock and Spindle' volcanic vent, St Andrews, Fife. *Geol. Mag.*, **8**, Dec. 5, 282.
—— 1919. Geology of Kinkell Ness, Fifeshire. *Geol. Mag.*, **56**, 498–506.
—— 1920a. Descriptions of some volcanic vents near St Andrews. *Trans. Edinb. geol. Soc.*, **11**, 69–80.
—— 1920b. Descriptions of some new volcanic vents in east Fife. *Trans. Edinb. geol. Soc.*, **11**, 81–85.
—— 1923. Further observations on the volcanic geology of east Fife. *Geol. Mag.*, **60**, 530–542.
—— 1927. Contemporaneous volcanic activity in east Fife. *Geol. Mag.*, **64**, 481–494.
CARMICHAEL, I. S. E. and MACKENZIE, W. S. 1964. The lattice parameters of high temperature triclinic sodic feldspars. *Mineralog. Mag.*, **33**, 949–962.
CHAPMAN, N. A. 1974. Ultrabasic inclusions from the Coalyard Hill vent, Fife. *Scott. Jnl Geol.*, **10**, 223–227.
CLOOS, H. 1941. Bau und Tätigkeit von Tuffschloten. *Geol. Rdsch.*, **32**, 709–800.
CLOUGH, C. T., BARROW, G., CRAMPTON, C. B., MAUFE, H. B., BAILEY, E. B. and ANDERSON, E. M. 1910. The geology of East Lothian. 2nd edit. *Mem. geol. Surv. Gt Br.*
CLOUGH, C. T., HINXMAN, L. W., WILSON, J. S. G., CRAMPTON, C. B., WRIGHT, W. B., BAILEY, E. B., ANDERSON, E. M. and CARRUTHERS, R. G. 1911. The geology of the Glasgow district. *Mem. geol. Surv. Gt Br.*
COLVINE, R. J. L. 1968. Pyrope from Elie, Fife. *Scott. Jnl Geol.*, **4**, 283–286.
COOMBS, D. S. and WILKINSON, J. F. G. 1969. Lineages and fractionation trends in undersaturated volcanic rocks from the East Otago volcanic province (New Zealand) and related rocks. *Jnl Petrol.*, **10**, 440–501.
CRAIG, R. M. 1912. Additions to the volcanic geology of east Fife. *Trans. Edinb. geol. Soc.*, **10**, 83–89.
CUMMING, G. A. 1928. The Lower Limestones and associated volcanic rocks of a section of the Fife coast. *Trans. Edinb. geol. Soc.*, **12**, 124–140.
—— 1936. The structural and volcanic geology of the Elie–St Monance district. *Trans. Edinb. geol. Soc.*, **13**, 340–365.
DAWSON, J. B. 1971. Advances in kimberlite geology. *Earth Sci. Rev.*, **7**, No. 4, 187–214.
FITCH, F. J., MILLER, J. A. and WILLIAMS, S. C. 1970. Isotopic ages of British Carboniferous rocks. *C. R. 6th Int. Congr. Carb. Strat. Geol.*, **2**, 771–789.
FORSYTH, I. H. and CHISHOLM, J. I. 1968. Geological Survey boreholes in the Carboniferous of east Fife, 1963–4. *Bull. geol. Surv. Gt Br.*, No. 28, 61–101.
—— and RUNDLE, C. C. *In press*. The age of the volcanic and hypabyssal rocks of east Fife. *Bull. geol. Surv. Gt Br.*, No. 60.

REFERENCES

FRANCIS, E. H. 1960. Intrusive tuffs related to the Firth of Forth volcanoes. *Trans. Edinb. geol. Soc.*, **18**, 32–50.
—— 1962. Volcanic neck emplacement and subsidence structures at Dunbar, southeast Scotland. *Trans. R. Soc. Edinb.*, **65**, 41–58.
—— 1968a. Pyroclastic and related rocks of the Geological Survey boreholes in east Fife, 1963–4. *Bull. geol. Surv. Gt Br.*, No. 28, 121–135.
—— 1968b. Effect of sedimentation on volcanic processes, including neck–sill relationships, in the British Carboniferous. *23rd Sess. Int. geol. Congr., Czechoslovakia 1968*, Sect. 2, 163–174.
—— 1969. In *Field excursion guide to the Carboniferous volcanic rocks of the Midland Valley of Scotland*, 23–30. Edinburgh: Scottish Academic Press.
—— 1970. Bedding in Scottish (Fifeshire) tuff-pipes and its relevance to maars and calderas. *Bull. volc.*, **34**, 697–712.
—— and EWING, C. J. C. 1961. Coal Measures and volcanism off the Fife coast. *Geol. Mag.*, **98**, 501–510.
—— and HOPGOOD, A. M. 1970. Volcanism and the Ardross Fault, Fife, Scotland. *Scott. Jnl Geol.*, **6**, 162–185.
—— and WOODLAND, A. W. 1964. The Carboniferous period. *In* The Phanerozoic time-scale. *Spec. Pub. geol. Soc. Lond.*, **1**, 221–232.
GEIKIE, A. 1879. On the Carboniferous volcanic rocks of the basin of the Firth of Forth—their structure in the field and under the microscope. *Trans. R. Soc. Edinb.*, **29**, 437–518.
—— 1897. *The ancient volcanoes of Great Britain*. Vol. 1. London: Macmillan.
—— 1902. The geology of eastern Fife. *Mem. geol. Surv. Gt Br.*
GREEN, D. H. and RINGWOOD, A. E. 1967. The genesis of basaltic magmas. *Contrib. Miner. Petrol.*, **15**, 103–190.
HATCH, F. H. 1892. The Lower Carboniferous volcanic rocks of East Lothian (Garleton Hills). *Trans. R. Soc. Edinb.*, **37**, 115–126.
HEDDLE, M. F. 1901. *The mineralogy of Scotland*. Vol. 2. Edinburgh: Douglas.
KIRK, S. R. 1925. The geology of the coast between Kinkell Ness and Kingask, Fifeshire. *Trans. Edinb. geol. Soc.*, **11**, 366–382.
KNOX, J. 1954. The economic geology of the Fife coalfields, Area III. *Mem. geol. Surv. Gt Br.*
KYNASTON, H. and HILL, J. B. 1908. The geology of the country near Oban and Dalmally. *Mem. geol. Surv. Gt Br.*
LORENZ, V. 1973. On the formation of maars. *Bull. volc.*, **37**, 183–204.
LUNN, J. W. 1932. The intrusion of Binny Craig, West Lothian. *Trans. Edinb. geol. Soc.*, **12**, 74–79.
MCBIRNEY, A. R. 1959. Factors governing the emplacement of volcanic necks. *Amer. Jnl Sci.*, **257**, 431–448.
MACDONALD, G. A. and KATSURA, T. 1964. Chemical composition of Hawaiian lavas. *Jnl Petrol.*, **5**, 82–133.
MACGREGOR, A. G. 1939. Exhibit of specimens of Scottish monchiquitic rocks containing large apatites. *Q. Jnl geol. Soc. Lond.*, **95**, cxviii–cxix.
MACGREGOR, A. R. 1968. *Fife and Angus geology: an excursion guide*. Edinburgh and London: Blackwood.
MACGREGOR, M. and MACGREGOR, A. G. 1936. The Midland Valley of Scotland. *Br. reg. Geol.*
MITCHELL, G. H., WALTON, E. K. and GRANT, D. 1960. *Edinburgh geology, an excursion guide*. Edinburgh: Oliver and Boyd.
REYNOLDS, D. L. 1954. Fluidization as a geological process, and its bearing on the problem of intrusive granites. *Amer. Jnl Sci.*, **252**, 577–613.
TOMKEIEFF, S. I. 1952. The nepheline-basanite of Southdean, Roxburghshire. *Trans. Edinb. geol. Soc.*, **14**, 349–359.

WALKER, F. and IRVING, J. 1928. The igneous intrusions between St Andrews and Loch Leven. *Trans. R. Soc. Edinb.*, **56,** 1–17.
WALKER, G. P. L. and CROASDALE, R. 1972. Characteristics of some basaltic pyroclastics. *Bull. volc.*, **35,** 303–317.
WALLACE, I. F. 1916. Notes on the petrology of the agglomerates and hypabyssal intrusions between Largo and St Monans. *Trans. Edinb. geol. Soc.*, **10,** 348–362.
WOOD, W. 1887. *The East Neuk of Fife: its history and antiquities*. 2nd edit. Edinburgh: Douglas.

Chapter 13
STRUCTURE

INTRODUCTION

EAST Fife lies at the north-eastern end of the Fife coalfield area, a structural region in which rocks of mainly Carboniferous age have been folded about relatively short northerly to north-easterly axes. To the north and north-west the coalfield area is separated by a well-marked structural line from the Strathmore region (Armstrong and Paterson 1970), in which rocks of mainly Old Red Sandstone age have been folded about elongate north-easterly axes. The line is not a simple fault, although there are zones of faulting associated with it; in central Fife it coincides with the easterly extension of the Ochil Fault but near Glenrothes it swings round into a north-easterly direction and reaches the coast, as the Dura Den Fault, at St Andrews Bay. Most of the folds in the coalfield region are gentle open structures, with dips of less than 30°. Dips greater than 70° are local features associated either with faulting or with igneous intrusions. Evidence from the trends of isopachytes (Knox 1954, pp. 13, 35, 67, 116; figs. 19–28; Knox *in* Allan and Knox 1934, pp. 153–154; Francis and others 1961, p. 123) suggests that some of the folds may already have been developing during the deposition of Upper Carboniferous sediments. Faults with easterly to southeasterly trends are a marked feature of the coalfield region (Knox 1954; Francis and others 1961). Most of them are normal faults, at least in areas where they have been proved by mining (Knox 1954). North-easterly-trending belts of complex structure, which include tight folding about northerly to north-easterly axes as well as strong normal and transcurrent faulting, also affect the region. They are believed to overlie deep-seated crustal fractures (Anderson 1951, pp. 94–100).

East Fife contains examples of all the structural features which characterise the Fife coalfield region as a whole, and the coast section provides good exposures of many of them. Folds are mainly shallow, with axes trending between north and north-east, while most of the faults belong to the easterly and southeasterly system. Two of the north-easterly-trending belts of complex structure enter the area. A consideration of the trends of the folds and the geometry of the north-easterly belts of complex structure (see p. 227 and Fig. 26) suggests that all developed in response to east–west compression: the easterly to southeasterly normal faults may be related to the same stress system. The age of the movements, particularly in relation to episodes of igneous activity, is not clear. The available evidence (pp. 136, 171 and Anderson 1951, p. 36) suggests that the igneous and tectonic activity were broadly contemporaneous and took place at intervals during Upper Carboniferous times.

FOLDS

The distribution of outcrops in the western half of east Fife is controlled by the presence there of the Largo Syncline, a large but shallow southward-plunging structure. Its axis runs in a northerly direction from the coast at Lundin Links

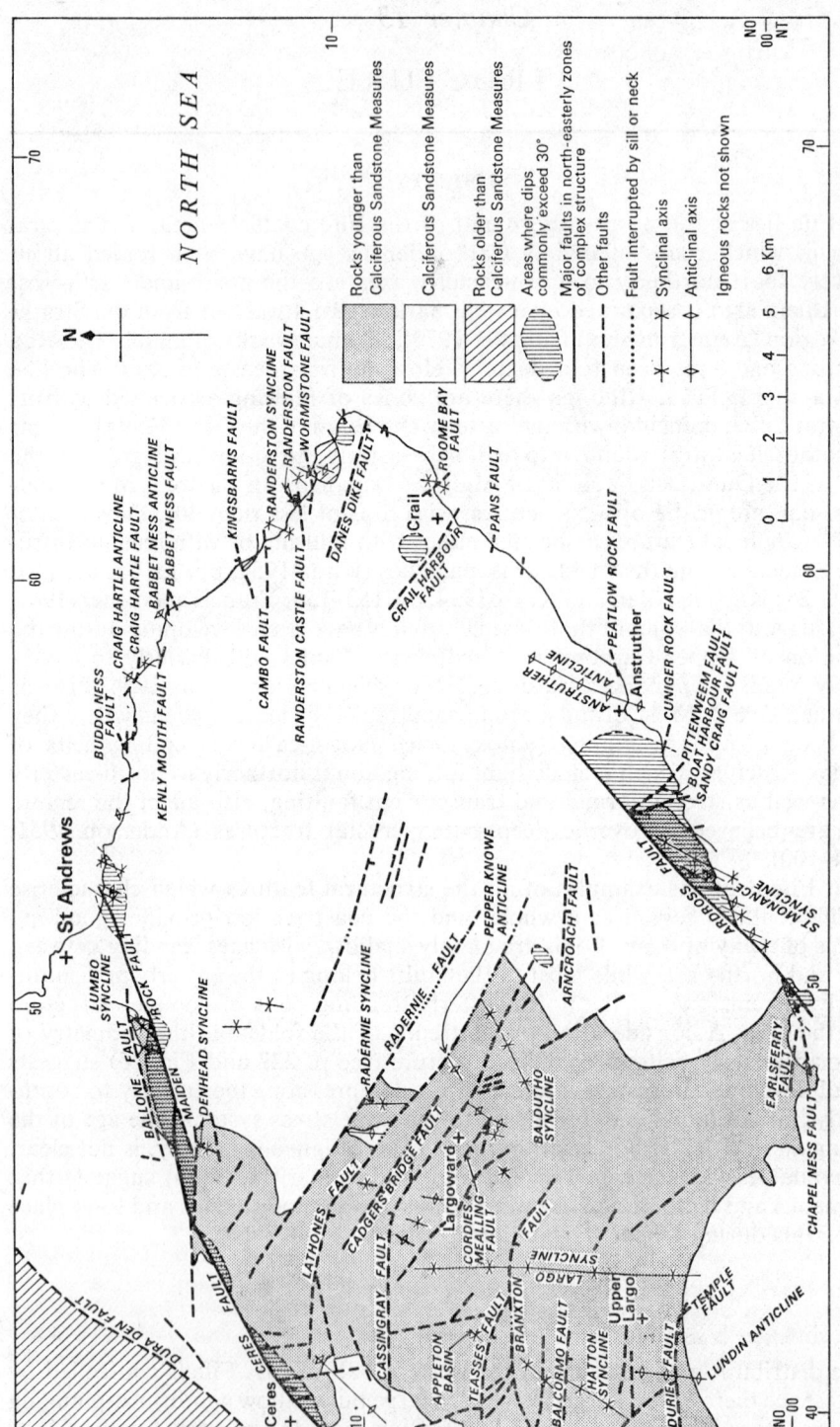

Fig. 26. Sketch-map showing structural features in east Fife

and can be recognised as far as Backmuir of New Gilston; further north the synclinal structure is less clearly marked, but it continues to control the broad disposition of outcrops as far as the Ceres Fault. Minor folds, mostly short periclines and short elliptical basins, are present throughout the area. As in adjoining areas to the west (Knox 1954; Francis and others 1961), fold axes generally trend in directions between north and north-east. A distinction can be drawn between shallow structures, with dips up to about 30°, and more strongly folded structures, with dips between 30° and 90°. The former are ubiquitous but the latter are found only along the two major north-easterly fault zones (Fig. 26) where they appear to pre-date the main fault movements. Both types of fold probably owe their origin ultimately to a phase of east–west compression. Some of the minor folds have been named (Fig. 26).

Details

Minor folds are best exposed at the coast, and good examples of shallow structures can be seen between Fife Ness [NO 638097] and Cambo Ness [NO 609118], and between Babbet Ness [NO 592142] and Kinkell Ness [NO 538158]. The axes of most of the folds trend between north and north-east, but one anticlinal fold, about 700 m south-east of Kinkell Ness, is an almost circular dome (Kirk 1925). The Anstruther Anticline is a somewhat larger structure than those exposed on the north coast. It determines the north-easterly trend of outcrops between Anstruther and Crail, and the axial region of the fold, plunging gently south-westwards, is exposed on the shore at Anstruther Wester.

Tight folds associated with the Ardross and Maiden Rock faults are also well exposed on the shore, between Ardross Castle [NO 50840070] and Pittenweem and between Kinkell Braes [NO 528157] and Kinkell Ness. In both areas the folding is more extensive on the south-east side of the fault than it is on the north-west, although near the inland extension of the Maiden Rock Fault around Lumbo [NO 489148] the most extensive folding lies on the north-west side of the fault. The folds nearest the Ardross Fault are described in detail by Francis and Hopgood (1970).

The Denhead Syncline, a north-north-easterly structure that underlies the village of Denhead [NO 468137], is known only from old bores and records of mining in the Largoward Splint and Black seams (pp. 77,78). It lies in the belt of complex structure associated with the Ceres and Maiden Rock faults.

The Radernie Syncline, a north-trending structure through March of Lathones [NO 475093], is probably the best authenticated of the inland structures, being proved by boreholes, by workings in the Radernie coals (p. 71) and by quarries in the St Monance Brecciated Limestone. To the west the presence of a flanking anticline is suggested by the trend of old limestone quarries but to the east the detailed structure is unknown.

The presence of the Baldutho Syncline, a shallow structure trending east-north-eastwards through Cunner Law [NO 491063], is suggested by boreholes, underground information and scattered exposures. To the north the Pepper Knowe Anticline, a shallow structure of similar trend, is indicated entirely by boreholes and mining information. The Appleton Basin, a small structure centred near Appleton [NO 423076], is known mainly from boreholes and workings in the Appleton Ell Coal.

The Hatton Syncline, a north-easterly-trending structure, is known from outcrops of the Castlecary Limestone in the Hatton Burn [NO 40100437 and 40410425]. It is flanked to the south-east by the Lundin Anticline, a northerly-trending structure exposed on the shore at Lundin Links. This fold here involves seams high in the Coal Measures. Inland, across the Durie Fault (p. 224), the persistence of the fold as far north as the region of Hattonlaw [NO 408039] is inferred partly from surface exposures, partly from borehole information and partly from workings in the Carhurlie Coal (p. 108).

EASTERLY TO SOUTH-EASTERLY NORMAL FAULTS

A group of major normal dip-slip faults can be recognised in the coast section, where they trend in directions between east-north-east and south-east, and dip at moderate to steep angles in the direction of downthrow. Slickensides, where visible, indicate dip-slip movements, and zones of shear-jointing are generally restricted to the immediate vicinity of the fault-planes. These faults are thought to belong to a widespread suite of generally eastward-trending normal faults well known in other parts of the Fife coalfield area (Knox 1954; Francis and others 1961). A group of easterly to south-easterly faults in the inland coalfield area of east Fife north of Largo is thought to belong to the same suite, while the Dura Den Fault, although of north-easterly trend, appears to be a normal fault and is therefore included here also.

DETAILS

COAST SECTION

The Durie Fault enters the area about 2 km west of Lundin Mill and runs eastwards towards Lower Largo, forming a northern limit to the outcrop of Coal Measures rocks. To the west of the present district the fault has a southerly downthrow of between 275 and 365 m (Knox 1954, p. 67) and this increases eastwards, probably reaching at least 600 m near Lower Largo. A south-easterly-trending fault on the shore at Temple [NO 425026], called the Temple Fault by Knox (1954, p. 67), is now taken to be the eastward continuation of the Durie Fault, which must therefore swing round from its more usual easterly trend somewhere in the region of Lundin Mill. As seen on the shore the fault brings red sandstones belonging to the Upper Coal Measures against a sequence of sandstone and mottled seatearth with marine bands, believed to lie in the Passage Group (p. 115; Fig. 14), so that the whole of the Lower and Middle Coal Measures have been cut out at this point. In the red sandstones on the downthrow side of the fault, shear-zones and minor faults close to the main fault-plane dip south-westwards at angles of 50° to 60°, but no slickensides are visible, perhaps because of the friable nature of the sandstones. The strata on the upthrow side of the fault are not well exposed near the fault-plane, but appear to contain a large number of irregular joints. A small east-south-easterly fault (marked 'transcurrent fault' on Fig. 14) cuts the Passage Group strata and can be traced for several hundred metres along the shore until it joins the Durie Fault. It shows horizontal slickensides and it presumably belongs to the widespread group of minor easterly-trending wrench-faults (p. 230) but because of poor exposure the age relationship between it and the Durie Fault cannot be established.

The Chapel Ness Fault (Cumming 1928, p. 131) is exposed at Earlsferry, running eastwards across the foreshore north of Chapel Ness. The fault-plane dips consistently northwards at angles between 45° and 60°, and a small dyke of bleached basalt has been intruded into it. Along most of its length shear-jointing is absent and no slickensides have been observed. The displacement of the outcrop of the Upper Kinniny Limestone shows that the fault has a northerly, probably 'normal', downthrow of about 45 m. A second fault runs across the shore about 300 m to the north, but is less well exposed. It also runs in a roughly easterly direction and its downthrow, which is towards the north, has been given as 55 m (Geikie 1902, p. 164). It is here named the

Earlsferry Fault in preference to the name 'Thirty-fathom fault' used by Cumming (1928).

The Sandy Craig Fault trends in a direction slightly east of south at its outcrop on the foreshore at the western end of Pittenweem. The beds affected by the fault are mainly argillaceous, at least on the higher parts of the shore, so that no shear-jointing or slickensiding is visible, and the dip of the fault-plane cannot be measured. The strata are locally disturbed against the fault. The displacement of the outcrops on the shore indicates that the fault throws down in a westerly direction by an amount exceeding 120 m.

The Pittenweem Fault, which has a south-easterly trend, is exposed in the cliffs and on the shore a few metres east of Pittenweem Harbour. The fault-plane dips at about 45° to the south-west, and the displacement of outcrops suggests a downthrow in that direction of 90 m at the very least. No slickensides are visible, and there is little shear-jointing.

The Peatlow Rock Fault runs in a roughly east-north-easterly direction across the foreshore at Cellardyke [NO 577037], where the fault-plane can be clearly seen. In some places it dips steeply southwards, and in others it is vertical. A very narrow zone of slickensided shear-joints is present where the fault-plane cuts sandstone; the orientation of the slickensides indicates a dip-slip movement. A southerly downthrow of about 60 m is estimated from the displacement of outcrops on the shore.

The Pans Fault, a vertical east-south-easterly fracture at The Pans [NO 605065], south of Crail, may belong to the group of normal faults. Slickensides indicating vertical movement are present and stratigraphical considerations suggest that the fault has a downthrow to the north, but the amount of displacement cannot be assessed.

The Crail Harbour Fault is exposed on the shore just to the east of Crail Harbour [NO 612074]. The fault-plane dips in a roughly south-westerly direction at a moderate angle and is associated with a belt of rather weak shear-jointing about 0·6 m wide. Stratigraphical considerations suggest that the fault has a southerly downthrow of at least 200 m.

The Danes Dike Fault trends in a direction slightly south of east and is well exposed on the shore at the south-east end of Danes Dike [NO 636094]. The fault-plane dips southwards at an angle of about 60°. Slickensides indicating dip-slip movement are present in a belt of shear-jointing 1·5 m wide in sandstones immediately adjacent to the fault-plane, and zones of minor faulting and shear-jointing are present in a belt about 150 m wide on the south side of the fault. General stratigraphical considerations suggest that the fault has a southerly downthrow of at least 300 m.

The Cambo Fault cuts off the rocks involved in the Randerston Syncline at the northern end of that structure. It is exposed at Cambo Ness [NO 609118] where it trends due east and dips very steeply southwards. No slickensides are seen but there is very little shear-jointing associated with the fault and it may therefore belong to the group of normal faults. Stratigraphical considerations show that it has an effective downthrow to the south of over 300 m.

The Kingsbarns Fault causes the juxtaposition of two very different sedimentary facies on the shore [NO 604124] about 200 m south of Kingsbarns Harbour. The main fault-plane is not exposed but a small subsidiary fault cuts nodular cherty carbonate rocks (p. 8) a few metres to the south of it. The subsidiary fault dips northwards at about 45°, with slickensides indicating dip-slip movement. The difference in stratigraphical level between the groups of strata separated by the main fault suggests that it has a northerly downthrow possibly as great as 600 m.

The Kenly Mouth Fault is exposed on the shore on both sides of the mouth of the Kenly Water [NO 581144], where it runs in an easterly direction. At the more westerly exposure the fault-plane is seen in a low cliff, and dips at about 45° towards the south. Minor faults on the north side of the main fault dip northwards, also at angles of about 45°, and throw down to the north. At the easterly exposure, fault-planes dipping both to north and south at angles of about 45° may be seen, and slickensides indicating dip-slip movement are present on some of them. The fault is believed to throw down towards the south and the amount of displacement may be relatively small, but this cannot be confirmed since no correlation of marker bands has been established across it.

The Craig Hartle Fault, exposed [NO 579148] near Craig Hartle, trends in an easterly direction. The fault-plane can be seen at one point to dip at about 50° towards the south, and traces of slickensides suggest a dip-slip movement. The amount and direction of the displacement cannot be assessed although the similarity of the facies on the two sides of the fault suggests that its throw is not very large. The fault has been linked with one with an apparent southerly downthrow which crosses the coast about half a kilometre to the west.
J.I.C., I.H.F.

INLAND AREAS

Several easterly-trending faults are present in the area to the north of Largo. The most important of these is the Branxton Fault, which throws down to the south and separates the Lower Limestone Group strata around Teasses from the Upper Limestone Group and Passage Group strata in the Hatton district. The identification of the Orchard Beds in the Hatton Burn near Houndshead Bridge (see p. 107), and of strata high in the Upper Limestone Group in the Balcormo Wood Borehole, show that this fault must lie rather further north than had previously been supposed. It is now thought to pass close to Branxton [NO 408065], where its throw must exceed 300 m. Its eastward continuation is largely hypothetical but it is believed to extend through unexposed country north and north-east of Largo Law as far as the Rires neck near Cathrie [NO 455056]. In the area to the north-east of the Branxton Fault, around Peat Inn, Lathones and Largoward, a number of faults trending between south-east and east-south-east have been proved during mining operations.

The Cordies Mealling Fault has a south-westerly downthrow of about 90 m at Cordies Mealling but the fault cannot be traced for more than a kilometre from there in either direction. The line of the Cassingray Fault is fairly accurately known around West Cassingray [NO 475072] where it has a south-westerly downthrow of about 45 m. Towards the south-east it can probably be traced as far as the gully between the two quarries at Baldutho Craig [NO 500064]. In the Falfield Bank district it is taken to separate the northward-dipping strata to the north from the folded beds around Falfield House [NO 445088]. Like the Lathones Fault, the Cassingray Fault is believed to change direction towards the west and to run almost due west to near Bankhead [NO 407092], cutting off the Gathercauld sill on the north side and the Teasses sill on the south side.

The Cadger's Bridge Fault, although known to have a downthrow to the south-west of about 120 m at Cadger's Bridge [NO 47270835], has not been traced towards the south-east for more than 700 m, and to the north-west it seems to be transgressed by the Higham sill. It is thought to reappear on the north-west side of the sill and to cause duplication of the outcrops of the Largoward Splint and Black coals south of Peat Inn [NO 453099] before dying out near Bankhead [NO 440105].

The Lathones Fault has a south-westerly downthrow estimated at 270 m near Lathones Hotel [NO 475088]. Towards the north-west it is apparently transgressed by the Higham sill, beyond which it serves to separate the outcrop of the Radernie coals at Nether Radernie from those of the higher seams around Peat Inn, with a throw of some 90 m. Its continuation further west is problematical, but on general grounds there appears to be a need for an east–west fault with southerly downthrow in the Greigston–Baldinnie district, to separate the basin around Bruntshiels [NO 435102] from the area of northerly dips around Baldinnie. The line of the Lathones Fault has therefore been continued for about one kilometre west of South Callange [NO 427113]. South-east of the Lathones Hotel the fault probably cuts off the Pepper Knowe Anticline and the Baldutho Syncline to the north-east. South-east of Kellie Law it is believed to affect the outcrops of the Back and Fore coals but it cannot be traced further in that direction.

The Radernie Fault cuts off the Radernie Syncline to the north near Brewsterwells [NO 481098] where its throw must be of the order of 300 m down to the south-west. It appears to run approximately E.35°S.; towards the north-west it cuts off the outcrops of the Radernie coals near Nether Radernie [NO 458105] and of the St Monance Brecciated Limestone south of Wilkieston [NO 450121]. The fault probably continues north-westwards as far as the Kinninmonth sill of quartz-dolerite: south-east of Brewsterwells its extent in poorly exposed ground is unknown, but it does not appear to affect significantly the outcrops of the Back and Fore coals.

To the south of the group of faults just described lies the Arncroach Fault, first reported by Landale (1837, p. 296) as a large east–west fault, which displaces the Back and Fore coals between Arncroach and Kellie Castle. He estimated it to have a downthrow of 165 m to the south, but in later unpublished reports amended this value to about 145 m. Its lateral extent west of Arncroach and east of Kellie Castle is unknown.   I.H.F.

The Dura Den Fault trends in a north-easterly direction but is described here since there is no strong folding, or evidence of transcurrent movement, associated with it. As already stated (p. 221), it defines the north-western boundary of the major structural region to which east Fife belongs. The fault-plane appears to be vertical, with a downthrow to the south-east. It is exposed in the bed of the Ceres Burn at a point [NO 41491442] near The Laurels, where a 5-m belt of shattered material separates flat-lying Upper Old Red Sandstone, on the north-west side, from south-eastward-dipping Carboniferous rocks. The south-easterly dip is a local feature and does not extend far from the fault. A quartz-dolerite sill exposed by the roadside [NO 41581437] near the fault appears to have been deformed along with the country rocks, and may therefore have been intruded prior to the fault movement, although the Kemback quartz-dolerite dyke (p. 167) seems to cross the fault without interruption. The fault is exposed again about 200 m north-east of the stream-bed locality, and can be traced accurately up the side of Dura Den and along the lower slopes of Kemback Hill as far as a point [NO 42741652] near Durdum. Its further continuation towards the north-east is largely conjectural. Its course cannot be traced south-west of Dura Den, and it is believed to die out among quartz-dolerite intrusions north of Ceres.   J.I.C.

North-Easterly Zones of Complex Structure

Two north-easterly to east-north-easterly belts of complex structure, about 13 km apart, affect the Carboniferous rocks in east Fife. Both contain strong north-easterly to easterly faults and narrow zones of tight folding about north-easterly to northerly axes, and in both there is evidence for transcurrent fault movements. The more northerly belt, termed the Ceres–Maiden Rock Fault Zone, can be traced over a distance of about 15 km, while the more southerly, the Ardross Fault, is known to extend for a distance of about 8 km. The Ceres–Maiden Rock Fault Zone may be a continuation, *en échelon*, of a similar belt of complex structure associated with the Ochil Fault between Ballingry and Cadham (Francis and others 1961, pp. 121–123; fig. 1). The Ardross Fault may be linked in some way with the Pentland Fault (Anderson 1951, pp. 99–100).

The structure of both belts suggests that they are the surface manifestations of dextral transcurrent movements along deep-seated crustal fractures (Moody and Hill 1956, p. 1215). The transcurrent nature of north-easterly fault zones in central Scotland, and their probable relationship to wrench-movements along deep-seated fractures, was first suggested by Anderson (1951, pp. 92–111). This author believed, however, that the movement along the fractures had been in a sinistral direction, as appears to have been the case in the Highlands, whereas the evidence now suggests that, in east Fife at least, the movements were dextral.

The age-relations of the north-easterly faults in central Scotland are not clearly defined, perhaps because intermittent movements took place along them over a long period of time. Some of them appear to have been active from Lower Carboniferous times onwards, and to have had an important influence on sedimentation (Francis *in* Craig 1965, pp. 309–357). In east Fife, the Ardross Fault was the site of tight folding of Lower Carboniferous sediments before the emplacement of the necks now exposed on the coast between Elie and St Monance (Francis and Hopgood 1970). Some of the intrusions in these necks have now been tentatively dated as Stephanian (pp. 177–180). The main fault movements at Ardross, however, post-date the emplacement of these necks.

The stress system responsible for early Carboniferous movements along the north-easterly faults cannot be worked out, but the pattern of later folding and wrench-faulting suggests that, during Upper Carboniferous times at least, there was a period of east–west compression.

The Dura Den Fault has a north-easterly trend but appears to be a normal fault; it has therefore been described in the previous section (p. 227).

DETAILS

The Ardross Fault is exposed on the shore over a distance of 2 km between East Links [NO 502000] and Newark Castle [NO 518112] where it disappears inland. As was first suggested by Cumming (1936, pp. 362–363) a north-eastward continuation of the fault is taken to mark the northern limit of the St Monance Coalfield. The fault may continue further in that direction, but it does not reach the coast, unless it runs through the area of cryptovolcanic structures at Balcomie Links [NO 631102] (p. 198). Towards the south-west, the fault probably runs out into the Firth of Forth at Wood Haven [NT 497996] (Cumming 1936, pp. 352–353; Francis and Hopgood 1970, p. 169).

The transcurrent nature of the fault was first recognised by Cumming (1936). He believed that the volcanic necks along this stretch of coast were emplaced before the faulting, and that they had then been broken up and displaced by a dextral transcurrent movement of about 1200 m. Francis and Hopgood (1970) confirmed that a dextral transcurrent movement has taken place along the fault, but pointed out that there is also evidence for a strong vertical component of movement, with a downthrow towards the north-west. Though agreeing with Cumming that the main fault movements were later than the emplacement of the necks, these authors were unable, from the structural evidence, to accept his correlation of necks on opposite sides of the fault and, in consequence, his estimate for the amount of lateral movement. Minor faulting, mainly along easterly to south-easterly lines, took place both before and after the period of neck emplacement and major faulting; the movements had vertical as well as horizontal components (Francis and Hopgood 1970, p. 183).

Strong folding around northerly to north-easterly axes affects the country rocks on the south-east side of the fault and, to a lesser degree, on the north-west side also. The folds die out rapidly away from the fault zone, and there is no reason to doubt that they were produced by movements along a deep-seated fracture which later gave rise to the Ardross Fault. Both folding and faulting can be attributed to a period of roughly east–west compression.

The Ceres Fault was first recognised (Geikie 1902, p. 175) between Ceres and North Callange [NO 420122] where it separates the Ceres Coalfield from an area of older rocks to the north. The coal-bearing strata dip very steeply in immediate proximity to the fault, but level off rapidly away from it. South-westwards the fault can be traced as far as a small gorge [NO 39371043] on the course of the Glassy How Burn, but it is then lost in a quartz-dolerite intrusion. Towards the north-east the fault can be recognised on the hillside south of Backfield of Ladeddie [NO 440138], where it separates an area of vertical strata belonging to the Limestone Coal Group, on the south, from flat-lying Lower Limestone Group strata, on the north. At Denork [NO 45421385] the fault appears to split but the main fault, throwing strata high in the Calciferous Sandstone Measures against coal-bearing strata in the Limestone Coal Group, can be mapped with fair accuracy between Denork and a point [NO 46651427] not far north of Denhead. Throughout its length the fault has the effect of a southerly downthrow, although the movement along it may have been partly transcurrent.

A fault associated with strong local folding runs east-north-eastwards from Denhead to the coast at Maiden Rock [NO 52641579] and may be a continuation of the Ceres Fault. Since the direction of effective downthrow across it is northwards, however, this stretch of fault is distinguished as the Maiden Rock Fault. The fault is well exposed at the coast (Balsillie 1920) where it consists of two vertical sub-parallel fractures about 60 m apart. The more easterly fracture runs across the foreshore just north of Maiden Rock, and to the east of it the strata are strongly folded about north-north-easterly axes. The intensity of folding decreases away from the fault. A small south-easterly fault cuts across an anticlinal axis [NO 52831581] east of Maiden Rock, effecting a sinistral lateral displacement of about 5 m.

The more westerly fracture forms a narrow furrow across the shore [NO 52551585] west of Maiden Rock. It contains a poorly exposed intrusion of bleached basalt breccia, as do some of the associated minor faults farther west (Balsillie 1920). There is a belt of strong folding, about north-easterly axes, along the north-western side of the fault, but this zone of disturbance is much narrower than that which affects the rocks east of the fault. An easterly-trending vertical fault, poorly exposed along the foot of the cliff [NO 52441581], appears to offset a synclinal axis, perhaps by means of a dextral transcurrent throw, but the effect of this cross-fault on the Maiden Rock Fault itself cannot be determined.

South-westwards from the coast the line of the fault cannot be traced accurately until it crosses the Lumbo and Cairnsmill burns, where it throws strongly folded Lower Limestone Group strata, on the north side, against less disturbed strata high in the Calciferous Sandstone Measures. It probably crosses the Cairnsmill Burn near a point [NO 49571467] where small intrusions of bleached basalt cut the country rock, and is fairly accurately located in the Lumbo Burn near a point [NO 48751449] where a sudden reversal of dip takes place. Farther west the fault joins the Ceres Fault, and separates an area of Limestone Coal Group strata, with the Mount Melville sill, from Calciferous Sandstone Measures south of Craigtoun Hospital [NO 48041441].

The style of folding associated with the Ceres and Maiden Rock faults (cf. Moody and Hill 1956), and its similarity to the folding associated with the Ardross Fault, suggest that they too are partly transcurrent, dextral faults which developed over a deep-seated fracture during a period of east–west compression. J.I.C.

Minor Wrench-Faults, Oblique Faults and Thrusts

Small wrench-faults are exposed in many parts of the coastal section, and are dominantly of east–west trend. Typical members of the group are vertical or nearly vertical, cause only slight displacement of the strata, and show near-horizontal slickensides. Many of them lie within zones of shear-jointing. Their transcurrent nature was first recognised by Kirk (1925, pp. 366–367) in the area of Kinkell Ness. The wrench-faults grade into faults with oblique slickensides and among this group both vertical fault-planes and east–west orientation are less strongly developed. Thrust-faults (low-angle reversed faults) are rare.

Details

WRENCH-FAULTS

A group of small south-easterly-trending vertical faults is exposed on the shore south of Caiplie [NO 590052]. They lie in a well defined belt about 400 m wide, in which zones of shear-jointing are also present. On some of the fault-planes slickensides are visible, dipping at low angles towards the south-east. A similar group of minor faults is exposed on the coast, over a distance of about a kilometre, south-westwards from Barns Mill [NO 601060]. This group includes at least six small near-vertical faults and several zones of shear-jointing, all of which trend in directions between east and south-east. Horizontal slickensides can be seen on some of the faults, and are best exposed on the most northerly fault [NO 601059], which lies in a belt of parallel shear-jointing over 15 m wide. The throw of this fault is small, and in a dextral sense. Another example is well exposed in the cliffs and on the shore towards the eastern end of Witch Lake [NO 51131706] at St Andrews. The fault-plane is nearly vertical and can be traced in an east-north-easterly direction for a distance of about 200 m across the shore. In the cliffs, horizontal slickensides can be seen on large joint-surfaces parallel to the fault-plane, as well as on complementary joints.

Minor wrench-faults of this type are also to be seen at Temple [NO 425026] (p. 224; Fig. 14); at the Boat Harbour [NO 547023] at Pittenweem; just south of Peatlow Rock [NO 577037] near Cellardyke; near the Mermaid's Cradle [NO 616076] at Crail; on the shore [NO 620078] east of Roome Harbour; and at a point [NO 63350894] on the shore north of Kilminning Castle. A fault on the shore [NO 637099] about 100 m north of Fife Ness Harbour also belongs to this group, and may be associated with the belt of oblique faulting at Craighead Quarry (p. 231). Another example is visible on the shore [NO 554151] at the mouth of Kittock's Den, while between this point and Kinkell Ness [NO 538158] the several examples noted by Kirk (1925, pp. 366–367) are to be found.

In addition to the minor wrench-faults described above, a number of larger easterly faults seem to have had important components of strike-slip movement along them. These are the Cuniger Rock Fault, the Wormistone Fault and the Randerston Castle Fault. The Cuniger Rock Fault trends in a generally easterly direction from Cuniger Rock [NO 55660271] and runs along the intertidal zone almost parallel to high water mark for a distance of over 500 m before disappearing below low water mark near Billow Ness. At the Billow Ness end the fault is trending somewhat south of east and appears to have a considerable effect on the strata through which it runs. Several zones of strong shear-jointing, with some apparently minor faults, diverge from the main fault on its northern side and run inland in roughly west-north-westerly directions. The displacement of the strata effected by these shear-zones cannot easily be made out, and it may be that a major fault runs through them. Slickensides at various angles suggesting oblique-slip and dip-slip (perhaps reversed) movements are present. Farther to the west, the Cuniger Rock Fault loses throw rapidly and dies out near Cuniger Rock. At various points in this part of its course the fault-plane can be seen, dipping southwards at angles of about 60° and showing oblique to horizontal slickensides. The displacement of outcrops shows that the transcurrent component of movement has been in a sinistral sense.

The Wormistone Fault is exposed at Wormistone Hind [NO 620107] and trends in an easterly direction, throwing strata of Carboniferous and Upper Old Red Sandstone facies together. The fault-plane is almost vertical, and slickensides are visible on it at one point, dipping at low angles towards the east. The displacement of the strata can be interpreted in terms of a slightly oblique dextral strike-slip movement of several hundred metres. As in the case of the Randerston Castle Fault, a zone of minor open folds, about northerly to north-easterly axes, is present to the north of the main fault, and minor faults of various types are associated with it.

The Randerston Castle Fault resembles the Wormistone Fault and lies a few hundred metres to the north of it. It is exposed on the shore and in the cliffs at Chincough Well [NO 617110]. Its effect on the strata suggests that a dip-slip component of movement, with a downthrow to the north of at least 90 m, has taken place but where the fault-plane is best exposed, on the shore, it dips very steeply and shows almost horizontal slickensides. Small near-vertical splay-faults are also visible in the cliffs and on the shore, and some of these are associated with zones of shear-jointing. A belt about 200 m wide containing minor faults is present on the northern side of the main fault, and minor open folding about north-easterly axes is also present here.

OBLIQUE FAULTS

Movements in directions intermediate between dip-slip and strike-slip are recorded in oblique slickensides exposed in the coast sections, and faults transitional between 'normal', 'transcurrent' and 'reversed' types are found. Most of these oblique faults are minor fractures and many are associated with a larger fault showing dominantly dip-slip or strike-slip movement.

A small splay-fault with oblique slickensides diverges from the more westerly of a pair of minor wrench-faults (the Boat Harbour Fault) [NO 547023] at Pittenweem, and another example of this relationship can be seen on the shore at Cellardyke, where a pair of minor south-easterly wrench-faults runs out to sea just south of Peatlow Rock [NO 577037]. Here a small splay-fault with oblique slickensides runs out from the more northerly of the two wrench-faults, and a third south-easterly fault a few metres to the south of these [at NO 57520356] dips north-eastwards and has oblique southward-dipping slickensides. A group of small faults which trend somewhat north of east occurs on the shore near Craighead Quarry [NO 634101]; the fault-planes dip southwards at angles of about 45° and oblique slickensides dip in south-westerly directions. Oblique slickensides are also visible in a belt of strong shear-jointing along a small easterly-trending vertical fault at Babbet Ness [NO 592142]. The fault is not obviously associated with any other fracture, although the larger northward-dipping Babbet Ness Fault (of unknown type) cuts across the shore about 200 m to the south. The Cuniger Rock Fault also contains evidence of oblique movement but it is not clear whether the main movement along this complex fracture has been transcurrent or oblique (p. 230). Oblique faults are also associated with the major north-easterly fault zones at the Maiden Rock and Ardross Castle (pp. 227–229) and an example of an oblique fault associated with a major normal fault can be seen [NO 63600930] on the south side of the Danes Dike Fault.

REVERSED FAULTS

Small reversed dip-slip faults have been noticed at a few places in the coast sections, but they are by no means as common as the other types described. A group of small low-angle faults with slickensides showing dip-slip movement is exposed on the shore [NO 61610775] at Roome Bay near Crail. The two complementary sets of fault-planes dip at low angles towards the east and the west. Geikie (1902, p. 285) illustrated two examples of small low-angle thrusts in limestones, one from the centre of the Randerston Syncline [NO 611116] and one below the Dovecot [NO 51930122] of Newark Castle. Thrusting may also have been involved in the movements along the Cuniger

Rock Fault (see above). It is possible that the dip-slip component of movement along some of the oblique faults may have taken place in a 'reversed' direction.

SHEAR-JOINTS

Zones of shear-jointing are associated with almost all the faults exposed on the coast, whether these are of normal, oblique, transcurrent or reversed type. The widest zones are developed along wrench-faults and oblique faults, whereas along even the largest normal faults the jointing is in general limited to zones of only a few centimetres thickness. The joints are best seen in sandstones where they appear as thin anastomosing stringers, paler in colour than the surrounding rock. The joint surfaces usually show slickensides. In thin section (S 54302, 54298) the paler stringers can be seen to consist of broken quartz and feldspar fragments of all sizes, packed together with a minimum of pore space.

J.I.C.

REFERENCES

ALLAN, J. K. and KNOX, J. 1934. The economic geology of the Fife Coalfields, Area II. 1st edit. *Mem. geol. Surv. Gt Br.*

ANDERSON, E. M. 1951. *The dynamics of faulting.* 2nd edit. Edinburgh and London: Oliver and Boyd.

ARMSTRONG, M. and PATERSON, I. B. 1970. The Lower Old Red Sandstone of the Strathmore region. *Rep. Inst. geol. Sci.*, No. 70/12. 23 pp.

BALSILLIE, D. 1920. Descriptions of some volcanic vents near St Andrews. *Trans. Edinb. geol. Soc.*, **11**, 69–80.

CRAIG, G. Y. (Editor) 1965. *The geology of Scotland.* Edinburgh and London: Oliver and Boyd.

CUMMING, G. A. 1928. The Lower Limestones and associated volcanic rocks of a section of the Fifeshire coast. *Trans. Edinb. geol. Soc.*, **12**, 124–140.

—— 1936. The structural and volcanic geology of the Elie–St Monance district, Fife. *Trans. Edinb. geol. Soc.*, **13**, 340–365.

FITCH, F. J., MILLER, J. A. and WILLIAMS, S. C. 1970. Isotopic ages of British Carboniferous rocks. *C. R. 6th Int. Congr. Carb. Strat. Geol.*, **2**, 771–789.

FRANCIS, E. H., ALLAN, J. K. and KNOX, J. 1961. The economic geology of the Fife Coalfields, Area II. 2nd edit. *Mem. geol. Surv. Gt Br.*

—— and HOPGOOD, A. M. 1970. Volcanism and the Ardross Fault, Fife, Scotland. *Scott. Jnl Geol.*, **6**, 162–185.

GEIKIE, A. 1902. The geology of eastern Fife. *Mem. geol. Surv. Gt Br.*

KIRK, S. R. 1925. Geology of the coast between Kinkell Ness and Kingask, Fifeshire. *Trans. Edinb. geol. Soc.*, **11**, 366–382.

KNOX, J. 1954. The economic geology of the Fife Coalfields, Area III. *Mem. geol. Surv. Gt Br.*

LANDALE, D. 1837. Report on the geology of the East of Fife Coalfield. *Trans. Highl. agric. Soc. Scotl.*, **11**, (vol. 5, new series), 265–348.

MOODY, J. D. and HILL, M. J. 1956. Wrench-fault tectonics. *Bull. geol. Soc. Amer.*, **67**, 1207–1246.

Chapter 14

PLEISTOCENE AND RECENT

GLACIAL EROSION FEATURES AND TILL DEPOSITS

THE ice-sheet of the last glaciation covered the whole area and till, consisting of a variety of stony and sandy clays, was deposited over most of the lower ground. Layers of sand and gravel have been recorded within or beneath the till at a few localities but there is still no good evidence for more than one period of till deposition in the region. The till is buried beneath younger drift deposits along a coastal strip, particularly in the south, and along some stream courses but crops out at the surface over most of east Fife, forming a smooth, rather featureless landscape broken in the west by the upstanding outcrops of igneous intrusions and volcanic necks. Locally the till surface has been moulded into ridges and drumlins; the direction of ice movement suggested by these (Fig. 27) agrees with that deduced from other evidence (see below). In parts of east Fife, smooth terrace-like platforms are a feature of the landscape up to heights of about 120 m O.D. Some of the terraces have been interpreted as marine erosion features (Chambers 1848, pp. 59–63, 325; Geikie 1902, p. 307) but as no deposit other than till has ever been found on them, it seems more likely that they are simply a part of the sub-glacial topography. The till reaches 20 m in thickness locally but is generally much thinner than this, and streams in many areas have cut through into the Carboniferous sediments beneath. The till has been removed, exposing bedrock, along much of the coastline also. In colour the till is generally grey or brownish-grey although there are some localised patches of red material in the south, while along the southern edge of Stratheden the reddish-brown till characteristic of the neighbouring Old Red Sandstone areas to the north and west has been carried over on to the Carboniferous outcrop for several kilometres (Fig. 27). Besides local material the till contains rock fragments derived from the Highlands and the Ochil and Sidlaw Hills, indicating that the ice moved across the area from west to east. Erratic blocks on the surface (Geikie 1902, pp. 292–295) support the same conclusion.

The rock surface beneath the till has been variously affected by the ice movement. Where the country rocks are soft, as in areas underlain by sedimentary rocks, a zone of disturbed and broken rock often separates till from solid rock, but hard igneous rocks are generally smoothed off and, in places, striated. Striations are rare, however, except where the rock surface has recently been stripped of till. The striae are orientated in an easterly direction (Geikie 1902, pp. 289–290), as are the local larger-scale groovings and channellings of the rock surface (Fig. 27).

DETAILS

Reddish-brown till is exposed in Dura Den opposite Charlie's Rock [NO 41731392] and at Ladeddie Limeworks [NO 44001361]. Farther east, up to 8 m of the same material is exposed in old quarries [NO 454164] west of Strathkinness while in the Kinness Burn [NO 49481570] near St Andrews over 9 m of it can be seen resting on rock. In most of this area the till varies in thickness up to 8 m, but locally, between Strathkinness,

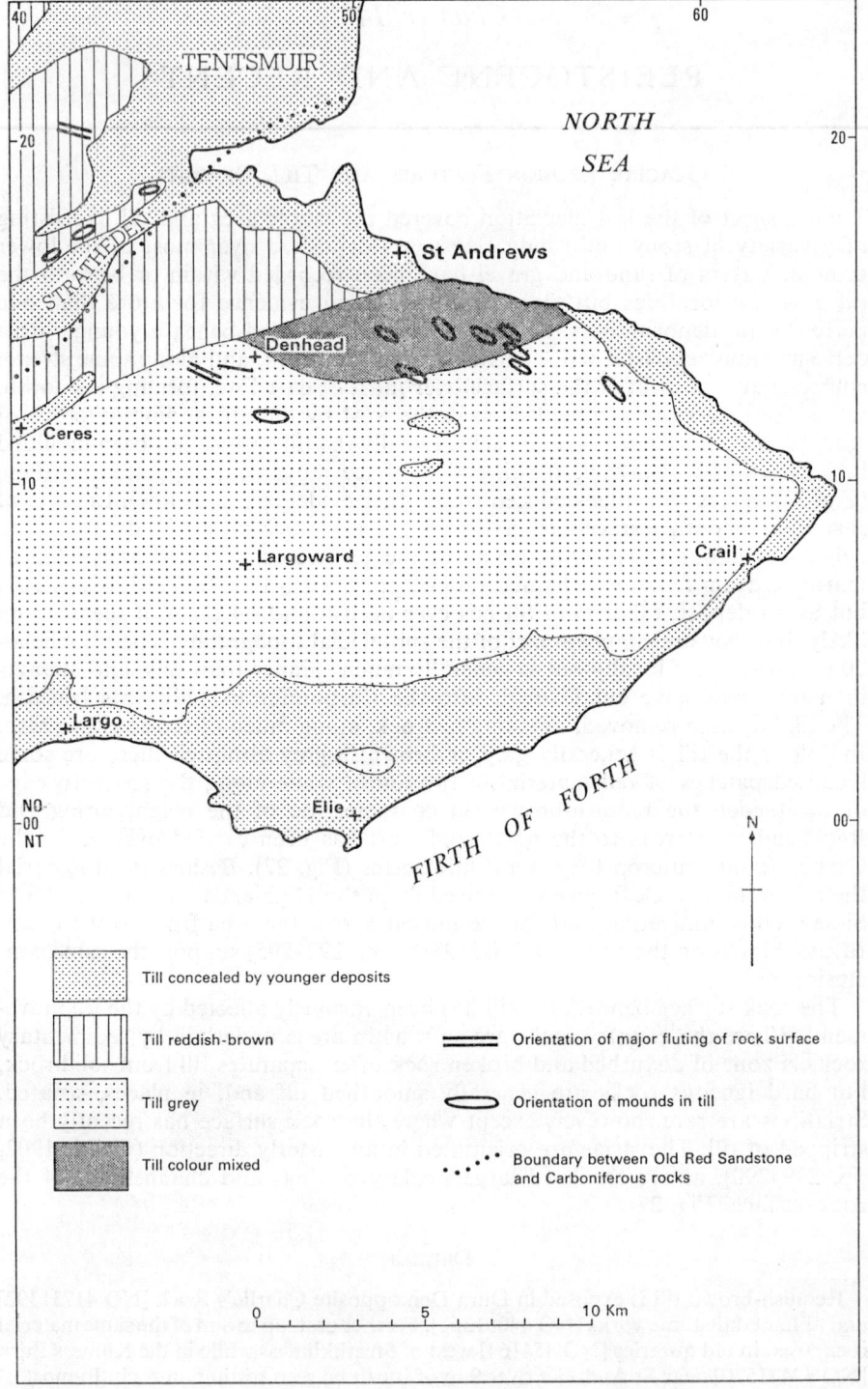

FIG. 27. *Sketch-map showing distribution of till deposits and glacial erosion features*

Kincaple and St Andrews, it reaches 12 to 15 m as in bores at Cauldside [NO 47451695] and at Balnacarron House [NO 49301590].

The boundary between reddish-brown till and grey till is fairly sharp in the western part of the area, and runs quite close to the line of the Ceres Fault from Ceres [NO 402117] to Denhead [NO 468137], but farther east the boundary becomes less well-defined and an area of till of intermediate character has been mapped (Fig. 27). Till of mixed coloration is exposed at Catcraig Quarry [NO 49131446] and in old quarries [NO 52961332] at Prior Muir.   J.I.C.

Grey till, which often weathers to a brownish colour at the surface, is present over most of the area south of that just described. It is exposed in numerous watercourses and old quarries. Boreholes in the coalfields show that generally it varies in thickness up to 8 m but have proved the existence of localised areas of greater thickness, up to 12 or 15 m, as at Ladeddie [NO 443129], Bandirran [NO 407103], Teuchats Toll [NO 405074], Little Pilmuir [NO 407038] and March of Lathones [NO 475093]. Geikie (1902, p. 291) recorded a thickness of 21 m at Gillingshill Reservoir [NO 510062]. Layers of sand and gravel are recorded within or beneath the till around Baldastard [NO 421069], around Lathallan [NO 461062], and north of East Cassingray [NO 491 070]. Isolated patches of reddish till have been recorded in bores and surface sections around Little Pilmuir and in a trench section [NO 58910588] near Caiplie, and they doubtless occur elsewhere; the red coloration probably derives from reddened rocks in the Carboniferous nearby.

Along the southern coastal strip of east Fife till is generally concealed beneath younger deposits (Fig. 27), but is exposed near Incharvie [NO 480028] and around Balbuthie [NO 502020]. Along the coastline itself the till cover has in many places been stripped off during late-Glacial and post-Glacial times and the raised beach deposits often rest directly on rock. Inland, in boreholes at Muircambus [NO 468024], the till is likewise thin or absent but in bores around Grangemuir [NO 539041] up to 9 m of till containing bands of sand and gravel were recorded beneath late-Glacial sands and clays.   I.H.F., J.I.C.

Glacial Meltwater Deposits

During the retreat stages of the final glaciation meltwaters deposited clastic sediments at and near the margins of the melting ice-sheet in several parts of east Fife. Some of the deposits accumulated in areas occupied by masses of inactive or dead ice and such deposits, although in some cases they may originally have been terrace-like, have collapsed into a hummocky topography of kames and kettleholes. They are shown on the geological map as glacial sand and gravel. In other areas little or no dead ice was present at the site of deposition and the original flat or terraced landforms are preserved. Deposits of this type are distinguished as fluvio-glacial sand and gravel.

The deposition of some of the sands and gravels took place during the period of high late-Glacial sea-level and many deposits lying at levels lower than about 36 m O.D. were laid down in the sea or were subsequently reworked by it. There is consequently no sharp distinction between glacial and fluvio-glacial deposits laid down below ice or on land by meltwater, and similar deposits laid down in or reworked by the sea and classified as late-Glacial marine deposits. In areas where the two types of deposit are present together the boundary mapped between them is often an arbitrary one.   J.I.C.

Details

Ceres–Pitscottie. Between these two villages a hollow containing dead ice received meltwater flowing from the south and west, as is shown by the presence of a mass of moundy gravel along the south side of the hollow between Ceres and a point [NO 41551234] about 500 m west of North Callange. Terraced gravel just above 60 m O.D. around Newbigging of Ceres [NO 40571186] and along the north side of the hollow from Ceres Mill [NO 40551238] to Pitscottie lies at a slightly lower level than the moundy gravel and presumably dates from a slightly later period. The meltwaters drained into Stratheden through a system of channels at the north-eastern end of the hollow. Since the inlet to the more westerly channel, at Pitscottie Vale [NO 417133], lies at a slightly lower level (about 58 m O.D.) than that to the more easterly channel (about 73 m O.D.) at Blebo Hole [NO 42381345], the westerly channel probably came into use relatively late, during and after the phase of terrace gravel deposition.

Dunino–Boarhills. Several patches of sand and gravel have been mapped at levels between 30 and 137 m O.D. along the course of the Kenly Water and its tributaries between Lathockar [NO 493110] and Boarhills. The largest patch, which is about two kilometres long by half a kilometre wide, with a flattish surface between 75 and 100 m O.D., lies beside the Cameron Burn between Brigton [NO 513116] and Stravithie [NO 535116]. A short distance to the south, a long narrow patch of sand and gravel with a subdued moundy surface between 100 and 130 m O.D. extends alongside the Kinaldy Burn for about two kilometres from Kinaldy [NO 512104] to Tosh [NO 524105]. The remaining patches are all small. Those at Bonnytown [NO 546126] and Winchester [NO 554140] have a subdued moundy topography between 60 and 80 m O.D. but a patch by the Kenly Water above Boarhills has a terrace-like form at about 30 m O.D. Its height was probably determined by the sea-level at the time of deposition.   I.H.F.

South coast of east Fife. Along the south coast of east Fife from Leven, just beyond the western limit of the district, to Kilrenny there is a low-lying tract of sandy ground in which both glacial sand and gravel and late-Glacial marine deposits have been recognised (Geikie 1902, pp. 296–313; Knox 1954, pp. 118–120; Cullingford and Smith 1966). There are few well-marked ice-contact features in this area, however, and in many places it is difficult to distinguish unmodified glacial sand and gravel from deposits reworked by the sea. Most of the deposits which lie below about 30 m O.D. are gently undulating or almost flat and have probably been subjected to marine action. Sand and gravel occur at higher levels, between about 30 and 75 m O.D., in two areas, one between Hatton and Lundin Mill and the other around Colinsburgh (see below); these are more moundy and are regarded as unmodified glacial meltwater deposits. Although the distinction between the two types of deposit is not so clear-cut in this area as it is around St Fort (Chisholm 1966) the general relationship between them is probably the same; material transported southwards or south-eastwards from higher ground was at first deposited at the margins of dead ice masses in the low ground around the Firth of Forth, but as the glacial remnants melted, the late-Glacial sea entered the area, reworking the deposits below about 30 m O.D. and perhaps spreading the sand eastwards along the strandline.   I.H.F., J.I.C.

The patchy deposits of glacial sand and gravel between Hatton and Lundin Mill have their upper limit at about 75 m O.D. between Pitcruvie [NO 414046] and Upper Largo; their lower limit, around Lundin Mill, lies at about 30 m O.D. In the flat coastal area, isolated mounds of gravel project above the surface of the late-Glacial raised beach, especially around Toll Cottage [NO 41640297]. In a borehole [NO 41410311] sited on the largest of these mounds, the gravel is 11·6 m thick, and rests directly on rock; another borehole [NO 41030399] on higher ground to the north proved 7·9 m of gravel resting on till.

In the Colinsburgh area, the sand and gravel deposits form an elongated spread with subdued moundy topography which extends between Dumbarnie [NO 449031] and Balcarres Mill Bridge [NO 48630359]. On the north side it rises to about 75 m O.D. near Balcarres House [NO 474044] and along the southern margin a feature, indefinite in places, separates it from the late-Glacial marine deposits at around 30 m O.D.

There are no sections through the sand and gravel, whose thickness is therefore unknown.
I.H.F.

Late-Glacial Marine Deposits and Shoreline Features

The late-Glacial deposits described in this section were laid down under marine or estuarine conditions during the final deglaciation, up to the end of Upper Dryas (pollen Zone III) times. They undoubtedly span different periods of time in different areas. The sea lay at a higher level relative to the land than it does at present, but was falling rapidly, probably as a result of an isostatic adjustment to the melting of the ice. Sediments then laid down in the sea are now exposed on dry land, therefore, and lie at levels up to about 36 m O.D. (Fig. 28). Some of the earlier deposits are thought to be contemporaneous with meltwater deposits laid down around decaying masses of glacial ice, and as some of these masses lay in the sea, the two types of deposit may be closely associated in space as well as in time. In the St Fort area, just to the north (Chisholm 1966), there is evidence that when the ice melted the sea entered areas previously occupied by ice, depositing new sediment and reworking existing meltwater deposits. Masses of buried glacial ice that survived the period of marine reworking, and persisted until the area became dry land, melted to produce moundy topography, so that there are apparently anomalous patches of 'glacial' topography among the marine deposits; some similar features in the east Fife deposits may be explained on the same basis.

The deposits range from plastic clays to sands and gravels. The clays are brown or reddish-brown in colour, often well-laminated, with estuarine and shallow-marine foraminifera and, more rarely, marine bivalves of 'arctic' affinities. They also contain isolated stones, some of distant origin. Brown laminated silts, yellow and red sands, and various interlaminated deposits also occur. Gravel is found mainly in marginal situations, on raised beaches or close to sources of gravel supply. The clays and associated interlaminated materials were probably laid down in subtidal and intertidal environments; the isolated stones suggest that floating ice was common. Sands probably deposited in tidal channels are also known. Beach gravels and sands are found only in areas that were exposed to wave action, but occur at all levels from the marine limit at about 36 m O.D. down to the margin (at about 9 m O.D.) of the deposits laid down during the post-Glacial marine transgression.

A stratigraphical sequence has been established in the St Andrews area but this is not necessarily applicable everywhere; and it has not proved possible to work out a chronology based on pollen-dating, radiocarbon dating or fossil successions anywhere in east Fife. A time sequence based on the heights of successive raised beaches is unlikely to be satisfactory, for several reasons. Firstly, raised beach terraces cannot always be identified as such with certainty, and may be confused with fluvio-glacial terraces. Secondly, the late-Glacial sea did not accomplish much erosion, except in soft materials, and the landward limit of a raised beach is therefore likely to have been determined by features of the pre-existing land surface; it may thus coincide locally with an abrupt rise in the ground surface which may simulate a coastal erosion feature, but this will not necessarily be related to any widespread stillstand of the sea. This means that individual fragments of raised beach are unlikely to be isochronous, and also that the slopes of back-features may reflect original till or bedrock

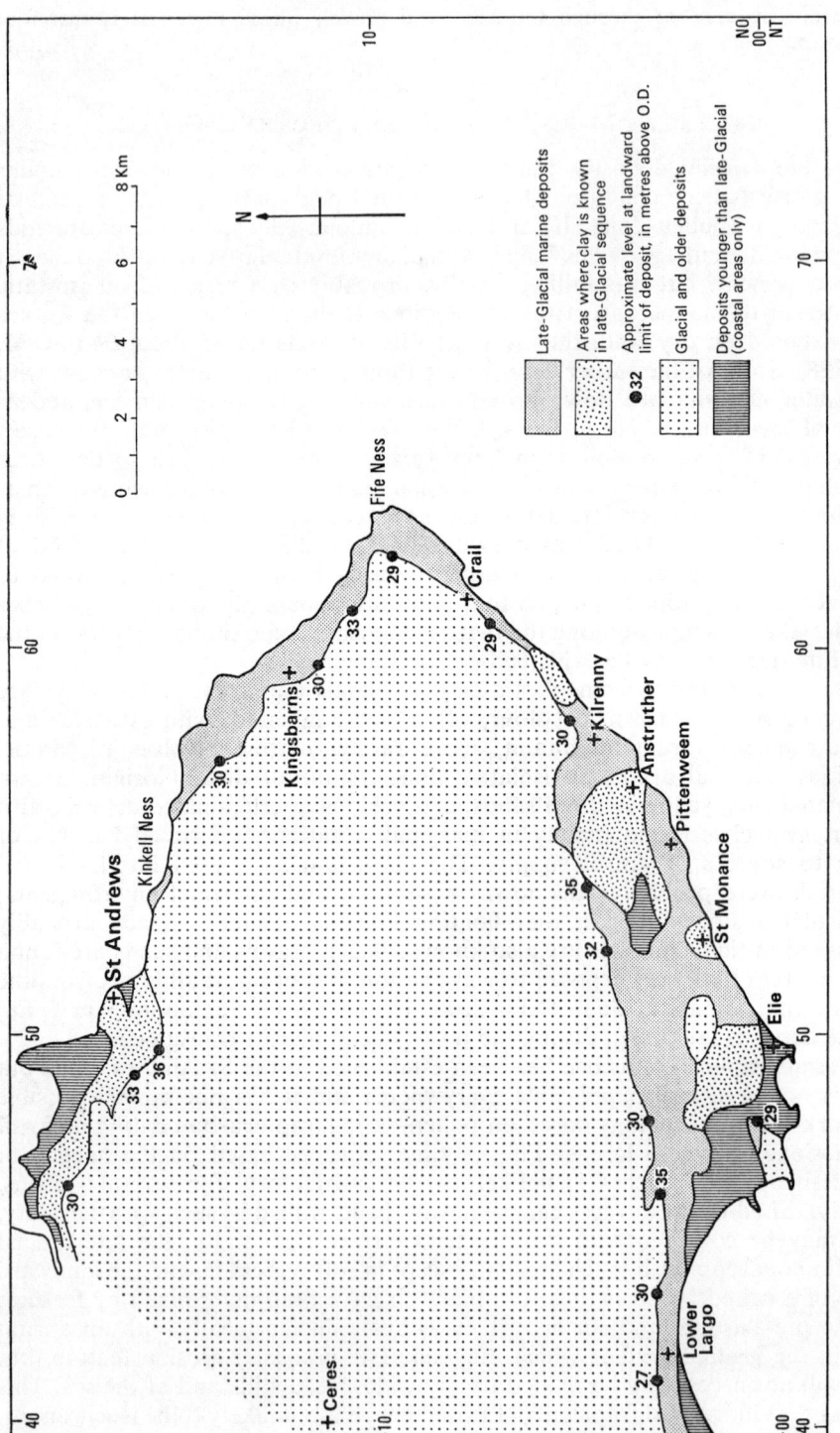

FIG. 28. *Sketch-map showing distribution of late-Glacial marine deposits*

slopes rather than tilting of shorelines. Thirdly, suitable material for radiocarbon or pollen dating is generally absent from gravel deposits, so that disconnected fragments of raised beach cannot be correlated by these means. Fourthly, the relationship between a raised beach deposit and fossiliferous clays nearby is generally disconformable, making a correlation between the fossiliferous deposits, which might eventually be dated, and a particular raised beach impossible.

It is clear from the foregoing that individual fragments of raised beach cannot be related with any certainty to the actual shorelines of the period. Indeed, it is doubtful if individual shorelines can be identified in this area, and it is unlikely that the easterly tilt, which must affect them, can successfully be measured (cf. Cullingford and Smith 1966). However, the level of the marine limit, which may be diachronous, can be identified with more certainty than can individual shorelines. It varies between 30 and 36 m O.D. in the north, between 29 and 33 m O.D. in the east, and between 27 and 35 m O.D. in the south (Fig. 28).

J.I.C., I.H.F.

LUNDIN LINKS TO KILRENNY

From the western margin of the district near Lundin Links [NO 409025] as far east as Dumbarnie [NO 449031] the mapped limit of late-Glacial marine deposits lies up to about one kilometre inland from the present shoreline, but east of Dumbarnie the coastal plain widens out and for about 13 kilometres, as far as Kilrenny [NO 575048], the deposit margin lies between two and four kilometres back from the present shore (Fig. 28).

The deposit in most areas consists of no more than two or three metres of sand, with fine gravel in places, but locally thicker sequences are known, and in some areas sequences containing interbedded clay occur (Fig. 28). At Elie an 'arctic' marine fauna has been obtained from the clay (Wood 1863; Brown 1867; Geikie 1902, pp. 302–303, 383) but there are no well authenticated fossil records from the sand deposits. The stratigraphy of the deposits is very poorly known, and beds of clay lie among the sands at more than one level. The clay with arctic shells at Elie appears to be among the oldest deposits, and rests on till; it has been disturbed, perhaps by ice-pushing prior to the deposition of overlying sands. In the western part of the area, between Lundin Mill and Colinsburgh, the deposits pass into contemporaneous or near-contemporaneous glacial sand and gravel, and the limit of late-Glacial marine material is not readily apparent. The less moundy material, however, lies at lower levels, mainly below 30 m O.D., and has been mapped as late-Glacial marine deposits, whereas the more moundy material, at higher levels, is regarded as glacial sand and gravel. Elsewhere the limit of deposit generally lies against gently sloping or undulating till-covered ground and is rarely marked by good features. Seaward-sloping terraces interpreted as raised beaches are present at some localities but there are also many gently undulating areas with no raised beach topography. Cullingford and Smith (1966, p. 43) believed that the lack of good beach features in parts of the area is due to the presence there of moundy glacial deposits but except in areas where dunes have developed, the ground is much less moundy than these authors suggest and cannot be described as typical ice-contact topography. It is more likely that outwash material in the area has been reworked and that the deposit now consists mainly of late-Glacial marine sediment, the surface form of which has been modified by the melting of buried glacial ice remnants (cf. Chisholm 1966).   J.I.C., I.H.F.

Details

Lundin Mill to Dumbarnie. Around Lundin Mill sandy deposits forming flattish areas between mounds of glacial gravel and sand have been mapped up to a level of about 27 m O.D. and from Lundin Mill eastwards to Dumbarnie there is a well-marked seaward-sloping terrace about half a kilometre wide with a back-feature varying from about 27 to 30 m O.D. Boreholes around Toll Cottage [NO 41640297] and Upper Largo suggest that the deposit is generally thin.

Dumbarnie to Kilrenny. Farther east, in the coastal plain between Dumbarnie, Elie and Kilrenny, the area occupied by late-Glacial marine deposits is much wider and consists of gently rolling ground mainly between about 9 and 35 m O.D. Raised beach topography is generally poorly developed in this tract although broad terraces have been mapped north of Balbuthie [NO 502020] and west of Kilrenny. The prevailing rolling topography may be due in part to the melting of buried glacial ice (see above) but is certainly due in places to the development of dune topography, as around Muircambus [NO 468024].

Within the coastal plain there are upstanding areas where older deposits come to the surface, for example at Kincraig [NO 467002] and Balbuthie. Kincraig, a large mass of vent agglomerate, rises to over 60 m O.D., and a narrow strip of late-Glacial sand has been mapped, at about 30 m O.D., on its northern face. At the western end of the hill (Plate I, frontispiece), three benches cut into the agglomerate at about 11·5, 22 and 24·5 m O.D. (Cullingford and Smith 1966, p. 50) are believed to mark short periods of stillstand during the withdrawal of the late-Glacial sea. At Balbuthie a low till-covered ridge rises to about 36 m O.D.

The late-Glacial marine deposits vary considerably both in thickness and lithology within the coastal plain area but the surface material is almost universally sandy. Boreholes around Muircambus show that in that area the whole sequence consists of sand, up to 9 m in thickness. Plastic clay is known to form part of the late-Glacial sequence around Kilconquhar, Elie, St Monance and north of Anstruther, and may lie hidden beneath the surface sand in other areas where there is no subsurface information; Fig. 28 shows only the areas where clay has been recorded.

At a disused clay pit [NO 480017] near Kilconquhar Station 'iron-grey brick clay rudely jointed, covered with yellow sand' at about 15 m O.D. was recorded during the primary geological survey, and in the banks of the Cocklemill Burn nearby [NO 47680150] about 2 m of chocolate-brown laminated clay with scarce foraminifera and small isolated stones can be seen. Clay was recorded at the station itself during the construction of the railway line (Wood 1863, p. 126).

Around Elie the clay contains 'arctic' marine shells, and two good sections are recorded in the literature. Both are now obscured. Wood (1863, pp. 126–127) first described a section in the railway cutting [NO 493003] west of Elie Station, where 'unctuous blue or reddish-blue clay' containing marine shells rests on top of till at about 12 m O.D. Brown (1867, pp. 620–621) described the cutting section in more detail and described the famous shore section [NT 49609972] at Wood Haven, where a similar sequence was then exposed at high water mark. The sections recorded by Brown are summarised below; the present interpretation of each bed is indicated in brackets.

Elie Inland Section (railway cutting)

	Thickness m
Sand, with peaty layers up to 1·8 m thick, containing freshwater and land shells [blown sand and marsh deposits at landward limit of post-Glacial raised beach]	variable, increases eastwards to 6 or 9
Peat, with roots of *Equisetum* [Sub-Carse Peat]	0·15 to 0·25
Gravel and sand with some clay; bedded, with depositional dips at low angles conformable with contours of present ground surface [late-Glacial marine deposits]	not stated
Unconformity	
Sand, passing down into clay with arctic marine shells; slightly folded; boulders present [late-Glacial marine deposits]	not stated
Till, black and brown	not stated

Elie Shore Section (Wood Haven)

Blown sand	1·2 to 1·8
Shingle with marine shells [post-Glacial marine deposits]	'thin'
Peat, with abundant remains of *Phragmites*, seeds and wood [Sub-Carse Peat]	0·12 to 0·25
Unconformity	
Sand with root traces at top, passing down into clay with arctic marine shells; gently folded [late-Glacial marine deposits]	not stated

Brown listed the fauna obtained in the clay and, from a consideration of the present-day habitats of the species concerned, concluded that when the clay was deposited the sea lay at least 45 to 60 m higher than at present. A later, more comprehensive list of the fauna was compiled by Tait (*in* Geikie 1902, p. 383). The clay probably extends beneath most of the town of Elie (Wood 1863, p. 127). The presence of scattered stones, including pieces of chalk and flint, in the clay at Wood Haven was noted by Geikie (1902, pp. 302–303); he believed that they had been dropped from floating ice.

At St Monance Station [NO 524019] Geikie (1902, pp. 310–311) recorded a section of clay and sand with contorted lamination at a level of about 18 m O.D.; Wood (1863, p. 126) also recorded clay at St Monance Station during the construction of the railway line. Clay is known at several places to the north and west of Anstruther. At Anstruther Station [NO 562036] it was recorded, at about 15 m O.D., during the construction of the railway (Wood 1863, p. 126) and to the north of the town dark-brown clay has been recorded beneath the surface sand at levels between 23 and 30 m O.D. in ditches between Clephanton [NO 550050], Easter Grangemuir [NO 547041] and Kilrenny. Clay formerly recorded at a clay pit [NO 551051] near Clephanton proves, however, to be till. Between Clephanton, Easter Grangemuir, Inch [NO 534037] and the Dreel Burn a series of boreholes have proved up to 14 m of interbedded clay, sand and gravel overlying till and rock. North of a line joining Wester Grangemuir [NO 538041] and Easter Grangemuir several boreholes encountered up to 5·5 m of clay with scattered stones, which lies at or near the surface at a level of about 30 m O.D.; farther south the clay is almost everywhere buried beneath up to 8 m of sand. In the bed of the Dreel Burn, at locality NO 53730339, dark-brown clay can be seen resting on till.

J.I.C., I.H.F.

KILRENNY TO KINKELL NESS

Seaward-sloping benches interpreted as raised beaches are common along this exposed stretch of coast; thin spreads of sand and fine gravel are usually present on them. The landward limit of the deposit is in general between 29 and 33 m O.D. (Fig. 28) but pronounced back-features are rare.

Details

From Kilrenny to Crail a thin deposit consisting mainly of sand and gravel rests on a well-marked seaward-sloping terrace interpreted as a raised beach. The landward limit of the deposit is defined locally by a back-feature at about 30 m O.D. An area of moundy topography on the terrace above Caiplie [NO 590052] may mark a residual mass of glacial material, or a dune deposit. A thin bed of clay among the sands north-east of Caiplie was formerly worked at a small clay pit [NO 591054].

Between Crail and Fife Ness the raised beach broadens out to about a kilometre and the ill-defined landward limit of the deposit lies below 30 m O.D. A thin deposit of sand and fine gravel can be seen resting on till and rock at the top of the cliffs at Roome Bay [NO 620079].

From Fife Ness to Buddo Ness the landward limit of deposition varies irregularly between about 21 and 33 m O.D. The limit is marked by features locally, as at Balcomie [NO 625100], around Cambo House [NO 604114] and from Boarhills to Buddo Ness [NO 559154]. The deposit, of sand and fine gravel, is generally thin. It is exposed locally in cliffs and stream sections, as at Buddo Ness and in the banks of the Kenly Water between Boarhills and the sea shore. At Pitmilly Law [NO 575136] the raised-beach topography is interrupted by some low mounds at about 30 m O.D.; they may be the remnants of glacial deposits reworked by the late-Glacial sea. Small features have been mapped within the raised beach at various levels down to about 15 m O.D., but these are fragmentary and probably mark short periods of erosive activity when local topographical features were revealed as the coast emerged from the sea bed. It is unlikely that individual lengths of feature formed in this way can be correlated with one another, and the existence of the series of eastward-sloping shorelines proposed by Cullingford and Smith (1966) cannot, therefore, be proved. West of Buddo Ness as far as the eastern outskirts of St Andrews the cliffs along the present shoreline rise to about 30 m O.D. and late-Glacial marine deposits are preserved only in one small patch, around Kinkell Ness [NO 538158].   I.H.F., J.I.C.

KINKELL NESS TO EDEN ESTUARY

The stratigraphy of the late-Glacial marine deposits in the immediate area of St Andrews (Fig. 29) has been pieced together from temporary sections and boreholes examined during the years 1961–67. The general succession is as follows:

		Thickness m
4	Terrace gravel and sand	0–3
	Erosion surface	
3	Channel sand	0–2·5
	Erosion surface	
2	Plastic clay	0–16
1	Basal gravel	0–3
	Surface of till or bedrock	

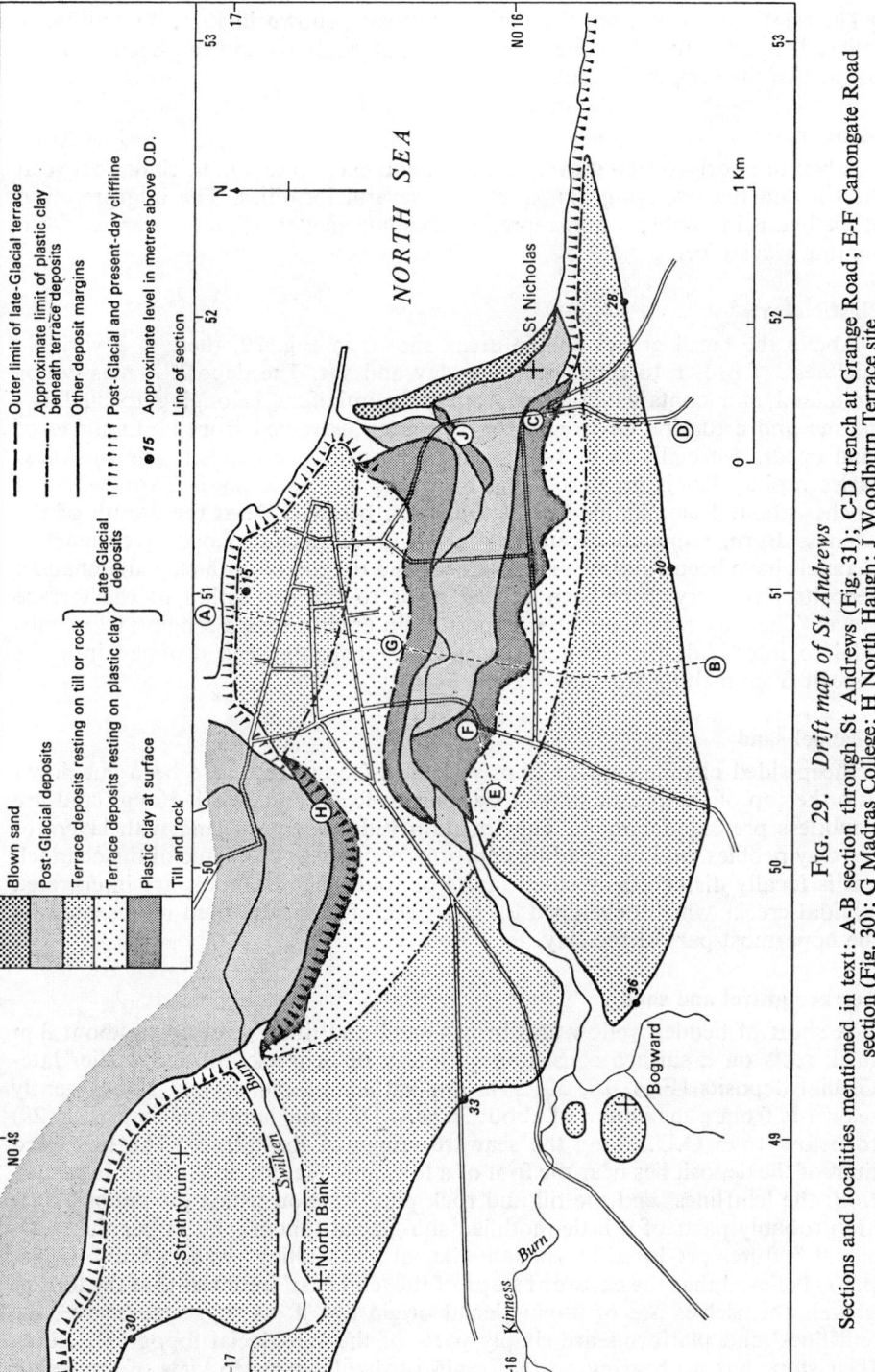

Fig. 29. *Drift map of St Andrews*

Sections and localities mentioned in text: A-B section through St Andrews (Fig. 31); C-D trench at Grange Road; E-F Canongate Road section (Fig. 30); G Madras College; H North Haugh; J Woodburn Terrace site

The relationship between the various deposits, shown in Figs. 30 and 31, is believed to apply to the whole area between St Andrews and the Eden estuary, not just to the area of Fig. 29.

Basal gravel

A bed of poorly-sorted gravel, sand, silt and clay up to 3 m thick lies between the till and the overlying plastic clay at several localities. The interpretation of the bed is in doubt; it may represent a submarine 'ablation till' derived from floating glacial ice.

Plastic clay

Above the basal gravel, in the areas shown in Fig. 29, there is a variable thickness of brown to reddish-brown clay and silt. The deposit is massive or laminated and contains isolated stones; foraminifera belonging to shallow-marine and estuarine environments have been extracted from it. Laminae of sand occur, especially near the top and bottom of the deposit. The thickness varies rapidly but is greatest along the valley of the Kinness Burn, where it reaches about 16 m. Its base lies about 15 m below O.D. at the mouth of the Kinness Burn, rising to about 21 m O.D. at Canongate Road. Well-marked channels have been cut down into the clay and the top of both clay and channel deposits has everywhere been eroded prior to the deposition of the terrace gravel. The clay may have been deposited in environments ranging from sub-tidal to intertidal; the isolated stones have probably been dropped into the deposit from melting sea-ice.

Channel sand

Steep-sided channels up to 12 m wide and 2·5 m deep have been cut down into the top of the plastic clay at Canongate Road and North Haugh and are doubtless present elsewhere. They contain yellow or grey sand with layers of red clay pebbles, and the bedding conforms roughly to the shape of the channel, but is locally disturbed by small faults or folds. The channels are interpreted as tidal creeks which developed on a surface of intertidal mud represented by the uppermost part of the clay.

Terrace gravel and sand

A sheet of bedded yellow and brown sand and fine gravel up to about 3 m thick rests on a surface consisting variously of bedrock, till and earlier late-Glacial deposits (Figs. 29, 30, 31). The surface of the deposit slopes gently seawards from a maximum of about 36 m O.D. at the landward limit (Fig. 29) to below 15 m O.D. along the seaward fringe of the terrace. The landward limit of the deposit lies near the foot of a low back-feature or 'cliffline' in places. Both the 'clifflines' and the till and rock platforms on which the deposit rests are probably parts of a little-modified sub-glacial topography rather than erosional features produced by the late-Glacial sea. Cullingford and Smith (1966, p. 37) believed that the eastward slope of the 'clifflines' indicates that the terrace gravels themselves are of fluvio-glacial origin but if, as suggested above, the 'clifflines' and platforms are simply parts of the sub-glacial topography, then their slope has no bearing on the origin of the deposits. In view of the stratigraphical position of the terrace gravels, above the foraminiferal plastic clays, the interpretation favoured here is the traditional one, namely that the terrace gravels are raised beaches—diachronous strand-line deposits laid down on platforms consisting partly of pre-existing and little-modified surfaces of rock and till and partly of the eroded surface of earlier late-Glacial deposits.

Details

St Andrews. Up to 3 m of gravel and sand resting on rock were proved in boreholes and exposed in foundations [around NO 510155] on the raised beach south of Lamond Drive, and a similar sequence is exposed in the cliffs [NO 507170 to NO 515167] north of the town. A trench (Fig. 29, C–D) dug alongside Grange Road from New Grange House [NO 51641540] to Dunolly Cottage [NO 51661598] revealed bedded sand and gravel resting on rock along the landward fringe of the raised beach, but to the north red plastic clay, with layers of sand at the top, comes in beneath the terrace gravel and thickens northwards.

Temporary sections along Canongate Road (Fig. 29, E–F) westwards from the junction with Largo Road showed up to about a metre of terrace gravel resting on an erosion surface cut across the top of red plastic clay with sand-filled channels (Fig. 30; Chisholm 1964, p. 55). Foraminifera were found in the clay about 40 m west of Largo Road, at a level of about 10 m O.D.: the single species represented here, *Elphidium selseyense* (Heron-Allen and Earland) was identified by Dr. J. R. Haynes. The clay increases in thickness towards the valley of the Kinness Burn. The following representative section was measured between two channels at a point about 180 m west of Largo Road:

	Thickness m
Gravel, fine, grading up into soil at top	1·0
Erosion surface	
Clay, brown, weathering red-brown, silty, massive; rare isolated stones	1·3
Silt and fine sand, pale-brown; flat lamination	0·2
Clay as above but well laminated; sandy bands near base; rare isolated stones	1·2
Gravel with silty sand matrix	0·1
Till ..(seen)	1·0

South and south-west of the site just described the clay thins rapidly to zero; in trenches at South Haugh [NO 504160] the clay varies from 0·2 to 1·0 m in thickness, and in trenches over a large area west of the abattoir [NO 50351580] terrace gravel lies directly on till. A short distance to the east a borehole [NO 50741597] proved clay 3·7 m, on gravel 0·3 m, on clay 3 m, on sandstone.

On the north side of the Kinness Burn, at the site [NO 50791638] of an extension to Madras College (Fig. 29, G and Fig. 31), boreholes and pileholes proved the following sequence:

	Thickness m	Approximate level above O.D. m
Terrace top		17
Sand and gravel (with soil and made ground at top)	2 to 10	15 to 7
Plastic clay and silt, brown, some parts laminated	13	2
Bound gravel ..(seen)	3	

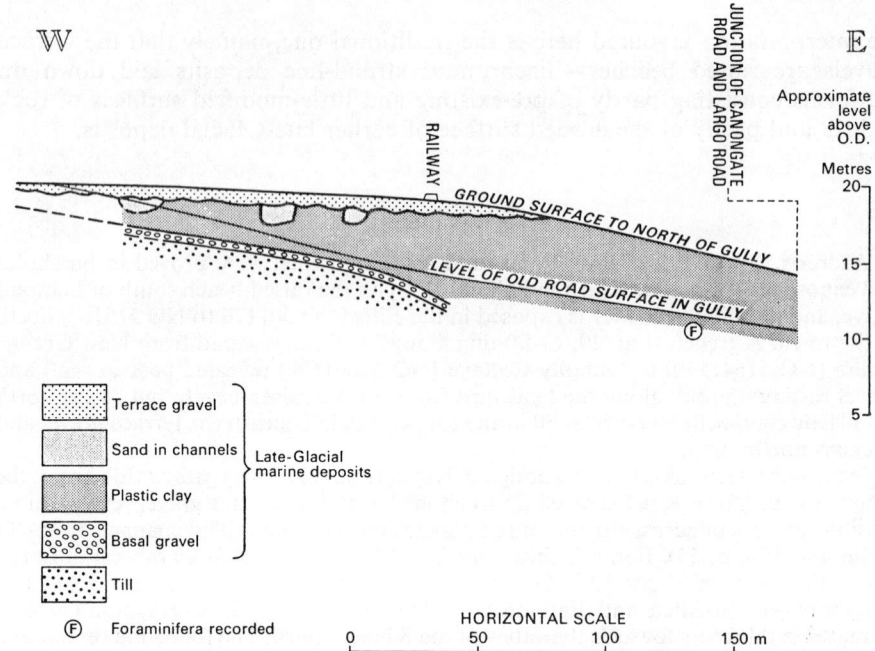

FIG. 30. *Diagrammatic horizontal section along Canongate Road west of junction with Largo Road, St Andrews*
(Based on road and sewer excavations, 1963)
Line of section shown on Fig. 29, E-F

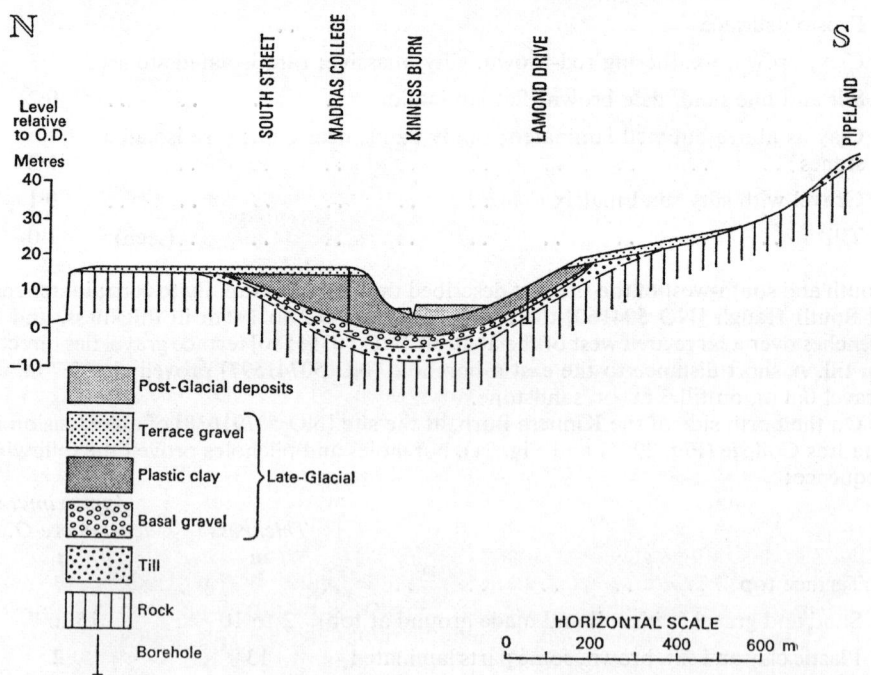

FIG. 31. *Diagrammatic horizontal section through drift deposits at St Andrews*
Line of section shown on Fig. 29, A-B

The top gravel is less than 3 m thick in most of the boreholes and this layer must represent the terrace deposit; an extra 7 m or so proved at the north end of the site may represent a channel deposit. Specimens of brown silty clay and silt obtained during piling operations contained isolated stones and showed massive and laminated textures. They also contained shallow-marine to estuarine foraminifera, among which Dr. J. R. Haynes identified *Elphidium clavatum* Cushman, *E. incertum* (Williamson) and *Protelphidium anglicum* Murray, and the ostracod *Rabilimis mirabilis* (Brady), identified by Dr. R. C. Whatley.

Along the outer margin of the high raised beach terrace at North Haugh [NO 500166] (Fig. 29, H) the following sequence was proved in boreholes:

	Thickness m	Approximate level above O.D. m
Terrace top		19
Sand with fine gravel bands	up to 3	16
Plastic clay	4	12
Surface of till or bedrock		

Subsequent excavations revealed the existence of sand-filled channels up to 3 m wide by 2 m deep, cut into the top of the plastic clay. The bedding in the channel deposits is disturbed by small faults and folds. The clay is reddish-brown in colour, poorly laminated, and yielded the following foraminifera (identified by Dr. J. R. Haynes): *Elphidium selseyense*, *E. sp.* and *Guttulina dawsoni* Cushman and Ozawa, and the ostracod *Rabilimis mirabilis*, identified by Dr. R. C. Whatley. Differential thermal analysis shows that the clay consists of about 80 per cent. illitic or micaceous material. The plastic clay is exposed at a point [NO 51521619] on the south bank of the Kinness Burn, near Woodburn Terrace, beneath a thin sequence of post-Glacial deposits (Fig. 29, J; p. 249), and clay was said to have been encountered to a depth of at least 23 m (i.e. to about 15 m below sea-level) in a borehole nearby (Walker 1864). The fauna recorded from this area by Walker consists mainly of recent temperate-zone forms, and these were probably collected from the post-Glacial deposits; the arctic bivalve species, recorded as *Tellina calcarea* Gmelin, is more likely to have been collected from the top part of the underlying late-Glacial clays.

St Andrews to Guard Bridge. The raised beach, with its deposit of terrace gravel, can be traced westwards from St Andrews as far as Kincaple. The landward limit of the deposit is marked by cliff-like features at levels between 30 and 33 m O.D. Reddish-brown plastic clays have been mapped to seaward of the raised beach platform between Easter Kincaple [NO 474179] and the River Eden [NO 453188]. The stratigraphical relationship between these deposits and the terrace gravels is probably the same as that found at St Andrews.

In an old clay pit [NO 45321878] at Guard Bridge the clay is about 3 m thick, with sand-filled channels cut down into it. The base of the clay can be seen resting on till in the river bank nearby. At Seafield clay pit [NO 470184], which lies at about 15 m O.D., 8 m of stratified clay with isolated stones were formerly seen resting on till, and starfish remains were recorded (Geikie 1902, p. 305) but only 2 m of clay can be seen now. The clay contains shallow-marine to estuarine foraminifera, identified by Dr. J. R. Haynes as *Elphidium clavatum* and *Pyrulina sp.*, and the ostracods *Cyamocytheridea punctillata* Brady and *Cytheropteron montrosiense* Brady, Crosskey and Robertson, which were identified by Dr. R. C. Whatley. J.I.C.

Deposits of the Post-Glacial Marine Transgression and Regression

The post-Glacial deposits described in this section are those laid down in marine environments or in areas close to the sea between the end of the Upper Dryas Period (pollen Zone III) and the present day, but deposits still accumulating or being reworked are dealt with separately (pp. 250–251). During late-Glacial and early post-Glacial times in eastern Scotland (Sissons, Smith and Cullingford 1966) the level of the sea was falling relative to the land surface, and reached a minimum during the Boreal period. The post-Glacial (or Flandrian) transgression, which followed, reached its maximum extent during the Atlantic period, and the sea then withdrew gradually towards the position it now occupies.

The most complete post-Glacial sequences recorded in Fife (Chisholm 1971) lie just outside the boundary of the present district, in Stratheden and around Tentsmuir. These sequences consist of two parts separated by an erosion surface. The deposits below the erosion surface were laid down before the transgression and during its early stages, and consist of the Sub-Carse Peat with overlying grey planty clays. Radiocarbon dates show that the Sub-Carse Peat at a site on the margin of Tentsmuir accumulated between about 9945 B.P. and 7605 B.P. (Chisholm 1971, p. 103). The erosion surface was formed during the transgression. In exposed coastal areas it was cut as the shoreline advanced over the land surface and in places a cliffline marks the landward limit of erosion; in estuaries and other protected inlets, it was produced by scour in tidal channels. The deposits above the erosion surface were laid down in near-shore, intertidal and supratidal environments during the transgression and regression and have been accumulating up to the present day. In some areas the deposits below the erosion surface are absent, and the upper deposits rest directly on older strata.

In east Fife, deposits of the lower subdivision are known only in the St Andrews area and near Elie, but beach deposits belonging to the upper subdivision are widespread. Along the most exposed parts of the coast these consist of up to about 3 m of shelly gravel and sand resting on a narrow platform cut into rock and backed by a cliff—the familiar Low or '25-foot' Raised Beach (Geikie 1902, pp. 313–316). Around parts of Largo Bay and St Andrews Bay, however, the deposit consists of up to 12 m or more of sand with a surface topography of dunes and this tends to obscure the form of the raised beach.

The maximum level reached by the post-Glacial sea is indicated fairly accurately by the height of the deposit margin in areas protected from wave action and, more roughly, by the height of the foot of the cliff in exposed areas. In the Stratheden–Tentsmuir area (Chisholm 1971) the sea reached a level of about 8·9 m O.D., and this figure fits reasonably well on the 'Main Post-Glacial Shoreline' graph for the Forth–Tay area (Sissons, Smith and Cullingford 1966, fig. i). Along most of the exposed east Fife shoreline, however, the level at the foot of the post-Glacial cliff varies irregularly between about 6 and 9 m O.D., and the exact level of maximum submergence is difficult to determine.

Details

Largo to Elie. Around Largo Bay the surface of the marine deposits is predominantly sandy. Dune topography is widely developed and this obscures the form of the raised beach, especially around Earlsferry and Elie. A section formerly exposed on the shore at Wood Haven [NT 49609972] (Brown 1867, pp. 619–620) provided the most complete sequence of post-Glacial deposits in the area. The section was as follows:

	Thickness m
Blown sand	1·2 to 1·8
Shelly shingle	'thin'
Peat with wood and *Phragmites* remains	0·12 to 0·25
Late-Glacial deposits, with root traces at top	not stated

The peat bed was recorded (Brown 1867, p. 620) below blown sand in Elie railway cutting [NO 495002] and beneath sand in the town of Elie itself (Wood 1863, p. 127); it is believed to be the same as the 'submerged forest' peat of Largo Bay (see below). Both probably correlate with the Sub-Carse Peat of early post-Glacial age, and should be distinguished from a younger shelly peat formerly exposed near Elie Station (p. 250).

The area south-west of Kilconquhar Station provides a discontinuous section through some of the post-Glacial deposits (Brown 1867, pp. 621–626). The lowest bed is the 'submerged forest' peat (Fleming 1830; Brown 1867, pp. 622, 625–626; Etheridge 1881, pp. 110–111) that lies beneath a thin cover of modern beach material on the foreshore in Largo Bay and Shell Bay [NO 460003]. The peat varies up to 1·2 m thick and contains the remains of tree stumps in position of growth. Farther inland, in the banks of the Cocklemill Burn [NO 461009 to NO 467012], and at a higher stratigraphical level than the peat bed, raised beach deposits consisting of bedded shelly sand and gravel appear to overlie interbedded clays, silts and sands with well-preserved marine shells. Both deposits are overlain by blown sand (Brown 1867, pp. 621–622; Etheridge 1881, pp. 107–110; Bell 1890, 1893). The finer deposits presumably accumulated in a sheltered intertidal environment, perhaps to landward of a beach bar which is now represented by the overlying raised beach deposit.

Elie to St Andrews. Between Elie and St Andrews the marine deposits occupy a narrow discontinuous raised beach platform cut mainly into rock and the old cliff line can be recognised in many places, complete with sea stacks and caves. The deposit consists of up to about 2 m of bedded gravel and sand containing littoral shells (Geikie 1902, pp. 313–315).
J.I.C., I.H.F.

St Andrews to Guard Bridge. In the valley of the Kinness Burn on the south side of St Andrews a small carse-like terrace of silt and clay has been laid down behind a protective bar of shell-sand and dunes. A section [NO 51521619] near Woodburn Terrace (Fig. 29, J) shows 2 m of soil and recent alluvium resting on 1·8 to 2·1 m of grey silty clay with planty layers, overlying reddish-brown late-Glacial clay. A section recorded in the same area by Walker (1864) showed about 1·5 m of shelly raised beach material resting on over 1·4 m of planty clay with shells. The nature of the fossils suggests that the deposits are of post-Glacial age, although a thick bed of clay encountered in a well nearby, and thought by Walker to belong to the same sequence, is more probably late-Glacial (p. 247).

Immediately west of St Andrews, the post-Glacial deposits are mainly sandy, forming a raised beach backed by a well-marked cliff. In boreholes and excavations at North Haugh [NO 501167] shelly sand up to 3 m thick rests on a rock platform so that, in this area at least, the earlier post-Glacial deposits have been overlapped by the beach sand. A short distance to the north, at Pilmour Links, a spit of dune sand (p. 251) marks the eastern limit of the Eden estuary.

Farther west, beyond Coble Shore [NO 468194] the surface deposit becomes silty and the raised beach passes laterally into the carse-like terrace of Stratheden (Chisholm 1971, pp. 92–94). The eastern limit of the estuary at the time of maximum submergence may have been defined by a small spit of sand dunes at Coble Shore. On the north side of the estuary (Chisholm 1971, p. 100) a full sequence of post-Glacial deposits is developed, but on the south side the sequence is thinner, and may be incomplete. A section [NO 46411897] west of Coble Shore shows the following strata:

	Thickness m
Sand, erosion surface at base	1·0
Clay, grey, silty and sandy	0·4
Clay, dark-brown, peaty	0·1
Sand, muddy, with pebbles at base	0·2
Clay, red-brown (late-Glacial) (seen)	0·9

The peaty clay may represent the Sub-Carse Peat, or a stratigraphically higher peaty horizon such as that recorded near Moonzie Mill (Chisholm 1971, pp. 94–97). The grey clays are also present beneath the beach sand on the east side of Coble Shore [NO 469191] and form the floor of the river channel [NO 480194] not far away. J.I.C.

Freshwater Alluvium and Peat

Freshwater alluvium is found along the courses of rivers and streams and at the sites of former lochs (Geikie 1902, pp. 332–335). It consists of brown and grey mud, sand, and loam, with lenses of gravel in stream alluvium and rare marly beds among the lake deposits. Peat forms a part of the alluvial sequence in some areas, especially at the sites of former lochs, but is only mapped separately from the alluvium where it predominates over other lithologies.

Details

Narrow discontinuous strips of alluvium lie along the courses of many small burns in east Fife, and there are larger patches along the Dreel Burn east of Balcaskie [NO 52670365] and along the Ceres Burn between Ceres and Pitscottie. Patches of alluvium that probably mark the sites of former lochs are to be found south-west of Branxton [NO 40830652], north and west of Clockmadron [NO 42700874], north of Larennie [NO 44400977], at the west end [NO 465113] of Cameron Reservoir, north of Cassindonald [NO 46471219], east of Newbigging of Blebo [NO 430135], south of Lochend [NO 51391394], west of Priorletham [NO 497127], east of Lathockar Mill [NO 49480952], and west of Kippo Plantation [NO 564107]. At Clockmadron, Larennie, Cassindonald and Priorletham the alluvium is overlain in places by peat. Peat containing land and freshwater shells lies near the limit of the post-Glacial raised beach around Elie Station [NO 496001] (Scott 1890; Bennie and Scott 1893). At the eastern end of the railway cutting nearby the peat was found to be interbedded with dune sand (Wood 1863, p. 127; Brown 1867, p. 620). I.H.F., J.I.C.

Blown Sand

The largest areas of blown sand lie around the shores of Largo Bay where dunes cover most of the post-Glacial raised beach between Lundin Links and

Kincraig [NO 465002]. There are smaller deposits at Earlsferry and at Elie. Inland, dunes have developed on the surface of sandy late-Glacial deposits between Muircambus [NO 468024] and Dunotter [NO 461030]. On the rocky coast between Elie and St Andrews, only small patches of blown sand occur, at Balcomie Golf Links [NO 630103], Cambo Sands [NO 604123] and St Nicholas [NO 51811595]. Beyond St Andrews, however, larger areas of dune sand occur. At Pilmour Links [NO 496185] there are lines of dunes parallel to successive, relatively recent, shorelines, and there are also 'blowout' dunes orientated parallel to the prevailing south-westerly wind direction. I.H.F., J.I.C.

References

BELL, A. 1890. Notes on the marine accumulations in Largo Bay, Fife. *Proc. R. Phys. Soc. Edinb.*, **10**, 290–297.

—— 1893. On a deposit in Largo Bay. *Proc. R. Phys. Soc. Edinb.*, **12**, 22–26.

BENNIE, J. and SCOTT, A. 1893. The ancient lake of Elie. *Proc. R. Phys. Soc. Edinb.*, **12**, 148–170.

BROWN, T. 1867. On the arctic shell-clay of Elie and Errol. *Trans. R. Soc. Edinb.*, **24**, 617–633.

CHAMBERS, R. 1848. *Ancient sea-margins*. Edinburgh: Chambers.

CHISHOLM, J. I. 1964. In *Summ. Prog. geol. Surv. Gt Br.* for 1963.

—— 1966. An association of raised beaches with glacial deposits near Leuchars, Fife. *Bull. geol. Surv. Gt Br.*, No. 24, 163–174.

—— 1971. The stratigraphy of the post-Glacial marine transgression in north-east Fife. *Bull. geol. Surv. Gt Br.*, No. 37, 91–107.

CULLINGFORD, R. A. and SMITH, D. E. 1966. Late-Glacial shorelines in eastern Fife. *Trans. Inst. Br. Geogr.*, **39**, 31–51.

ETHERIDGE JUN., R. 1881. Notes on the post-Tertiary deposits of Elie and Largo Bay, Fife. *Proc. R. Phys. Soc. Edinb.*, **6**, 105–112.

FLEMING, J. 1830. Notice of a submarine forest in Largo Bay, in the Firth of Forth. *Q. Jnl Sci.*, **1**, 21–29.

GEIKIE, A. 1902. The geology of eastern Fife. *Mem. geol. Surv. Gt Br.*

KNOX, J. 1954. The economic geology of the Fife Coalfields, Area III. *Mem. geol. Surv. Gt Br.*

SCOTT, T. 1890. Preliminary notes on a post-Tertiary freshwater deposit at Kirkland, Leven, and Elie, Fifeshire. *Proc. R. Phys. Soc. Edinb.*, **10**, 334–345.

SISSONS, J. B., SMITH, D. E. and CULLINGFORD, R. A. 1966. Late-Glacial and Post-Glacial shorelines in south-east Scotland. *Trans. Inst. Br. Geogr.*, **39**, 9–18.

WALKER, R. 1864. On clays, containing fossils, near St Andrews, with remarks on some of the latter. *Ann. Mag. Nat. Hist.*, (3) **14**, 200–209.

WOOD, W. 1863. On the strata discovered in making the east of Fife extension railway. *Proc. R. Phys. Soc. Edinb.*, **3**, 125–128.

Chapter 15
ECONOMIC GEOLOGY

COAL

No coal has been mined in east Fife since 1945 and there is at present no likelihood of a resumption of operations. The coal seams are subject, more than is usual in Scotland, to the occurrence of steep dips, sharp folds, faults, igneous intrusions and volcanic necks, all of which make mining more difficult. The most easily obtainable coal has already been extracted, generally on a small scale and in rather a haphazard manner. Most of the workings are very old (those at St Monance go back at least to the 17th century) and consequently such information as is available is incomplete, lacking in detail and liable to be inaccurate. Some of the seams are undoubtedly thick however and considerable reserves of coal still exist; the possibility of opencast operations or of a resumption of mining cannot therefore be totally discounted. Seams have been worked in all the major subdivisions of the Carboniferous, except possibly the Passage Group, and one of the notable features of the coals in east Fife is their good development in the Lower Limestone Group, which generally in Scotland has few if any workable coals. Indeed the most extensively worked seam in east Fife is the Largoward Splint Coal, which is now known to lie within this group. Details of the development of all the important seams are given in the relevant stratigraphical accounts.

LIMESTONE

Quarrying of limestone in east Fife was formerly extensive and in some cases mining was resorted to as the amount of overburden increased; all activity however has long since ceased. Most of the quarries were in the St Monance Brecciated Limestone (the Hurlet Limestone of Geikie 1902), which is generally between 2·5 m and 3·5 m thick. The Castlecary Limestone was quarried at Hatton, where it is more than 4·5 m thick, and the Mid Kinniny Limestone was quarried at Balcarres, where it shows local thickening to 3 m. Some of the 'limestones' (probably bedded dolomites, see p. 15) in the Calciferous Sandstone Measures near Crail have also been worked. There is no definite evidence of quarrying of the St Monance White Limestone, though Cumming (1936, p. 349) thought that some of the coral limestone presumed to be this bed, which is exposed on the foreshore [NT 50179960] near the Wadeslea neck, had been worked. The St Monance Brecciated Limestone is still partly visible in several of the quarries and samples of the other two limestones are also still obtainable. Details of these quarries are given in Chapters 4 and 6. Three chemical analyses from surface exposures of the St Monance Brecciated Limestone are quoted, under the name of Charlestown Station Limestone, by Muir and others (1956, pp. 47–48, 112–114: sample nos. SL 37, 47 and 236). These authors also provide one analysis (SL 237) of the St Monance White Limestone and another (SL 34) which is probably of the Mid Kinniny Limestone, although they refer to it as the Charlestown Station Limestone (see Fig. 8 for correlation of these limestones). All these analyses show that the limestones are dolomitised to varying degrees. Another three analyses were made of the St Monance Brecciated Limestone in the Higham Borehole (Forsyth and Chisholm 1968, p. 73) which is less than 500 m from the old quarries at Radernie (for which no analysis is available). They too indicate varying degrees of dolomitisation, with the percentage of magnesia varying from 1·5 to 15·4, and less than 2 per cent. of acid-insoluble material. The bedded dolomites formerly quarried near Crail are no longer visible and no analyses are known, but many similar, though generally thinner, beds are exposed in the coast section.

Ironstone

Both blackband and clayband ironstones are known to occur in the Carboniferous rocks, but the only seam known to have been mined is the Denhead Blackband Ironstone, which is found in the Lower Limestone Group between the St Monance Little and Charlestown Main limestones (see p. 72). It is of local occurrence only and was regarded as exhausted by about 1866.

Oil-Shale

No attempt to use any of the shales in the Calciferous Sandstone Measures as oil-shales has been made during this century and the only record of working in east Fife is that at Pitcorthie, 4 km west of Crail (Geikie 1902, p. 119).

Building Stone

The sandstones in the Calciferous Sandstone Measures have been quarried, mainly for local building stone, at many places in east Fife. The quarries are now all disused and many of the smaller ones especially have been partly or totally obscured. The largest are situated west of St Andrews in the Kemback, Nydie, Knock Hill and Strathkinness areas and these quarries are mostly still open. They clearly provided large amounts of good stone, much more than enough for local requirements. They are also known to have been worked for several hundred years: Nydie Quarry was already in existence in the 13th century when it provided stone for Balmerino Abbey (Smith and Johnson 1951). At Kemback Wood and Knock Hill the quarries were extended southwards in the form of mines as the amount of overburden increased. The sandstones which were worked are in the Pittenweem and Sandy Craig beds. A few sandstones from higher in the Carboniferous have also been quarried on a small scale. Various doleritic, basaltic and basanitic rocks have been worked, probably mainly for roadstone (see below) or kerb stones, but in some places, e.g. Colinsburgh, these rocks have also been used for buildings.

Roadstone

The Kilbrackmont–Baldutho sill of olivine-basalt of Dalmeny type is quarried for roadstone at Baldutho [NO 500063] and was formerly quarried at Kilbrackmont. These are the two largest roadstone quarries in east Fife but numerous smaller disused ones exist in other sills of olivine-dolerite and allied types, in both sills and dykes of quartz-dolerite and in some of the neck intrusions. Most of these quarries are still open so that samples could easily be obtained for testing. The total reserves are very large, but are dispersed over a large number of discrete intrusions.

Mineral Veins

A vein of lead ore was described by McDonald (1795, pp. 305–306) from Myreton (now Morton) of Blebo. Apparently large masses of the ore were found on the surface in 1722. A vein was opened up nearby and found to be about 60 cm thick 'containing spar and other vein stuff, mixed with large spots or flowers of fine ore'. About one kilometre to the west 'a nest of the purest lead ore', containing lumps up to 152 kg in weight, was subsequently discovered, with a 30-cm vein a little below it. The latter was said to have contained 'a rib of pure metal' 7 cm to 18 cm thick. Working was, however, abandoned because of the hardness of the rock. According to Anderson (1841, p. 421) various unsuccessful attempts were made later to work the ore but were abandoned because of the 'disturbed and ruptured state of the strata' and the occurrence of igneous intrusions or necks. Anderson identified the ore as galena and placed the second vein north, not west, of the first one. He regarded it as a continuation of the latter but he also gave the trend as north-easterly; the two statements do not appear to be compatible. A small shaft [NO 43221472], now partly filled in, has been noted near Morton of Blebo. It is believed to mark the site of one of these old lead mines. The tip beside it contains baryte as well as baked sandstone and shale.

Sand and Gravel

The glacial and fluvio-glacial sand and gravel deposits in east Fife have not hitherto been exploited to any significant extent. In general they are probably rather thin for quarrying operations on a large scale, but thicknesses of up to 11·5 m have been encountered in boreholes in the Lundin Links area. Sufficient amounts for local requirements could be obtained from the deposits north of Lundin Links, north-east of Ceres, around Colinsburgh and west or north of Dunino. Some gravel and a good deal of sand are locally present near the coast in the late-Glacial marine and estuarine deposits, but the sand is mainly fine-grained and liable to be associated with silt and clay. The dune-like deposits west of Muircambus [NO 468024], however, could provide considerable quantities of sand. All these deposits are discussed in greater detail in Chapter 14. Some of the less well-cemented thick white sandstones in the Calciferous Sandstone Measures might be soft enough to crush for sand. No analyses are available, but their appearance suggests that the average silica content may well be high. The belt with little or no superficial cover that runs parallel to the Kenly Water on its south-eastern side between Boarhills Station and Dunino, and an area 1 to 2 km east of Dunino on either side of the Kenly Burn both appear to offer ample scope for quarrying with little overburden and mainly sandy sequences. The disused sandstone quarries in the Kemback–Strathkinness area are another possible source of silica sand.

Brick-Clay

The principal source of brick-clay in east Fife has been the clays in the late-Glacial marine and estuarine deposits, which have been worked at various places including Seafield (west of St Andrews), Caiplie (between Crail and Anstruther) and around Elie. The deposit formerly worked at Clephanton (near Anstruther) is however thought to have been at least partly till. None of these deposits has been worked for a considerable time and most of the pits are now obscured, although red laminated clay with stones and gravelly layers is still visible at Seafield [NO 470184], where 7·6 m of stratified brick-clay with scattered stones was formerly seen to rest on till. Boreholes around Easter Grangemuir [NO 547041] suggest that these clays can be up to 5·5 m thick in that area.

Semi-Precious Stones

Pyrope garnets occur in parts of the agglomerate of the Elie Ness neck and are known locally as 'Elie rubies'. Heddle (1901, p. 48) provided two chemical analyses and reported that pyrope garnets also occur in one of the basaltic intrusions in the Ruddons Point neck and in two basaltic dykes about 1½ km east of Elie Ness. The latter observation was confirmed by Cumming (1936, p. 353). Balsillie (1927, p. 489) divided Heddle's collection of these garnets into two groups by colour and refractive index. Colvine (1968) carried out another analysis and found that the Elie garnets are a rare type of pyrope, low in calcium and chromium but relatively rich in titanium.

Water Supply

The Calciferous Sandstone Measures contain numerous thick, rather poorly cemented sandstones. They are commonest in the Fife Ness and Sandy Craig beds but occur in the other subdivisions as well. These sandstones appear likely to be reasonably good aquifers. A number of bores have already been put down and have yielded up to 10 litres per second but it is probable that larger supplies of potable, but possibly rather hard, water could be obtained from deeper boreholes of larger diameter. Not much is known about the water-yielding capacity of the rest of the Carboniferous. The succession contains a high proportion of sandstone and appreciable quantities of water should be available locally. A few bores which were mainly in Lower Limestone Group strata have been tested: the largest yield, at least 49 litres per second, may however be exceptional as the bore may have tapped water from either old workings or shattered rock in a fault zone. Thick sandstones occur locally in the Limestone Coal Group (one such having proved capable of yielding 21·5 litres per second) but as old mine workings are widespread the water may be highly mineralised. Drilling in the areas where necks or dolerite sills are numerous would be highly speculative: the sills appear to be very irregular and are unlikely themselves to produce much water unless a fissured zone is encountered. For details reference should be made to Jackson (1967). I.H.F.

References

ANDERSON, J. 1841. On the geology of Fifeshire. *Trans. Highl. agric. Soc. Scotl.*, **7** (new series), 376–431.
BALSILLIE, D. 1927. Contemporaneous volcanic activity in east Fife. *Geol. Mag.*, **64**, 481–494.
COLVINE, R. J. L. 1968. Pyrope from Elie, Fife. *Scott. Jnl Geol.*, **4**, 283–286.
CUMMING, G. A. 1936. The structural and volcanic geology of the Elie–St Monance district, Fife. *Trans. Edinb. geol. Soc.*, **13**, 340–365.
FORSYTH, I. H. and CHISHOLM, J. I. 1968. Geological Survey boreholes in the Carboniferous of east Fife, 1963–4. *Bull. geol. Surv. Gt Br.*, No. 28, 61–101.
GEIKIE, A. 1902. The geology of eastern Fife. *Mem. geol. Surv. Gt Br.*
HEDDLE, M. F. 1901. *The mineralogy of Scotland.* Vol. 2. Edinburgh: Douglas.
JACKSON, N. P. D. 1967. Records of wells in the areas of Scottish one-inch geological sheets Kinross (40), North Berwick (41), Perth (48) and Arbroath (49). *Water Supply Papers geol. Surv. Gt Br., Well Catalogue Series.*
MCDONALD, J. 1795. Parish of Kemback. *The [old] statistical account of Scotland*, **14**, 297–310. Edinburgh.
MUIR, A., HARDIE, H. G. M., MITCHELL, R. L. and PHEMISTER, J. 1956. The limestones of Scotland. Chemical analyses and petrography. *Mem. geol. Surv. spec. Rep. Miner. Resour. Gt Br.*, 37.
SMITH, R. F. and JOHNSON, N. M. 1951. Quarry to abbey: an ancient Fife route. *Proc. Soc. Antiq. Scotl.*, **83**, 162–167.

Appendix 1

LIST OF CARBONIFEROUS FOSSILS

BY R. B. WILSON

THE following list of fossils is based mainly on specimens in the Geological Survey Collections which have been collected in recent years and for which accurate localities are known. Collections previously made in the area were also consulted both in the Geological Survey Collections and in other institutions, but in most cases the localisation of the material was found to be inadequate. Messrs. P. J. Brand and D. K. Graham have been responsible for the bulk of the collecting of the specimens listed and the identifications have been made by the writer. More detailed information of the exact provenance of the fossils is available in the records of the Palaeontology Department of the Institute of Geological Sciences, Edinburgh.

The following abbreviations are used for the stratigraphical units: CSM = Calciferous Sandstone Measures, LLG = Lower Limestone Group, LCG = Limestone Coal Group, ULG = Upper Limestone Group, PG = Passage Group, MCM = Middle Coal Measures, UCM = Upper Coal Measures. The numbers refer to the localities at which the fossils were found and are listed on pp. 263–268. Qualifications of the determinations such as cf., aff. or ? are given after the relevant locality number.

PLANTAE
Algae CSM 14, 41, 102, 120, 121.

PORIFERA
Hyalostelia parallela (McCoy) CSM 51; LLG 59?, 89, 138.
sponge? CSM 83, 117, 129.

SCYPHOZOA
'*Conularia*' *sp*. CSM 15, 86, 126, 127; PG 91.

ANTHOZOA
Aulophyllum cf. *fungites* (Fleming) LLG 89.
Caniniid CSM 113.
Chaetetes sp. CSM 15, 63?
Cladochonus sp. LLG 142?
Clisiophyllum sp. CSM 113; ULG 67.
Diphyphyllum cf. *lateseptatum* McCoy CSM 134.
Hexaphyllia sp. CSM 105.
Lithostrotion junceum (Fleming) CSM 15, 113, 131, 134.
Syringopora? CSM 15.
Zaphrentites sp. CSM 131; LLG 5, 132, 133, 134, 139, 142; ULG 71.
solitary coral indet. CSM 4, 22, 107, 113, 128, 131; LLG 38, 89, 104, 110, 131, 133, 134, 139, 141.

BRYOZOA
Fenestella spp. CSM 4, 15, 20, 22, 23, 26, 30, 45, 56, 63, 82, 83, 86, 101, 102, 103, 105, 106, 107, 113, 117, 118, 123, 124, 125, 126, 127, 128, 131, 134; LLG 31, 38, 59, 89, 92, 104, 110, 114, 130, 132, 134, 137, 138, 139, 141, 142.
Penniretepora sp. LLG 65.

Polypora dendroides McCoy LLG 38, 130, 138.
Rhabdomeson sp. CSM 4.
trepostomatous bryozoa CSM 4, 22, 113, 117, 120, 129, 131, 134; LLG 10, 31, 38, 89, 104, 133, 134, 138, 141, 142.
bryozoa indet. CSM 45, 82, 83, 102, 113, 120, 131, 134; LLG 59, 89, 133, 134.

ANNELIDA
Serpula sp. CSM 41?
Spirorbis sp. CSM 1, 3, 40, 41, 52, 53, 68, 120, 121, 127, 135; PG 91a.
Serpuloides carbonarius (McCoy) CSM 4, 15, 19, 20, 83?, 105, 106, 107, 113, 127, 131; LLG 5, 13, 130, 131, 138, 139, 142; PG 91a.

BRACHIOPODA
Antiquatonia insculpta (Muir-Wood) LLG 114 cf.
Avonia youngiana (Davidson) CSM 4, 134; LLG 59, 134.
A. sp. LLG 131.
Brachythyris ovalis (Phillips) LLG 59.
B. triradialis (Phillips) LLG 142.
B. sp. CSM 113?
Buxtonia sp. CSM 4?, 19, 20, 105, 107, 124?; LLG 38, 50, 110, 130, 134, 137, 138, 141.
Cleiothyridina cf. *fimbriata* (Phillips) LLG 89?
Composita cf. *ambigua* (J. Sowerby) CSM 45, 113, 131; LLG 10, 12, 89, 131, 134.
C. sp. CSM 15, 113?; LLG 50.
Crania sp. LLG 133.
Crurithyris urii (Fleming) CSM 19?, 30, 105, 113 cf.; LLG 48, 49, 110, 131, 134.
C. sp. CSM 22?, 25, 51?, 106?, 107; LLG 50?, 131?, 141?
Dielasma cf. *hastatum* (J. de C. Sowerby) CSM 30?, 51, 131; LLG 37, 50, 131, 132?
Echinoconchus elegans (McCoy) CSM 113.
E. punctatus (J. Sowerby) CSM 4?, 20?, 23?, 82, 107 cf., 113, 128, 129.
Eomarginifera lobata (J. Sowerby) LLG 5, 37?, 89 cf., 132 cf., 133 cf., 138?, 141 cf.
E. longispina (J. Sowerby) CSM 19 cf.; LLG 137?
E. praecursor (Muir-Wood) CSM 105?, 107 cf.
E. setosa (Phillips) CSM 105 cf.; LLG 37, 93, 131, 134?, 139?, 142.
E. sp. CSM 4, 20, 23, 30, 134?; LLG 31, 38, 64, 89, 104, 110, 119, 131, 132.
Gigantoproductus sp. giganteus (J. Sowerby) group LLG 131, 134.
Girtyella saccula (J. de C. Sowerby) CSM 37.
Hustedia sp. CSM 124?; LLG 131, 138?
Lingula mytilloides (J. Sowerby) CSM 4, 6, 8, 15, 19, 23, 29 cf., 30, 45, 51, 83, 101, 103, 105, 107, 113, 120, 121, 125, 127, 128; LLG 11, 38, 48, 49, 50, 60, 62, 89, 93, 131, 134, 137, 138, 142; LCG 80; ULG 66, 72, 73, 75; PG 91a.
L. squamiformis Phillips group CSM 4, 8, 18, 33, 34, 36, 51, 70, 100, 102, 105, 106, 107, 109, 113, 123, 127, 128, 134; LLG 5, 12, 13, 31, 37, 47, 59, 90, 115, 130, 131, 132, 133, 134; LCG 42, 43, 44, 80.
Linoprotonia sp. CSM 82, 106, 113?; LLG 131; PG 91a.
Orbiculoidea nitida (Phillips) CSM 4, 8, 33, 45, 51, 63, 82, 83, 88, 102, 103, 105, 113, 117, 118, 123, 125, 127, 131, 134; LLG 5, 38, 59, 89, 110, 131, 133, 134, 142; ULG 75.
Orthotetoid CSM 15, 45, 63, 82, 86, 95, 100, 117, 118, 125, 127, 128, 129; LLG 12, 49, 60, 131, 134, 141, 142; ULG 66, 71; PG 91a.
Phricodothyris sp. CSM 51?, 82, 113; LLG 59, 93, 110, 132?
Pleuropugnoides sp. CSM 36, 82, 83, 86, 95, 96, 113, 118, 131; LLG 10, 12, 37, 50, 114, 131.
Plicochonetes sp. LLG 60, 134.
Productus cf. *carbonarius* de Koninck LLG 131; ULG 66?; PG 91.
P. concinnus J. Sowerby CSM 63; LLG 37?

P. cf. *redesdalensis* Muir-Wood CSM 4?, 45, 82, 86?, 107?, 113, 117, 118?, 131; LLG 10, 50, 94, 104?, 131, 133, 134?
P. sp. CSM 4, 15, 21, 22, 26, 30, 51, 56, 79, 83, 95, 96, 102, 103, 105, 106, 113, 117, 123, 125, 127, 128, 129; LLG 31, 38, 59, 64, 132, 134, 137, 138.
Promarginifera trearnensis Shiells LLG 38, 89, 134, 138.
Pugilis cf. *pugilis* (Phillips) CSM 56, 82, 113, 131; LLG 12.
P. sp. CSM 4?, 127; PG 91a?
Pugnax cf. *pugnus* (Martin) CSM 45?, 63, 127?
Punctospirifer cf. *scabricosta* North CSM 117.
P. sp. CSM 45?
Rhipidomella cf. *michelini* (Léveillé) LLG 60, 138, 142.
Rhynchonelloid CSM 15, 25, 45, 56, 63, 120, 126, 127; LLG 132, 133, 134, 137.
Rhynchopora sp. CSM 131?; LLG 59.
Rugosochonetes sp. LLG 49, 59, 60, 64, 132, 133, 134, 139, 142.
Schizophoria cf. *resupinata* (Martin) CSM 18, 30?, 113?, 134; LLG 5, 37?, 38?, 50, 65, 89, 110, 119?, 131, 132, 133, 134, 139?, 141, 142?; ULG 71?
Spirifer sp. crassus de Koninck group CSM 117, 131.
S. cf. *trigonalis* (Martin) LLG 59, 89, 132, 134?, 142.
S. sp. CSM 56, 82, 129?; LLG 5, 12, 31, 38, 59, 64, 131, 134.
Spiriferellina cf. *insculpta* (Phillips) CSM 83.
S. cf. *perplicata* (North) CSM 15; LLG 59, 132?, 134.
Tornquistia cf. *polita* (McCoy) LLG 38, 110, 119, 131, 134, 138, 139, 141.

GASTROPODA

Bellerophon anthracophilus Frech ULG 66.
B. aff. *costatus* J. de C. Sowerby CSM 117.
B. randerstonensis Weir CSM 120, 122.
B. sp. CSM 85?; LLG 132.
Donaldina cf. *grantonensis* (Longstaff) CSM 127.
D. sp. CSM 45a, 51, 88?, 113, 117, 120, 127?; LLG 37, 110.
Euphemites urii (Fleming) CSM 51 cf., 102 cf., 117 cf., 123 cf.; LLG 13, 57, 59, 60, 65, 90, 119, 132, 139, 142; ULG 66 cf.
E. sp. CSM 4, 8, 19, 29, 30, 34, 45, 56, 86, 88, 100, 101, 102, 106, 117, 118, 123, 127; LLG 32, 38?, 48, 49, 59, 60, 93, 115, 133, 134, 138; ULG 72.
Girtyspira sp. CSM 1.
Glabrocingulum beggi E. G. Thomas CSM 105?
G. sp. CSM 23, 30; LLG 60, 142; ULG 66.
Hesperiella thomsoni (de Koninck) LLG 5, 132, 139, 141.
Hypergonia kirkbyi Donald CSM 120.
Latischisma globosa E. G. Thomas ULG 66.
Meekospira? CSM 30, 120.
Naticopsis? scotoburdigalensis (Etheridge jun.) CSM 81, 101, 117, 127.
N. cf. *variata* (Phillips) CSM 20, 23, 103, 128?; LLG 11?, 50, 92, 138?, 142.
Platyceras? CSM 103.
Pseudozygopleura robroystonensis (Longstaff) LLG 142.
P. cf. *rugifera* (Phillips) CSM 4, 19, 30, 105, 107; LLG 11?, 89.
P. sp. CSM 88, 127; LLG 110, 139.
Retispira concinna (Weir) CSM 102.
R. decussata (Fleming) CSM 4, 23, 30, 103, 117 cf.; LLG 5, 38, 59, 64, 94, 134, 137, 138, 142.
R. cf. *densistriata* (Weir) CSM 4, 19, 20.
R. striata (Fleming) CSM 21 cf., 30, 51, 103, 113; LLG 60, 104, 119, 134, 137, 138, 139.
R. undata (Etheridge jun.) CSM 45a, 85.

R. sp. CSM 7?, 15, 26?, 45, 102, 105, 107, 113, 117, 123, 125, 127; LLG 48, 49, 57, 60, 132, 141; ULG 72, 75.
Straparollus (*Euomphalus*) *carbonarius* (J. de C. Sowerby) CSM 23?; LLG 5, 59, 60, 119, 131, 138, 139, 141, 142.
Strobeus sp. ULG 71.
Tropidocyclus oldhami (Portlock) LLG 49.

SCAPHOPODA
Dentalium s.l. CSM 4, 117, 125, 127; LLG 5, 50, 59, 60, 114, 119, 132?, 134?, 138, 139, 141; ULG 73.

BIVALVIA
Acanthopecten sp. LLG 138.
Actinopteria persulcata (McCoy) CSM 4, 22, 29?, 30, 51?, 63, 105, 106, 107, 113; LLG 37, 50, 94, 134, 137.
Anthraconaia sp. MCM 98.
Anthraconauta cf. *phillipsii* (Williamson) UCM 97.
A. cf. *tenuis* (Davies and Trueman) UCM 97.
Anthracosia acutella (Wright) MCM 98.
A. aquilinoides (Tchernyshev) MCM 98.
A. atra (Trueman) MCM 98.
A. concinna (Wright) MCM 98.
A. fulva (Davies and Trueman) MCM 98.
Anthracosphaerium sp. MCM 98.
Aviculopecten aff. *interstitialis* (Phillips) CSM 128; LLG 110?, 138.
A? semicircularis (McCoy) LLG 59.
A. semicostatus (Portlock) LLG 110.
A. subconoideus Etheridge jun. CSM 4?, 7, 21, 33 cf., 45, 51 cf., 56 cf., 100, 102, 107?, 117, 123, 124, 125?, 127, 128?
A. subconoideus Hind *non* Etheridge jun. CSM 88, 102, 123, 127.
A. aff. *tabulatus* (McCoy) CSM 120.
A. sp. CSM 8, 15, 22, 30, 86, 101, 103, 105, 117, 120, 121, 126; LLG 38, 50, 59, 93, 104, 131, 134, 137, 141; LCG 57; ULG 72.
Aviculopinna mutica (McCoy) CSM 117; LLG 60, 119.
Caneyella? LLG 47.
Carbonicola antiqua Hind CSM 40, 55?, 68.
C. elegans (Kirkby) CSM 2, 9, 16, 77, 120.
C. sp. 54?, 111, 123?
Cardiomorpha hindi? Wilson CSM 4.
'*Ctenodonta*' *pentonensis* Hind CSM 22; LLG 137, 138?
Curvirimula scotica (Etheridge jun.) CSM 3, 24?, 47, 51, 52, 53, 84, 129.
Cypricardella cf. *acuticarinata* (Armstrong) CSM 4.
C. cf. *rectangularis* (McCoy) CSM 4, 19, 20, 22, 30, 103; LLG 5, 31, 38, 59, 89, 92, 93, 104, 114, 119, 131, 132, 137, 138, 139, 141, 142.
Dunbarella sp. CSM 107, 118; LLG 38?, 59?, 131?, 141.
Edmondia arcuata (Phillips) LLG 38.
E. maccoyi Hind CSM 86 cf., 113 cf.; LLG 89, 132, 142.
E. cf. *senilis* (Phillips) 4, 45, 51, 107, 113, 117; LLG 59, 138?
E. sulcata (Fleming) CSM 107?; LLG 38, 132.
E. cf. *unioniformis* CSM 4, 8, 45, 56, 79, 86, 102, 120?, 124, 127; LLG 115, 138.
E. sp. CSM 7, 15, 20, 22?, 30, 63, 95, 100?, 103, 105, 113?, 125; LLG 114, 141; PG 91a?
Euchondria neilsoni Wilson LLG 49.
Leiopteria hendersoni (Etheridge jun.) CSM 45a, 113, 118, 122.
L. thompsoni (Portlock) CSM 4 cf.; LLG 114.
L. sp. CSM 45, 63, 107?, 113, 117?, 129; LLG 31, 47, 50, 134.

Limipecten dissimilis (Fleming) CSM 4, 20, 86 cf., 113 cf., 129 cf.; LLG 50, 89, 94, 104, 110, 114, 138, 139.
L. sp. CSM 15, 22?, 26?, 45, 100?, 107?, 117, 123, 126?, 127?, 131?; LLG 38, 130, 134, 141, 142.
Lithophaga lingualis (Phillips) CSM 1, 4, 20, 30?, 40?, 82, 105, 106, 113, 129, 131; LLG 38, 59?, 89, 104, 133, 138, 139, 141.
Modiolus latus (Portlock) CSM 69, 124 cf.
M. sp. CSM 103?, 107, 131; LLG 47?, 115?; ULG 72?, 75.
Myalina sublamellosa (Etheridge jun.) CSM 100, 102, 117, 125?, 127.
M. sp. CSM 4, 8, 26, 27, 113, 131; LLG 38, 50, 115, 138; LCG 57; ULG 72?
Naiadites cf. *crassus* (Fleming) CSM 6, 7, 34, 39, 45, 45a, 88, 100, 102, 117?, 120?, 127, 136.
N. obesus (Etheridge jun.) CSM 1, 6, 9, 40, 46, 55, 81, 101, 102, 117, 120, 121, 123, 127, 135, 136, 140.
N. sp. CSM 14, 16, 51?, 52?, 70?, 111?, 122; LCG 80; MCM 98.
Nuculopsis gibbosa (Fleming) CSM 22, 23, 103, 117; LLG 5, 50, 132, 133, 134, 139.
Palaeolima cf. *simplex* (Phillips) 4, 20, 26?, 45, 82, 96?, 106, 107, 113, 118?; ULG 66?
Palaeoneilo laevirostrum (Portlock) CSM 26, 30, 86 cf., 103, 106 cf., 107 cf., 123 cf., 126 cf., 127 cf.; LLG 38, 59?, 89, 104, 131, 132, 133, 137 cf., 138, 139, 141.
P. luciniformis (Phillips) CSM 105, 113; LLG 48, 60, 61, 132.
P. mansoni Wilson CSM 21, 30, 51, 103, 105; LLG 12, 13, 31, 32, 38, 59, 61, 131, 132, 133, 134; ULG 71.
Paleyoldia macgregori Wilson CSM 30; LLG 38, 114.
Parallelodon elegans Hind non McCoy LLG 132.
P. cf. *semicostatus* (McCoy) CSM 103, 107, 113; LLG 5, 38, 59, 60, 89, 104, 110, 114, 130, 132, 137, 138, 139, 141, 142.
Pernopecten sowerbii (McCoy) CSM 26?, 51; LLG 37?, 131?, 137?, 138.
P. sp. CSM 4, 30, 86, 105, 106, 107, 113, 131; LLG 59, 114, 130; ULG 66.
Polidevcia attenuata (Fleming) CSM 15, 22, 23, 30, 45, 45a, 51, 56, 79, 86, 88, 100, 101, 102, 103, 105, 106, 113, 117, 123, 126, 127, 131; LLG 31, 38, 50, 59, 60, 61, 115, 119, 134, 139, 141, 142; ULG 66, 71.
P. attenuata traquairi (Etheridge jun.) CSM 88.
Posidonia becheri Bronn CSM 29, 107.
P. corrugata (Etheridge jun.) LLG 13, 32?, 48, 59, 134.
P. corrugata gigantea Yates LLG 48, 49, 60.
Prothyris cf. *scotica* Wilson CSM 105; LLG 132.
P. sp. CSM 101, 103?, 107?; LLG 114?
Pterinopectinella granosa (J. de C. Sowerby) CSM 107?
P. sp. CSM 19, 20, 22, 23?, 51?, 105, 106; LLG 110?, 131, 134, 138.
Pteronites angustatus McCoy CSM 56, 88 cf., 102 cf., 124, 127 cf.
P. sp. CSM 117, 127.
Sanguinolites cf. *abdenensis* Etheridge jun. CSM 4, 34?, 70, 105, 106, 107, 113; LLG 59, 104, 115.
S. clavatus (Etheridge jun.) CSM 4, 8, 15, 29, 45a, 51, 85, 88, 100, 101, 102, 103 aff., 105, 106, 107, 123, 125, 127; LLG 37 cf., 59?; ULG 75?
S. costellatus (McCoy) CSM 30, 102 cf., 113, 127; LLG 13, 32, 48?, 50, 57, 133.
S. cf. *plicatus* (Portlock) CSM 30, 45, 51, 105, 113, 131, 134?; LLG 60, 132, 133?, 134, 141.
S. striatolamellosus de Koninck CSM 4, 113; LLG 38, 50, 60?, 131, 132, 138 cf., 141.
S. striatus Hind CSM 4 cf., 51?, 88, 100, 102, 113?, 123, 127.
S. subplicatus Kirkby CSM 118, 120, 121.
S. tricostatus (Portlock) CSM 45, 127?; LLG 38?, 137.
S. sp. variabilis McCoy group CSM 25?, 30, 34, 127; LLG 115; ULG 72, 75.
S. sp. CSM 24, 26 sp. nov., 33 sp. nov., 85 sp. nov., 103, 113, 117.
Schizodus obliquus (McCoy) LLG 138, 142.

S. pentlandicus (Rhind) CSM 6, 7, 8, 39, 117, 120, 121, 125?
S. salteri Etheridge jun. CSM 4, 105, 107; LLG 131 cf.
S. sp. CSM 4, 15, 23, 30, 33, 45, 51, 88, 101, 103, 106, 113, 117, 123, 124, 127; LLG 32, 38, 57, 59, 119, 131, 132?, 133, 134, 137, 139; ULG 71.
Sedgwickia gigantea McCoy LLG 38, 132, 141.
S. suborbicularis Hind LLG 139, 141.
S. sp. CSM 4, 15, 56?, 63, 101?, 102, 107, 117?, 125, 138.
Solemya primaeva Phillips CSM 4 cf.; LLG 114.
Solenomorpha minor (McCoy) CSM 30, 106, 107; LLG 38, 57, 59, 132, 138, 139, 141; ULG 72.
S. sp. CSM 105; LLG 92.
Streblochondria cf. *anisota* (Phillips) CSM 4.
S. cf. *concentricolineata* (Hind) CSM 15, 86.
S. cf. *elliptica* (Phillips) CSM 30, 45?, 82; LLG 138.
S. sp. CSM 4, 19, 20, 45, 56, 63, 79, 96, 102?, 103, 105, 106, 107, 113, 117, 118, 123?, 127, 128, 129; LLG 38, 45, 49, 50, 59, 104, 110, 114, 119, 130, 131, 132, 137, 139, 141, 142; ULG 66.
Streblopteria ornata (Etheridge jun.) CSM 30, 34, 51, 103, 109, 113, 131; LLG 32, 37, 57, 59, 90, 115, 131, 132, 133, 134.
S.? redesdalensis (Hind) CSM 118, 127.
S. sp. CSM 105, 107; LLG 134, 139?
Sulcatopinna flabelliformis (Martin) CSM 4, 105?
Wilkingia elliptica (Phillips) CSM 8 cf., 21, 45, 56?, 117, 127; LLG 38?
W. maxima (Portlock) CSM 30.
W. sp. CSM 30, 45; LLG 60, 137.

CEPHALOPODA
Catastroboceras sp. CSM 127?; LLG 38, 50, 60, 104, 138.
Cyrtoceras sp. LLG 65.
Stroboceras sp. CSM 120?
Orthocone nautiloid CSM 4, 20, 21, 26, 30, 45, 45a, 51, 56, 63, 69, 77, 86, 88, 100, 101?, 102, 103, 105, 107, 113, 117, 120, 121, 122, 123, 125, 127; LLG 31, 48, 49, 57, 59, 60, 90, 92, 119, 132, 133, 134, 139; ULG 66, 71, 73; PG 91a.
Nautiloid indet. CSM 77, 105, 121, 122; LLG 134, 137, 139.
Beyrichoceratoides truncatus (Phillips) LLG 94.
B. sp. CSM 106.
goniatites indet. CSM 15, 19, 26, 29, 30, 51, 103, 105, 107, 113, 118, 127?; LLG 12, 48, 49, 59, 60, 93, 119, 130, 134?, 138; ULG 66, 71.

ARTHROPODA
trilobite fragments CSM 22, 30, 103; LLG 5, 38, 60, 62, 65, 89, 93, 110, 114, 119, 130, 131, 134, 138, 139, 141, 142.
Dithyrocaris sp. CSM 103.
'*Estheria*' *sp.* CSM 113, 127.
Leaia cf. *salteriana* Jones CSM 116.
L. sp. CSM 117.
Ostracods CSM 1, 2, 3, 6, 9, 18, 24, 27, 33, 39, 46, 51, 53, 69, 86, 100, 102, 103, 111, 117, 120, 121, 127, 128, 135; LLG 47, 59, 60, 134, 142.

CRINOIDEA
Crinoid fragments CSM 4, 7, 8, 15, 18, 19, 20, 21, 22, 23, 25, 26, 30, 33, 36, 39, 40, 45, 51, 56, 63, 79, 82, 86, 101, 102, 103, 105, 106, 107, 113, 117, 123, 124, 125, 126, 127, 129, 131, 134; LLG 5, 10, 11, 12, 37, 38, 47, 49, 50, 59, 60, 62, 64, 65, 89, 93, 104, 110, 130, 131, 132, 133, 134, 137, 138, 139, 141, 142; ULG 50, 66, 67, 71.

ECHINOIDEA
Archaeocidaris urii (Fleming) CSM 4, 19, 22, 39, 45, 51, 56, 102, 117, 125, 131; LLG 38, 50, 60, 65, 89, 119, 134, 137, 138, 139, 141, 142.

PISCES
Rhizodus hibberti (Agassiz) CSM 2.
R. sp. CSM 83.
fish fragments CSM 1, 6, 9, 24, 40, 41, 46, 51, 53, 103, 111, 117, 120, 121, 123, 127, 129, 135.

LOCALITY LIST

1. Anstruther, shore from West Haven to Anstruther Harbour, [NO 56300308 to 56540348]; CSM, including Anstruther Wester Marine Band, Anstruther Beds.
2. Anstruther, shore between Anstruther and Cellardyke harbours, [NO 56970339 to 57630373]; CSM, Anstruther Beds.
3. Ardross, shore 663 m E.25°N. of, [NO 51680107]; CSM, Pathhead Beds.
4. Ardross, shore 200 m S.41°W. of, [NO 50710060]; CSM, Ardross Limestone, Pathhead Beds.
5. Bankhead, quarry 180 m S.S.E. of, [NO 40760899], LLG, St Monance Brecciated Limestone.
6. Billow Ness, shore 550 m E.12°S. of Kirklatch to 860 m E.11°N. of Kirklatch (Anstruther Wester Haven), [NO 55900271 to 56200298]; CSM, including Billow Ness Marine Band, Anstruther Beds.
7. Brigton, temporary exposures in ditch 310 m E.5°N. of, [NO 51661160]; CSM, Brigton Marine Band, Pittenweem Beds?
8. Buddo Ness, shore west of, [NO 55811553]; CSM, West Sands Marine Band, Pittenweem Beds.
9. Caiplie, shore from Innergellie Haven to Hermit's Well, [NO 58150447 to 59940578]; CSM, Anstruther Beds.
10. Cairnsmill Den, 90 m W.44°S. of Cairnsmill, [NO 49591477]; LLG, St Monance Brecciated Limestone.
11. Cairnsmill Den, 585 m S.26°E. of Bogward, [NO 49411505]; LLG, as 10.
12. Cairnsmill Den, right bank immediately under Cairnsmill, old collection; LLG, as 10.
13. Callange No. 1 Borehole, 440 m north of Callange, [NO 42751174]; LLG, Mid Kinniny Limestone.
14. Cambo Burn, right bank, opposite East Newhall, [NO 60291080]; CSM, Randerston No. 3 Limestone?, Anstruther Beds.
15. Carnbee Den, 300 m S.29°W. of Gordonshall, [NO 53120667]; CSM, Carnbee Marine Band, Pathhead Beds?
16. Cellardyke, section from harbour to Innergellie Haven, [NO 58080425]; CSM, Anstruther Beds.
17. Ceres Burn, unnamed tributary of, 150 m S.21°W. of Newbigging of Blebo, [NO 42961331]; CSM, shale above St Monance White Limestone?
18. Ceres Burn, same tributary, 45 m downstream from 17; CSM, as 17.
19. Ceres Burn, same tributary, 395 m downstream from 17, [NO 42601314]; CSM, Newbigging Marine Band, = Upper Ardross Limestone?
20. Ceres Burn, exposure near same tributary, 740 m E.13°S. of Easter Pitscottie, [NO 42341318]; CSM, Blebo Hole Marine Band, = Lower Ardross Limestone?
21. Ceres Burn, exposure immediately south of same tributary, 780 m E.15°S. of Easter Pitscottie, [NO 42371314]; CSM, Blebo Hole Marine Band, = Lower Ardross Limestone?

APPENDIX 1

22. Claremont Burn, 70 m east of road from Colinsburgh to Strathkinness, [NO 46131479]; CSM, Claremont Cottage Marine Band, = Lower Ardross Limestone?
23. Claremont Burn, 510 m downstream from same road, [NO 46571484]; CSM, Claremont Cottage Marine Band, = Lower Ardross Limestone?
24. Claremont Burn, 250 m downstream from road to Denbrae Farm, [NO 4761 1521]; CSM, Pathhead Beds?
25. Claremont Burn, 155 m east of Denbrae House, [NO 47731547]; CSM, Denbrae House Marine Band, Sandy Craig Beds?
26. Claremont Burn, 670 m E.27°N. of Claremont Farm, [NO 46571485]; CSM, Claremont Cottage Marine Band, = Lower Ardross Limestone?
27. Claremont Burn, small tributary of, 55 m N.19°W. of Denbrae Farm, [NO 47541513]; CSM, Denbrae Farm Marine Band, = West Braes Marine Band?
28. Claremont Burn, 55 m N.22°W. of Denbrae Farm, high on bank, [NO 47531513]; CSM, as 27.
29. Claremont Burn, 305 m W.5°S. of Denbrae Farm [NO 47201503]; CSM, as 27.
30. Claremont Geological Survey Borehole (1966); 825 m W.27°S. of Claremont, [NO 45181419]; CSM, Pathhead Beds.
31. Cordies Mealling No. 8 Borehole, 230 m N.43°E. of Cordies Mealling, [NO 46430738]; LLG, upper part.
32. Craighall Burn, 850 m E.38°N. of Hall Teasses, [NO 42010959]; LLG, Lower Kinniny Limestone.
33. Craighall Burn, right bank, 220 m W.36°N. of Newbigging of Craighall, [NO 41421041]; CSM, St Monance White Limestone, Pathhead Beds.
34. Craighall Burn, right bank, 575 m S.33°W. of Craighall Mains, [NO 40511054]; CSM, as 33.
35. Craighall Burn, 64 m N.38°W. of Newbigging of Craighall, [NO 41521030]; CSM, above St Monance White Limestone, Pathhead Beds.
36. Craighall Burn, 90 m S.35°W. of Harleswynd, [NO 41171046]; CSM, below St Monance White Limestone, Pathhead Beds.
37. Craighall Burn, high on left bank, 310 m W.14°S. of Craighall, [NO 40271051]; CSM, just below St Monance Brecciated Limestone, Pathhead Beds.
38. Craighall Quarry, 350 m S.6°E. of Craighall Mains, [NO 40861067]; LLG, St Monance Brecciated Limestone.
39. Craig Hartle, shore north of fault at, [NO 57931497 to 57341489]; CSM, Craig Hartle North Marine Band and below, Anstruther Beds.
40. Craig Hartle, section between fault south of Craig Hartle and Pitmilly Burn mouth, [NO 58001480 to 57881478]; CSM, Craig Hartle South Marine Band and below, Anstruther Beds.
41. Craighead, shore at, [NO 63591005]; CSM, Fife Ness Beds.
42. Craigtoun Hospital Geological Survey Borehole (1966); LCG. 290 m E.44°N. of Craigtoun Hospital, [NO 48271463].
43. Craigtoun Park No. 1 Geological Survey Borehole (1966); LCG. 300 m W.26°S. of Craigtoun Hospital, [NO 47751426].
44. Craigtoun Park No. 2 Geological Survey Borehole (1966); LCG. 300 m W.6°N. of Craigtoun Hospital, [NO 47721442].
45. Crail, shore from The Pans to Crail Harbour, [NO 60240667 to 61080742]; CSM, Pans Marine Band to Crail Harbour Marine Band, Pittenweem Beds.
45a. Crail, shore 140 m S.15°E. of The Pans, [NO 60560641]; CSM, Barns Mill Marine Band, Anstruther Beds.
46. Crail, shore east of Roome Harbour [NO 62030787]; CSM, Anstruther Beds.
47. Den Burn, left bank, 550 m S.3°W. of Balmakin, [NO 48800436]; LLG, Seafield Marine Band.
48. Den Burn, right bank, 985 m S.24°W. of Balmakin, [NO 48430401]; LLG, Charlestown Main Limestone.

49. Den Burn, right bank, 1010 m S.22°W. of Balmakin, [NO 48450396]; LLG, Charlestown Main Limestone.
50. Den Burn, 1150 m S.15°W. of Balmakin, [NO 48540380]; LLG, St Monance Brecciated Limestone.
51. Denork Geological Survey Borehole (1966); 695 m W.42°S. of Claremont, [NO 45401409]; CSM, upper part, Pathhead Beds.
52. Dreel Burn, right bank, 385 m S.26°W. of Wester Kellie, [NO 50870565]; CSM, Pathhead Beds.
53. Dreel Burn, right bank, 940 m W.13°N. of Balcormo, [NO 50580447]; CSM, Pathhead Beds?
54. Dreel Burn, 715 m E.38°S. of Crawhill, [NO 56280351]; CSM, Anstruther Beds.
55. Dunino Burn, right bank, 155 m downstream from Limelands Avenue, [NO 53700967]; CSM, Pittenweem Beds.
56. Dunino Burn, 120 m downstream from Limelands Avenue, [NO 53660968]; CSM, Chesters Marine Band, Pittenweem Beds.
57. Earlsferry, shore 45 m E.37°N. of Chapel, near Chapel Ness, [NT 48159941]; LLG to LCG, Upper Kinniny Limestone and Johnstone Shell Bed.
58. Earlsferry, shore 415 m E.37°N. of Chapel, [NT 48439966]; LLG, Mid Kinniny to Upper Kinniny limestones.
59. Elie, shore 690 m E.37°N. of Chapel, [NT 48659982]; LLG, Neilson Shell Bed to Mid Kinniny Limestone.
60. Elie, Wood Haven, 1480 m E.10°N. of Chapel Ness, [NT 49569967]; LLG, St Monance Brecciated Limestone to Neilson Shell Bed.
61. Gibliston, unnamed stream 365 m N.12°W. of, [NO 49310588]; LLG, horizon of Upper Kinniny Limestone.
62. Gibliston, old quarry 240 m S.45°W. of, [NO 49210535]; LLG, St Monance Brecciated Limestone.
63. Gordonshall, old quarry 300 m W.17°S. of, [NO 53120664], CSM, Carnbee Marine Band, Pathhead Beds?
64. Greigston No. 4 Borehole, 275 m south of Greigston House, [NO 44741089]; LLG.
65. Hairies Hole, quarry 27 m east of road, 530 m S.37°W. of Bankhead, [NO 40350872]; LLG, St Monance Brecciated Limestone.
66. Hatton Burn, left bank, 675 m N.30°E. of Pilmuir, [NO 39900445]; ULG, Plean Limestones.
67. Hatton Burn, right bank, 330 m W.6°S. of Hatton, [NO 40100437]; ULG, Castlecary Limestone.
68. Kenly Water, shore at west side of mouth of, north of fault, 485 m N.31°E. of Hillhead, [NO 58101449 to 58061440]; CSM, Anstruther Beds.
69. Kenly Water, shore at west side of mouth of, south of fault, 285 m N.23°E. of Hillhead, [NO 58071440 to 58081430]; CSM, including Kenly Mouth Marine Band.
70. Kenly Water, 130 m downstream from railway viaduct at Boarhills Station, [NO 56701338]; CSM, Anstruther Beds?
71. Keil Burn, 140 m W.35°N. of Pitcruvie, [NO 41250474]; ULG, Plean Limestones.
72. Keil Burn, right bank, 80 m W.2°N. of Pitcruvie, [NO 41280466]; ULG, as 71.
73. Keil Burn, 80 m W.36°S. of Pitcruvie, [NO 41290456]; ULG, as 71.
74. Keil Burn, left bank, 285 m W.21°N. of Blindwells, [NO 41520407]; ULG, as 71.
75. Keil Burn, left bank, 730 m S.37°W. of Largo House (ruin), [NO 41590282]; ULG, as 71.
76. Keil Burn, left bank, 670 m E.8°N. of Balcormo, [NO 41710565]; ULG, as 71.
77. Kilminning, shore from The Goats 1536 m S.13°W. of Craighead to Kilminning Castle 1260 m S.2°E. of Craighead, [NO 62870836 to 63250858]; CSM, including Goats Marine Band, Anstruther Beds.

78. Kinaldy Burn, 370 m E. 17°S. of Cutty Hillock, [NO 49930960]; CSM, Sandy Craig Beds?
79. Kincaple, Den Quarry, [NO 45541771]; CSM, Kincaple Marine Band, = Witch Lake Marine Band?
80. Kincraig, shore 565 m S.32°E. of, [NT 47199985]; LCG.
81. Kingsbarns, shore at The Lecks, 1100 m north of Cambo, [NO 60301255]; CSM, Anstruther Beds.
82. Kinness Burn, right bank below New Mill, 800 m west of St Andrews, old collection, locality could not be positively found; CSM, probably the same as 83.
83. Kinness Burn, left bank, 80 m E.12°N. of New Mill, [NO 49841602]; CSM, New Mill Marine Band, Sandy Craig Beds?
84. Kinness Burn, left bank, immediately upstream from New Mill, [NO 49741592]; CSM, Anstruther Beds.
85. Kirklatch, shore 520 m E.14°S. of, [NO 55870270]; CSM, including Chain Road Marine Band; Anstruther Beds.
86. Kittock's Den, cliff on shore 145 m W.20°N. of mouth of, [NO 55281514]; CSM, Witch Lake Marine Band, Pittenweem Beds.
87. Knockhill Farm, pipe trench at, [NO 44211639]; CSM, Pittenweem Beds?
88. Knockhill Quarry, tip material, 720 m E.16°N. of Knockhill Farm, [NO 44821673]; CSM, Knockhill Marine Band, = St Andrews Castle Marine Band?
89. Ladeddie Limeworks, tip material at east end, [NO 44201376]; LLG, St Monance Brecciated Limestone?
90. Ladeddie No. 8 Borehole, 670 m E.1°N. of Drumcarro, [NO 45981295]; LLG, upper part.
91. Largo Burn, 770 m S.25°E. of Largo House (ruin), [NO 42340271]; PG.
91a. Largo Burn, shore section 485 m E.8°S. of, and 1006 m W.17°S. of Broomhall, [NO 42750252]; PG.
92. Largobeath No. 2 Borehole, 550 m E.6°S. of South Cassingray, [NO 48070730]; LLG, upper part.
93. Lathones March, old quarry near, 900 m W.7°S. of Lathockar Mains, [NO 47970933]; LLG, St Monance Brecciated Limestone.
94. Lathones No. 6 Borehole, 570 m E.16°S. of Higham, [NO 47430922]; LLG, lower part.
95. Lochty No. 1 Borehole, 275 m W.14°S. of Lochty, [NO 52280799]; CSM, Sandy Craig Beds?
96. Lochty No. 2 Borehole, 560 m S.21°W. of Lochty, [NO 52330753]; CSM, Sandy Craig Beds?
97. Lower Largo, shore section on east side of Harbour Pier (east of Keil Burn), [NO 41670244]; UCM.
98. Lower Largo, shore section on west side of Harbour Pier (west of Keil Burn), [NO 41360240 to 41300235]; MCM.
99. Lumbo Den, 18 m downstream from road, [NO 48811483]; LLG, Charlestown Main Limestone.
100. Maiden Rock, shore east of, 695 m N.9°W. of Brownhills, [NO 52661582]; CSM, St Andrews Castle Marine Band?
101. Maiden Rock, shore east of, 680 m N.3°W. of Brownhills, [NO 52751581]; CSM, Witch Lake Marine Band.
102. Maiden Rock, shore east of, 720 m N.21°E. of Brownhills, [NO 53031581]; CSM, Witch Lake Marine Band to St Andrews Castle Marine Band, Pittenweem Beds.
103. Mount Melville Cottages Geological Survey Borehole (1966); 565 m N.22°E. of Craigtoun Hospital, [NO 48281496]; CSM, Pathhead Beds.

104. Newbigging of Craighall, old quarry 540 m north of, [NO 41601079]; LLG, St Monance Brecciated Limestone.
105. Newark Castle, shore 210 m S.21°W. of Newark, [NO 51800120]; CSM, Ardross Limestones, Pathhead Beds.
106. Newark Castle, shore 200 m S.26°E. of Newark, [NO 51970122]; CSM, as 105.
107. Newark Castle, shore 295 m E.18°S. of Newark, [NO 52150132]; CSM, as 105.
108. North Bank, unnamed stream 400 m W.24°S. of, [NO 47811065]; CSM, Pathhead Beds?
109. North Bank, unnamed stream 400 m W.10°S. of, [NO 47811075]; CSM, Pathhead Beds?
110. North Cassingray, old quarry 605 m W.32°N. of, [NO 47780865]; LLG, St Monance Brecciated Limestone.
111. Nydie Wood, quarry in, 425 m E.38°N. of Knockhill Farm, [NO 44471680]; CSM, Nydie Marine Band, = Witch Lake Marine Band?
112. Ovenstone Convalescent Home, stream N.E. of, [NO 53830498]; CSM, Ovenstone Marine Band, Pittenweem Beds?
113. Pathhead, shore section from Pathhead to Sandy Craig, [NO 53770212 to 54420223]; CSM, West Braes Marine Band to St Monance White Limestone, Pathhead Beds.
114. Patieshill, unnamed stream 570 m N.6°W. of and 575 m N.3°W. of, [NO 43810894 and 43840894]; LLG, Seafield Marine Band.
115. Patieshill, unnamed stream 435 m E.39°N. of, [NO 44210864]; LLG, horizon of Lower Kinniny Limestone.
116. Pitcorthie No. 6 Borehole, 660 m S.10°W. of West Pitcorthie, [NO 56920632]; CSM, Anstruther Beds.
117. Pittenweem, shore east of, from fault N.E. of harbour to 455 m E.19°S. of Kirklatch, [NO 55220258 to 55780268]; CSM, Cuniger Rock Marine Band to Pittenweem Marine Band, Pittenweem Beds.
118. Pittenweem, shore west of harbour, [NO 55670234]; CSM, Boat Harbour Marine Band, Sandy Craig Beds.
119. Radernie Quarry, old quarry 400 m N.5°E. of Higham, [NO 46970980]; LLG, St Monance Brecciated Limestone.
120. Randerston, shore opposite, east limb of syncline from 765 m E.10°N. of Cambo to 1315 m E.9°S. of Cambo, [NO 61131159 to 61681117]; CSM, Anstruther Beds.
121. Randerston, shore on west limb of syncline from 765 m E.10°N. of Cambo to Cambo Ness, [NO 61121160 to 60861182]; CSM, Anstruther Beds.
122. Randerston, shore immediately west of fence to east of, [NO 61871091]; CSM, Anstruther Beds.
123. Rock and Spindle, shore east of, 1295 m E.25°N. of Brownhills, [NO 53961566]; CSM, Witch Lake Marine Band to St Andrews Castle Marine Band, Pittenweem Beds.
124. Rock and Spindle, in stream high on cliff 960 m E.S.E. of, [NO 54761513]; CSM, Witch Lake Marine Band, Pittenweem Beds.
125. Rock and Spindle, shore 685 m to 730 m E.S.E. of, [NO 54461527 to 54631526]; CSM, Witch Lake Marine Band, Pittenweem Beds.
126. Rock and Spindle, high on cliff 660 m E.S.E. of, [NO 54381522]; CSM, Witch Lake Marine Band, Pittenweem Beds.
127. St Andrews, shore north of harbour pier [NO 51601667]; CSM, West Sands Marine Band to St Andrews Castle Marine Band, Pittenweem Beds.
128. St Andrews, shore between St Nicholas and Maiden Rock, [NO 52361586 to 52311590]; CSM, St Nicholas Marine Band and above, Sandy Craig Beds?
129. St Andrews, shore between St Nicholas and Maiden Rock, [NO 52331593 to 52561587]; CSM, as 128.

130. St Andrews Wells, quarry at, [NO 43851457]; LLG, St Monance Brecciated Limestone?
131. St Monance, shore section between harbour and mouth of St Monance Burn, [NO 52550154 to 52300144]; CSM and LLG, St Monance White Limestone to Charlestown Main Limestone.
132. St Monance, section from core of syncline 65 m east of harbour to harbour wall, [NO 52790158 to 52710158]; LLG, Seafield Marine Band to Mid Kinniny Limestone.
133. St Monance, east limb of syncline east of harbour to swimming pool, [NO 52790158 to 53260176]; LLG, as 132.
134. St Monance, section from swimming pool to Pathhead, [NO 53260176 to 53730209]; CSM and LLG, St Monance White Limestone to Charlestown Main Limestone.
135. Salt Lake, section on shore west of Babbet Ness, [NO 58811423 to 58531440]; CSM, Anstruther Beds.
136. Sypsies Water Borehole, Crail, 455 m S.38°W. of Ribbonfield, [NO 59340829]; CSM, Anstruther Beds.
137. Teasses, old collection from quarry not seen now; LLG, St Monance Brecciated Limestone.
138. Teasses, quarry at, 580 m S.26°E. of Windygates, [NO 40660779]; LLG, St Monance Brecciated Limestone.
139. Wilkieston Quarry, [NO 44951195]; LLG, St Monance Brecciated Limestone.
140. Winchester, cliff 1145 m N.N.W. of, [NO 55021507]; CSM.
141. Windygates, old quarry 395 m N.11°W. of, [NO 40320872]; LLG, St Monance Brecciated Limestone.
142. Woodtop Quarry, 510 m S.31°W. of Windygates, [NO 40070793]; as 141.

Appendix 2

LIST OF GEOLOGICAL SURVEY PHOTOGRAPHS (EAST FIFE)

Taken by Messrs. W. D. Fisher, A. Christie, R. Anderson and I. Bowler

COPIES of these photographs are deposited for public reference in the library of the Institute of Geological Sciences, South Kensington, London SW7 2DE and in the library of the Scottish office, Murchison House, West Mains Road, Edinburgh. Prints and lantern slides are supplied at a fixed tariff on application to the Director: some of these subjects are in colour.

RECENT AND PLEISTOCENE

Post-Glacial raised beach platform and cliff. Easter Kincaple, St Andrews	D 583
Post-Glacial and late-Glacial raised beach platforms and cliffs. Kincraig, Elie	D 1685, 1766–7
Late-Glacial raised beach platform and cliff. Balgove, St Andrews	D 584

CARBONIFEROUS

Bioturbated shaly sandstone. St Monance Bathing Pool	D 541
Dipping sequence cut by small fault. Pathhead	D 543
Strata above Charlestown Main Limestone. Pathhead	D 1062
Strata cut by small faults. Pathhead	D 548
St Monance Brecciated Limestone. Pathhead	D 540
St Monance White Limestone. St Monance (west)	D 537
Folded strata below St Monance White Limestone. St Monance (west)	D 538
Trace-fossils in sandstone below St Monance White Limestone. Pathhead	D 1061
Fossil tree-trunk. Pathhead	D 544
Upper Ardross Limestone and shales, crumpled. Newark Castle, St Monance	D 623–5
Convolute sandstone. Pathhead (east)	D 536
Steeply dipping strata. Pittenweem Bathing Pool	D 542
Carbonate conglomerate. Pittenweem Bathing Pool	D 547
Sequence including nodular dolomitic seatbed. St Nicholas, St Andrews	D 581
Nodular dolomitic seatbed. St Nicholas, St Andrews	D 582
Steeply dipping Pittenweem Beds. Pittenweem (east)	D 535
Cyclic sediments in Pittenweem Beds. Crail (west)	D 552
Sandstone in Craigduff Dome. Coast east of St Andrews	D 563
Nodular dolomitic sandstone. Coast east of St Andrews	D 572
Steeply dipping strata. Coast east of St Andrews	D 578–9
Crail Coves, sandstone in Anstruther Beds. Crail (west)	D 545–6
Cyclic sediments in Anstruther Beds. Roome Bay, Crail	D 551
Kilrenny Mill Musselband. Kilrenny	D 539
Randerston Syncline. Randerston	D 549
Anticline to east of Randerston Syncline. Randerston	D 550

Structure

Folds near Maiden Rock Fault. St Andrews	D 573, 577
Vertical beds near Maiden Rock Fault. St Andrews	D 574
Maiden Rock Fault. Maiden Rock, St Andrews	D 575–6
Craigduff Dome in Pittenweem Beds. Coast east of St Andrews	D 560–1
Plunging folds on coast. St Andrews	D 580
St Monance Syncline. St Monance	D 533–4
Upper Ardross Limestone, crumpled. St Monance	D 623–4
Folding and faulting near margin of Kinkell Ness neck	D 553

Volcanic Necks

Largo Law	D 1097–8, 1765
Basalt passing through basaltic breccia to agglomerate. Ruddons Point	MNS 1742
General views of Kincraig neck, showing dipping tuffs and columnar basalt intrusion	D 1103, 1682–3
Columnar basalt intrusion in Kincraig neck	D 1100–2, 1684
Stacks of agglomerate in Kincraig neck	D 1099
Columnar basanite of Chapel Ness intrusion. Earlsferry	D 1104–5
Basaltic breccia, Elie Ness neck	D 1677
Basalt dykes cutting tuffs and agglomerates, Elie Ness neck	D 1106–7
Basalt bomb in tuffs, Elie Ness neck	MNS 1635
Collapsed bedding in tuffs, Elie Ness neck	MNS 1636
Cross-bedding and wedge-bedding in tuffs, Elie Ness neck	MNS 1637
Brecciation of basalt intrusion, Ardross neck	D 632
Basalt intrusion in agglomerate, Ardross neck	D 631
Tuffisite near margin of Ardross neck	D 630
Tuffisite invading sediments near margin of Ardross neck	D 628
Flow-banding in tuffisite in Coalyard Hill neck. St Monance	D 1680–1
Agglomerate cutting mudstone at margin of Dovecot neck. St Monance	D 629
Dovecot neck cutting Calciferous Sandstone Measures	D 626
Bleached basalt intrusion cutting shales with ironstone bands, near Davie's Rock. St Monance	D 627, 1770
Sandstone invaded by tuffisite at margin of St Monance neck	D 1768–9
Agglomerate and monchiquitic intrusions cutting steeply dipping sediments at margin of St Monance neck	D 1678–9, MNS 1638
General views of Kinkell Ness neck, including the Rock and Spindle	D 554–5, 1672
Rock and Spindle, Kinkell Ness neck	D 556–8, 1673, 1762–4
Basalt intrusion cutting agglomerate, Rock and Spindle	D 1675
Centroclinal bedding and vertical tuffs in Kinkell Ness neck	D 559
Tuffisite in sandstone block, Kinkell Ness neck	D 564
Basalt dyke cutting tuffs in Kinkell Ness neck	D 565
Tuffs and agglomerates, Kinkell Ness neck	D 566–71
Tuffisitic breccia near margin of Kinkell Ness neck	D 1676

INDEX

Accretionary lapilli, 105, 107
Aegirine, 158
—— -augite, 155, 158–160
Agglomerate, 4, 106–107, 109, 165, 167–168, 171–173, 175–177, 182–206, 213–215, 217, 240
Airbow Point, 32
ALLAN, J. K., iii
Alluvium, estuarine, 3, 237–250
——, freshwater, 3, 250
——, marine, 3, 237–250
Analcime-basalt, 206–209, 211, 213; see also monchiquite
—— -basanite, 137–140, **147–149**, 151, 153–156, 160, 206–207
Analcimite, olivine-, 207
Analyses, chemical: olivine-dolerite sill-complex, 150–153; neck intrusions, 208–209
——, modal: olivine-dolerite sill-complex, 155–157
ANDERSON, F. W., iii
——, J., 167, 254
——, R., iv, 269
Annie's Mine, 56
Anorthoclase, xenocrysts of, 207, **214–215**
Anstruther Anticline, 14, 22, 27, **222–223**
—— Beds, 10–12, 15–17, 19–20, **22–23**, 27–36, 38, 42, 55–56, 123–124
—— Borehole, 9, 11, 14, 16, **20–22**, 23, 27, 30, 124
—— —Pathhead coast section, 10, 13, **21–26**, 27–28, 34, 37–38, 44, 57
—— Station, 241
—— Wester, 223
—— —— marine band, 11, 20, **22–23**, 27, 29, 124
—— —— section, 20, **22–23**, 27, 56
Apatite, large xenocrysts of, 214
Appleton Basin, 77–80, 222–223
—— Ell Coal, 80
—— Splint Coal, 80
Arctic marine fauna, at Elie, 239–241
Ardross Castle, 57, 231
—— coast section, 56–57
—— Fault, 2, 12, 51, 56, 60, 171, 174, 181, 187, 190–197, 222–223, **227–228**
—— Limestones, 11, 26, 44–46, 56–57, 127–130, 197; see also Lower Ardross Limestone, Upper Ardross Limestone

Ardross *Lingula* Band, 11, 26, 47, 57
—— neck, 162, 173, 178, 181, **192–194**, 214
Arncroach, 52–54
—— Fault, 222, **227**
Arnsbergian Stage (E_2), 5, 103, 133, 136; see also Upper Limestone Group
Atlantic period, 248

Babbet Ness, 32, 231
—— —— Anticline, 32, **222**
—— —— Fault, 32, 222, **231**
Backbraes, 164
—— Quarry, 69, 70
Back Coal, 12–13, 16, 26, 34, 51, **52–54**, 227
—— —— (of St Monance), 87
Backfield of Ladeddie, 69, 72, 76
Backmuir of New Gilston, 165, 223
Balbuthie, 235, 240
Balcarres, 252
—— Craig, 1, 143–144
—— Den, 56, 63, 68, 70–71, 75–79, 205
—— House, 82, 236
—— Shaft, 92
—— sill, 144
Balchrystie, 93
Balcomie, 4, 7, 16–17, 30, 228, 242
—— Beds, 3, **7–8**, 20–21, 32
—— Golf Links, 251
—— Sands, 198, 207, 217
Balcormo Fault, 222
—— Mains Farm, 109
—— Wood, 109
—— —— Borehole, 103, 105–109, 133–134, 140–141, 164–165, 204
Baldastard, 78–79, 140, 235
—— Index Limestone, 79
—— sill, 139–140, 157–158
Baldinnie, 226
Balduthо Boreholes, 90, 92–93
—— Craig, 138, 143, 226
—— Quarry, 137, 140, 143, 149–150, 152, 156, 253
—— Syncline, 82, 93, 146, **222–223**, 226
Balhousie, 145
—— neck, 181, 204
Ballfield Coal, 94–96
Ballone Fault, 222
Balmain, 106, 204

271

T

Balmain neck, 106, 181, 204
Balmerino Abbey, 253
Balmonth, 55
Balmungo, 50
Balnacarron House, 235
Balniel, 143
—— Den, 68, 177, 205
BALSILLIE, D., 14, 40–41, 44, 46, 48–49, 112, 138–139, 143, 147–148, 162–163, 171, 177–178, 183–184, 190, 198, 200–201, 213–215, 229, 255
Bandirran, 51, 235
Bankhead, 164–165
—— Quarry, 69–70
Barkevikite, 160
Barncraig Coal, 118, 120
Barns Mill, 27, 230
—— —— Marine Band, 11, **28**, 124
Basalt, intrusions in Lundin Links neck, 179–180, 182–183, 207, **216–217;** see also olivine-basalt
Basanite, 4, 137–140, 147–149, 151, 153–156, 160, 172, 188, 196–198, 200–202, 204–209, 210–212, 215
Basanitic intrusions, 4, 113, 154, 179–180
—— volcanism, 179–180
Bell Craig, 35
Belliston, 68
BENNIE, J., 250
BENNISON, G. M., 14, 29, 122
Berryside Burn, 165
Billow Ness, 27, 230
—— —— bathing pool, 19, 23
—— —— Marine Band, 11, 20, **23**, 28, 124
—— —— section, 20, **23**, 27, 33
Biotite-basanite, 147–148, 156, 160
—— xenocrysts, 206, 214
Blackband Ironstone: see Denhead Blackband Ironstone
Black Metals Marine Bands, 83, 86, 93, 97, 99–100, 132–133
Blairhall Main Coal, 83, 87
Blebo, 163
Blebocraigs, 37, 40, 43–44
—— Quarry, 43
Blebo Hole, 236
—— —— Marine Band, 11, 46, **48**, 128
—— House, 166–167
—— Skellies, 166
Blindwells, 113, 178–179
Blinkbonny, 205
Blown sand, 3, 241, 243, 249, **250–251**
Boat Harbour (Pittenweem), 230
—— —— Fault, 20, 25, 222, **231**

Boat Harbour Marine Band, 11, 20, **24–25**, 54, 126–127, 129
Boghall Burn, 165, 204
—— Mine, 78
Bogward neck, 41, 201
Bonfield Quarry, 42
Bonnytown, 236
Boreal period, 6, 248
Boulder clay, 3; see also till
Bowanton Coal, 83, 95, 97
BOWLER, I., iv, 269
BRAND, P. J., iii, 122, 257
Branxton Fault, 163–164, 222, **226**
Brick-clay, 255
Bridgeton Den, 34
Brigton Marine Band, 11, **34,** 126
Brownhills, 168
—— dyke, 167, **168**
BROWN, T., 13, 24, 27–28, 31, 38–39, 57, 62, 122, 124, 239–241, 249–250
Bruntshiels, 82, 165, 226
—— neck, 162, 165, 181, **202**
Buddo Ness, 33, 200, 242
—— —— Fault, 33, 222
—— —— neck, 181, 200
—— —— section, 33–34, 37
Building stone, 253
Bungs of Cassingray, 202
Bush Coal, 119

Cadger's Bridge Fault, 77–78, 222, **226**
Caiplie, 27–28, 242, 255
—— *Lingula* Band, 11, 27–28
—— section, 27–28
Cairn, 145
Cairnhill, 202
Cairn Pit, 91–92
Cairns Den, 41–42, 63, 69
Cairnsmill, 50, 69
—— Burn, 50, 229
Calciferous Sandstone Measures, 2–4, **9–59,** 122–130, 178, 229, 252–255; conditions of deposition, 18; description of sections, 19–57; lithology and facies, 14–17; palaeontology, 122–130; stratigraphy, 9–14; trace-fossils, 19
Caliche, 4: see also cornstone
Callange, 79–80, 85
—— Borehole, 65, 78, 85–86, 94–97, 99, 101, 132–133, 165
Calmy Limestone, 104–106, **108,** 133
Cambo, 4, 32, 36, 242
—— Burn, 1, 36

Cambo Fault, 12, 31, 32, 222, **225**
—— Ness, 8, 21, 32, 225
—— Sands, 7–8, 32, 251
Cameron Burn, 34, 141, 201, 236
—— —— sill, 136, 140–141, 157
Canongate Road, St Andrews, 244–246
Capledrae Marine Band, 107
—— Parrot Coal, 104, 107
Carbonate conglomerate, 4, 7–9, **16**, 17, 24–25, 56
Carbonates, bedded, in Calciferous Sandstone Measures, **15–16**, 24
——, nodular, in Calciferous Sandstone Measures, **16**, 17, 25
Cardenden Smithy Coal, 83, 87–88, 98
Carhurlie Coal, 104–105, **108–109,** 112
Carhurly, 36
—— Burn, 36
Carnbee, 54–55
—— Den, 54
—— Law, 205
—— Marine Band, 11, **54–55,** 129
CARRUTHERS, R. G., iii
Carvenom, 56
Cassindonald, 68, 72, 250
—— Colliery, 70, 72
—— Quarry, 68, 137, 147–148, 151, 153–154, 156
Cassingray Colliery, 77, 80, 82
—— Fault, 77, 139, 143, 164, 222, **226**
—— Shaft, 79
Castlecary Limestone, 103–105, **109–110,** 111, 134, 252
Catcraig dolerite intrusion, 143
—— Quarry, 143, 235
Cathrie sill, 145, 158
Cauldside, 235
Cave Coal (of Denhead), 101
Cellardyke, 225, 231
—— Harbour, 27
—— section, 27
Ceres, 1, 82–84, 94–101, 163, 235–236, 250
—— Burn, 1, 166, 227, 250
—— Coalfield, 83–84, **94–98,** 228–229
—— Coals, 82–83, **94–98**
—— Fault, 2, 12, 60, 84, 94, 101, 163, 166, 181, 222–223, **228–229**
—— Five Foot Coal, 83, 95, **96–97**
—— intrusion, 165–166
—— Little Coal, 83, 95, **97**
—— Lower Four Foot Coal, 83, 95, **96**
—— –Maiden Rock Fault Zone, 36, 69, 101, 136, 142, 174, **227–229**
—— Mill, 236
—— Rum Coal, 83, 95, **96**

Ceres Six Foot Coal, 83, 95, **97,** 101
—— Thick Coal, 83, 95, **97,** 100–101
—— Two Foot Coal, 83, 95, 97–98
—— Upper Four Foot Coal, 83, 95, **97,** 101
——Whin Coal, 83, 95, **96**
Chain Road Marine Band, 11, 20, **23,** 28, 124
—— —— section, 20, **23, 27**
CHAMBERS, R., 233
Chapel Ness, 173
—— —— basanite intrusion, 173, 178–179, 181, **187–188,** 207–209, 211–212
—— —— Fault, 188–189, 222, **224**
CHAPMAN, N. A., 171, 213, 215
Charlestown Green Limestone, 64–65
—— Main Limestone, 61–62, 64–65, 67, 73–74, **75–76,** 131, 252
—— Station Limestone, 60, 63–65, 68
Charleton Borehole, 205
Charlie's Rock, 233
Chemical analyses: see analyses, chemical
Chemiss Coal, 118–120
Chert, 8
Chesters, 35
—— Marine Band, 11, **35,** 126
Chincough Well, 231
CHISHOLM, J. I., iii
Chlorophaeite, 168
Chonetes Limestone, 79
CHRISTIE, A., iv, 269
Chrome-diopside, 184
—— -spinel, 184; *see also* spinel
Clachreid Ha' Wood Quarry, 143
Claremont Borehole, 44–49
—— Burn, 46, 49
—— Cottage Marine Band, 11, 46–47, **49,** 128
Clatto Hill, 1, 44, 202
Clay (late-Glacial), 237–238, 240–247, 249–250
Clephanton, 241, 255
Clinopyroxenite, 196
Clockmadron, 77–78, 250
Coal Measures, 2–3, 5, 111–112, **117–121,** 134–135, 162–163, 178–180, 224; *see also* Westphalian
Coalyard Hill neck, 66, 173–176, 178–179, 181, **194–196,** 213
Coble Shore, 249–250
Cocklemill Burn, 240, 249
Colinsburgh, 236, 239, 253–254
COLVINE, R. J. L., 171, 190, 215, 255
Conglomerate: *see* carbonate conglomerate

Cordies Mealling, 91–92
—— —— Fault, 222, **226**
Cornceres, 29–30
Cornstone, concretionary, 7; *see also* carbonates, nodular
——, conglomeratic, 7; *see also* carbonate conglomerate
Cracoean Stage (B), 9, 130
Craig Coals, 104–105, 107
Craigforth neck, 173, 178, 181, **186–188**
Craighall, 50, 69
—— Burn, 50–51, 63, 78, 144
—— Colliery, 75
—— Den, 50, 69, 165
—— Quarry, 65, 70
—— sill, 144
Craig Hartle, 33
—— —— Anticline, 33, 222
—— —— Fault, 33, 222, **226**
—— —— Lower Marine Band, 11, **33**
—— —— North Marine Band, 11, **33**, 34, 124
—— —— north section, 33
—— —— South Marine Band, 11, **33**, 124
—— —— Syncline, 33
Craighead Quarry (Baldutho), 143
—— —— (Fife Ness), 230–231
Craiglumphart Quarry, 48, 166
CRAIG, R. M., 14, 41, 44, 46, 48–49, 171, 202
Craig Rock (Largo Law), 173, 178–179, **204**, 207–**209**, 211
Craigton Blackband Ironstone, 72–73
—— Common, 72
Craigtoun Hospital Borehole, 99, 101, 140, 142
—— Park No. 1 Borehole, 85, 99–101, 140, 142
—— —— No. 2 Borehole, 85, 101, 140, 142
Crail, 30, 122, 230, 242
—— Coves, 28
—— Harbour Fault, 12, 28, 222, **225**
—— —— Marine Band, 11, **28**, 37, 125–126
—— section, 28, 37
Cryptovolcanic ring-structures, 8, **173–174**, 183, 186, 188–189, 191, 194–195, 197–198, 200–201
CULLINGFORD, R. A., 236, 239–240, 242
CUMMING, G. A., 14, 56–57, 63, 66, 68, 71, 75, 79–80, 171, 174, 189–191, 193–194, 196, 214, 224–225, 228, 252, 255
Cuniger Rock, 23
—— —— Fault, 20, 23, 222, **230–232**

Cuniger Rock Marine Band, 11, 18, 20, 22–23, **24**, 28, 37–38, 125–126
—— —— section, 20, **23–24**, 27, 37
Cunner Law, 223
CURRIE, ETHEL D., 9, 60, 82, 103, 130, 132–133

Danes Dike Fault, 12, 21, 27, 29, 222, **225**, 231
—— —— section, 29
Davie's Rock, 178–179, 181, 195, **197**, 212–214
Dean Bridge, 206
Denbrae Farm, 49
—— —— Marine Band, 11, 46–47, **49**, 127
—— House Marine Band, 11, **49**, 127
Den Burn, 82, 91, 144
—— Coal, 118, 120
Denhead, 44, 47, 62, 65, 70, 72, 74, 76–77, 79, 82, 84, 94–101, 223, 229, 235
—— Blackband Ironstone, 62, 67–69, **72–74**, 253
—— Coalfield, 94, 101
—— No. 1 Borehole, 202
—— No. 2 Borehole, 202
—— Syncline, 77–78, 142, **222–223**
Denork, 48, 101, 138, 229
—— Borehole, 44–45
—— Coalfield, 94, 101
—— Craig, 101, 142
—— intrusion, 142, 156
—— Gas Coal, 101
—— House intrusion, 142
Den Quarry, 42
Den, The, 201
Dinantian, 179; *see also* Lower Carboniferous, Viséan, Tournaisian
Dolerite, tholeiitic, 165, 168; *see also* olivine-dolerite, quartz-dolerite
Donaldson Coal, 83, 95, **97**
Dovecot (of Newark Castle), 231
—— neck, 181, 195, **197**
Dreel Burn, 1, 56, 206, 241, 250
Drumcarro, 60, 80, 84–85, 138
—— Borehole, 14, 44–46, 52, 65, 67–74, 76–78, 101, 128, 132, 140, 144, 147
—— East Colliery, 94, 99
—— Eight Feet Coal, 100–101
—— Four Feet Coal, 100–101
—— intrusion or sill, 136, **141–142**, 149–150, 152, 156, 177, 202, 218
—— Main or Thick Coal, 100–101
—— Quarry, 142, 150, 152

Drumcarro Splint or Five Feet Coal, 100–101
—— West Colliery, 94, 100–101
Drumcarrow Craig, 1, 137, 141–142, 202
Drumeldrie, 204
Drumhead Pit, 77–78
Drumlins, 233–234
Drummochy Coalfield, 117–118
Dumbarnie, 236, 239–240
—— Borehole, 82, 93, 103, 105–109, 133–134
Dunfermline Splint Coal, 83, 87–88, 90, **92,** 95–96, 100
—— Under Coal, 83, 86–88, 90, **92,** 99
Dunicher Law, 1, 137, 140, 145, 206
—— —— sill, 136, 145, 148–150, 152, 154, 156, 158, 212
Dunino, 35
—— Burn, 35, 145, 203
—— neck, 181, 203, 216
Dunotter, 251
—— Borehole, 63, 65, 79, 83–86, 89–94, 98–99, 132–133, 138, 140, 144, 146, 156, 164–165, 168
—— sill, 138, 140, 144–146, 154, 156, 158–159
Dura Den, 44, 227, 233
—— —— Fault, 2, 12, 44, 163, 166–167, 221–222, 224, **227,** 228
Durie Fault, 111–112, 114, 117, 222–223, **224**
Dykes, basanitic, 145, 173; monchiquite, 136, 206; quartz-dolerite, 4–6, 162, 164, **167–168,** 198; *see also* neck intrusions

Earl David's Parrot Coal, 118–119
Earlsferry, 82, 86–89, 91–92, 224, 248, 251
—— Bay, 188, 197
—— Fault, 222, 224–225
Earl's Ferry seams, 86
EARP, J. R., iii
Easter Balrymonth Hill, 198
—— Grangemuir, 55, 255
—— Newburn, 204, 211
East Sands to Maiden Rock coast section, 41
Elderburn Farm, 101
Elie, 60, 71, 75–78, 80, 82, 84, 86–93, 239–241, 248–249, 251, 255
—— Coalfield, **86–89,** 93
—— Harbour neck, 173, 176, 178, 181, 187, **189**
—— Main Coal, 88–89, 92

Elie Ness neck, 173–178, 181, 187, **189–190,** 212, 214–215, 255
—— 'rubies', 171, 190, 215, 255
—— Salt Coal, **87–89,** 92
—— section, 77
—— Station, 250
—— –St Monance coast section, 6, 56–58, 188–198
—— Thick Coal, 88
ELLIOT, R. W., iii, 158, 163, 168, 183
'Encrinite-bed', 9, 13, 24, 38–39, 125, 129–130
Essexite, teschenitic, 139–140, **146,** 148, 151, 153–157, 159–160
'Extra' Coal, 118–119

Falfield, 79
—— Bank, 202, 226
—— Colliery, 76–77
—— House, 77
—— Mine, 78
Faults: normal, 224–227; north-easterly, 5, 227–229; oblique, 229–231; reversed, 231–232; thrust, 230; transcurrent, 114, 224, 227–231; wrench, 229–231
Feddinch, 201, 207
—— Borehole, 50, 140, 143
—— neck, 181, **201,** 217–218
Feldspathic patches and veinlets, in neck intrusions, **212–213,** 215
Fife Ness, 13, 21
—— —— Beds, 12–17, **20–21,** 30, 32, 123–124, 255
—— —— Sandstone, 21
—— —— section, 20–21
FISHER, W. D., iv, 269
Five Foot Coal, 83, 87–88, 90, **93,** 95, 97, 100
—— —— (of Ceres), 83, 95, **96–97**
—— —— Limestone, 68, 73, 76
Flagstaff Hill, 205, 208–209
—— —— intrusion, 205, 208–209, 211, 213
Flandrian marine transgression, 6, 248
Fleecefaulds, 164
—— sill, 164
FLEMING, J., 249
Flisk Quarry, 44
Fluke Dub, 8, 198
—— —— section, 20–21
Fluvio-glacial sand and gravel, 3, 235–236, 238, 254

Fore Coal, 12–13, 16, 26, 34, 51, **52–54,** 227
—— —— (of St Monance), 87
FORSYTH, I. H., iii
Foulhouse Coal, 87
Four Coals of St Monance, 87
'Fourteen-fathom' Pit, 91
FRANCIS, E. H., iii, 56, 68, 85, 87, 89, 91, 93, 111, 115, 119, 171, 174–177, 189–191, 223, 228
Freshwater alluvium, 3, 250

Galena, 254
Garnet, xenocrysts of, 111, 190, 215, 255
Gathercauld Craig, 139
—— sill, 138–139, 226
GEIKIE, A., iii
GEMMEL, J., 98–101
Geophysical investigations, 163, 167–168
Gibliston, 68
—— House, 56
Gillings Hill neck, 181, 206, 210
Gillingshill Reservoir, 235
Gilmerton, 34
Gils Law, 165, 204
Gilston, 156, 165
—— Burn, 204
—— Mains, 82, 157
—— sill, 165
Glacial drainage channels, 236
—— erosion features, 233–234
—— sand and gravel, 3, 233, **235–237,** 238–240, 254
Glassee Coal, 83, 87–88
Goats Marine Band, 11, **29,** 124
GOODLET, G. A., iii
Gordonshall, 54–55
—— Marine Band, 11, 55
GRAHAM, D. K., iii, 122, 126, 257
——, W. G. E., 122
Grange, 50
Grangemuir, 55, 235
—— House, 55
Grange Road, St Andrews, 167, 245
Gravel, 3, 6, 233, 235–249, 254
Greenside, 166
GREENSMITH, J. T., 14–16, 18, 27
Greigston House, 141
—— Mains, 141
—— Waterhole, 137, 141, 149–150, 152, 156
Grey facies (of Calciferous Sandstone Measures), 13, 17–18, 21–26
Ground-moraine, 6; *see also* till
Guard Bridge, 247

HALDANE, D., iii
Hall Teasses, 51, 164
Hatton, 103, 236, 252
—— Burn, 103, 106–107, 109, 113, 141, 223
—— Den, 109
—— Syncline, 109, 113, 222–223
Hattonlaw, 223
HAYNES, J. R., iii, 245, 247
HEDDLE, M. F., 171, 190, 214–215, 255
Higham Borehole, 14, 51–52, 65, 67–68, 70–72, 74–76, 129–131, 140, 145, 253
—— sill, 136, 138, 144–146, 154, 158–159, 226
'High Raised Beach' deposits: *see* late-Glacial marine and estuarine deposits
HOPGOOD, A. M., 56–57, 171, 189–191, 223, 228
Hornblende, kaersutitic, xenocrysts of, 212–214
HOWELL, H. H., iii
Humbie, The, 8
Huntershill Cement Limestone, 106
Hurlet Limestone, 9, 60, 63, 69, 252

Incharvie, 235
Index Limestone, 82, 103–104, **106,** 133
Innergellie Haven, 27
—— Marine Band, 11, 27
Intrusions: *see* olivine-dolerite sill-complex, quartz-dolerite sills and dykes, neck intrusions
IRVING, J., 138–139, 141, 145, 147–148, 162, 165, 168–169, 206
Isle of May sill, 136–138, **146,** 150, 152, 154, 156, 158–159

JACKSON, NAN P. D., iii
Jersey Coals, 83, 87–88, 97
John Barnet's Coal, 89
Johnny Dow's Pulpit, 23
JOHNSON, N. M., 253
JOHNSTONE, G. S., iii
Johnstone Shell Bed, 83, 86–87, 89–91, 94–96, 99, 132–133

Keddie's Pit, 89
Keil Burn, 1, 103, 106, 108–110, 113, 117, 134, 204, 206
Kellie, 52–54
—— Castle, 54
—— Law, 1, 52, 205–206, 208–209

Kellie Law neck, 52, 181, **205–206**, 212, 214–215
—— Mill, 56
Kelty Blackband Ironstone, 83
—— Main Coal, 83, 88, 97
Kemback, 37, 40, 42–44, 254
—— dyke, 163, **167,** 227
—— Hill, 42–43, 167, 227
—— Quarry, 43, 253
—— Wood, 42–43, 167, 253
Kenly Mouth Fault, 33, **225–226**
—— —— Marine Band, 11, **33,** 124
—— Water, 1, 33, **35,** 236, 242
—— ——, mouth of, 32–33
Kilbrackmont–Baldutho sill, 137, **143–144,** 149, 177, 253
—— Craig, 143
—— Knock, 205
—— Quarry, 137, 140, 149–150, 152, 156, 253
Kilconquhar, 240
—— Mill, 68
—— Station, 240
Kilduncan Burn, 36
Kilminning Castle, 29
—— —— Musselband, 29
—— —— section, 29
Kilrenny, 239–240, 242
—— Burn, 29–30
—— Mill Musselband, 27, 29
—— section, 27
Kinaldy, 236
—— Burn, 34–35, 203, 236
—— Den, 35, 145, 151, 153
—— neck, 145, 177–179, 181, **203,** 212
—— sill, 136, **144–146,** 149, 151, 153, 155–156, 158, 177
Kincaple, 42, 247
—— Den, 42
—— Marine Band, 11, **42**
Kincraig, 6, 175, 178, 240
—— neck, 176, 181, **185–186,** 210–211
—— Point, 186
Kingask, 136
—— Quarry, 146, 151, 153
—— sill, 138, **146,** 149, 151, 153–154, 156
Kingsbarns Fault, 8, 13, 32, 222, **225**
—— Harbour, 32
—— section, 32
Kinkell Cave, 39–40
—— Farm neck, 162, 168, 181, **198**
—— Ness, 37, 40, 229–230, 242
—— —— neck, 39–40, 181, **199–200,** 214, 216
Kinness Burn, 1, 41, 167, 233, 244–247, 249

Kinninmonth, 94
—— Burn, 147, 202
—— Den, 171
—— Hill, 162, 166, 168–169
—— sill, **166,** 227
Kippo, 36
—— Burn, 36
KIRK, S. R., 14, 39, 122, 171, 200, 223, 229–230
KIRKBY, J. W., 9, 13–15, 22–27, 29, 31–33, 39, 54, 62–63, 66, 70–71, 75, 111, 122, 129, 198
Kirklatch Marine Band, 11, 20, **24,** 28, 37–38, 126
Kittock's Den, 34, 168, 207
—— ——, mouth of, 33–34, 200, 217, 230
Knock Hill, 42–43
—— —— Farm, 167
Knockhill Marine Band, 11, 37, **42–43**
Knock Hill Quarry, 43, 253
—— —— ——–St Andrews dyke, 167–168
KNOX, J., iii

Ladebraes section, 37, 40, **41,** 42
Ladeddie, 70, 84–85, 98, 137, 235
——, Backfield of, 69, 72, 76
—— Backfields Colliery, 94, 101
—— Frontfields Colliery, 94, 98–99
—— Hill, 101, 141–142, 202
—— Limeworks, 69–71, 74, 101, 233
—— neck, 142, 177, 181, **201–202,** 211
—— Quarry, 70
Lady's Tower, 187, 190
Lambieletham neck, 181, 201
LANDALE, D., 13, 16, 34, 52, 56, 62, 68–72, 75–78, 80, 84–89, 91, 93–97, 100–101, 109, 112, 119–120, 227
Larennie, 250
Largo, 103, 119–120, 122, 134, 248
—— Bay, 117, 248–251
—— Burn, 115, 134
—— House, 113, 178–179
—— Law, 1–2, 5, 111–113, 115, 117, **203–204,** 211
—— —— neck, 112–113, 173, 175, 178–179, 181, **203–204,** 211, 214, 216
—— Syncline, 221–223
Largobeath Colliery, 76–77, 82
Largoward, 60, 62, 76–77, 79–80, 82, 84–85, 89–93
—— Black Coal, 61, 65, 67, **76–77,** 88
—— Colliery, 62, 76–77, 80
—— No. 1 Pit, 77
—— Parrot Coal, 92

Largoward Splint Coal, 61, 65, 67, 76, 77–78, 88, 252
—— Thick Coal, 87–88, 90, 91, 92, 96
Late-Glacial marine and estuarine deposits, 3, 235–236, 237–247, 249, 254
—— —— raised beaches, 236–237, 239–240, 242, 244–245
—— —— sea, 236–238, 244
—— —— shoreline features, 237–247
Lathallan, 60, 79–80, 82, 89, 91–92, 235
—— Borehole, 67, 70, 72, 75, 85, 205
—— Colliery, 76–77
—— Little Coal, 80
—— Mill, 143
Lathockar, 68
Lathones Fault, 68, 141, 144–145, 147, 222, 226–227
—— Quarry, 70
Lavas, olivine-basalt, 112–113, 178–179
Lawhead, 77
LAWRIE, T. R. M., iii
Laws Castle, 201
Lead ore, 254
Lecks, The, 32
LEITCH, D., 122
Lherzolite, 171, 196, 207, 213
Limburgite, 207, 218
Limestone Coal Group, 2–3, 5, 82–102, 132–133, 165, 178, 188, 229, 255; lithology, 84–85; palaeontology, 132–133; stratigraphy, 85–101; thickness, 82; volcanic rocks, 85
Lingo Burn, 146
—— Burnside, 145–146, 203
—— House, 146, 159
—— sill, 146, 159–160
Lingula bands, 11, 22, 83, 86, 89–91, 94, 96, 99–100, 107, 132
Little Coal (of Ceres), 83, 95, 97
—— Pilmuir, 103, 235
—— —— Borehole, 85, 103, 105–109, 133–134, 164–165
—— Splint Coal (of Central Fife), 83, 98
—— —— —— (of Ceres), 83, 95, 97
Lochgelly Blackband Ironstone, 83
—— Parrot Coal, 83, 87–88, 93, 97
—— Splint Coal, 83, 87–88, 93, 97
Lochore Marine Band, 107
Lochty, 52–54, 206
—— Marine Band, 11, 53–54
—— sill, 144–146, 149–150, 152, 154, 156, 159

Lower Ardross Limestone, 11, 20, 26, 45, 47–49, 57, 195, 197; *see also* Ardross Limestones
—— Baldinnie, 165
—— Bollandian Stage (P_1), 9, 130
—— Carboniferous, 3, 9–81, 228; palaeontology of, 122–132
—— Cardenden Smithy Coal, 83
—— Coal Measures, 3, 5, 117–119, 163; volcanism, 117–119, 163, 177–179; *see also* Westphalian A
—— Coxtool Coal, 118, 120
—— Jersey Coal, 83, 87–88, 97
—— Kenly, 36
—— Kinniny Limestone, horizon of, 61–62, 64, 67, 78, 132
—— Largo, 117, 134, 224
—— Limestone Group, 2–4, 55–56, 60–81, 86, 99, 123, 128, 130–132, 163, 178, 229, 252–253, 255; cycles, 61; lithology, 61–62; palaeontology, 130–132; stratigraphy, 62–80; thickness, 60
Low Lime Coal, 70
—— Little Coal, 70
—— Raised Beach, 248
Lumbo, 70, 72, 76–78, 223
—— Borehole, 140, 143
—— Bridge, 41, 49, 74, 76
—— Burn, 49–50, 142, 229
—— Den, 42, 63, 69, 74, 142
—— Farm, 69, 72, 74
—— sill, 74, 142
Luncart Coal, 83, 95, 98
Lundin Anticline, 117, 222–223
—— Coalfield, 117, 119–120
—— Links, 103, 105–106, 109, 112, 221, 223, 239, 254
—— —— neck, 162, 177–179, 180; intrusions in, 163, 178–180, 182–183, 207, 216–217
—— Mill, 224, 236, 239–240
Lyoncross Limestone, 133

MACGREGOR, A. R., 7–8, 14, 32, 39–40, 57, 171
Macgregor Marine Bands, 9, 130
MACLAREN, C., 62
MACNAIR, P., 63, 71
Madras College, St Andrews, 245–246
Maiden Rock, 37, 229, 231
—— —— Fault, 2, 12, 39–40, 50, 94, 143, 168, 181, 199, 222–223, 229

Maiden Rock–Kittock's Den section, 39–40
Main Coal (of Craighall), 75
—— —— (of Drumcarro), 100–101
—— —— (of Elie), 88–89, 92
—— —— (of Kelty), 83, 88, 97
—— —— (of Radernie), 62, 65, 67, **71**, 72–73, 75
—— —— (of Rires), 89
—— —— (of Teasses), 75
Make-Him-Rich Coal, 83, 95, 98
MANSON, W., iii, 52, 84
March of Lathones, 223, 235
Marine and estuarine deposits: *see* late-Glacial and post-Glacial marine and estuarine deposits
—— bands, 4–5, 85–86, 123–124, 129–130; in Calciferous Sandstone Measures, 11, 20, 22, 26, 123–124, 129–130
—— limit (late-Glacial), 237–239, 242
Marl Coal, 60–61, 63, 65, 67, 79, **80**, 90, 96
—— —— (of Craighall), 75
—— —— (of Radernie), 62, 65, 67, **71**, 72–73, 75
MCDONALD, J., 166–167, 254
McGowan's Harbour, 8
Meltwater channels, 236
Mid Coal of St Monance, 87
Middle Coal Measures, 3, 5, 112, **117–120**, 134, 162–163, 179–180; *see also* Westphalian B
Mid Kinniny Limestone, 61, 64–65, 67–68, **79**, 132, 252
Mill Hill Marine Band, 61–62, 64–67, **76**, 131–132
Millstone Wood, 42–43
Mineral veins, 254
Monchiquite, 5, 136, 172–173, 198, 203–204, 206–209, **210–212**, 213
Morton of Blebo, 254
—— Wood, 167
Mount Melville, 84, 142
—— —— Borehole, 50
—— —— Coalfield, 94
—— —— Cottages Borehole, 44–46, 48–49
—— —— Lodge, 76
—— —— sill, 100, 142, 229
Muircambus, 235, 240, 251, 254
—— Borehole, 65, 67, 76–77, 79, 131–132
Muiredge, 29
'*Myalina* Bed', 13, 38, 125–126
—— Limestone', 13, 38–39, 125–126

My Lord's Coal, 87
Mynheer Coal, 83, 87
Myreton of Blebo, 254

Namurian, 5, 82–116, 132–134, 136–137, 175, 177–179; *see also* Limestone Coal Group, Upper Limestone Group
Neck intrusions, 4, 6, 154, 173, **206–218**, 254
Necks, volcanic, 1–2, 4–6, 105, 136, 140, 145, 167, **171–220**
Neilson Shell Bed, 62, 65, 75–76, 131
Nepheline, 154, 207, 209–210, **212**
—— -basalt, 207, 211
—— -basanite, 197, 206–209, 211–213
—— -monchiquite, 206–209, 211–212
Nephelinite, olivine-, 207, 209
Nether Radernie, 72, 147, 226–227
NEVES, R., iii, 9, 14, 31, 94, 117, 119, 130, 163
Newark Castle, 57, 173, 228
—— —— Syncline, 57
—— neck, 178, 181, **195–197**, 213
Newbigging Marine Band, 11, 46, **48**, 128
—— of Blebo, 46, 48
—— —— Ceres, 236
—— —— Craighall, 50–51, 69, 75, 137
Newburn Church, 106
New Gilston, 165
New Mill, 41, 167
—— —— Marine Band, 11, 37, **41**, 127
Nodules, basic, 184, 190; ultrabasic, 184; *see also* ultrabasic inclusions, xenoliths
North Baldutho, 82, 146, 151, 153–154, 156, 159
—— Bank, 51
—— Bowhill, 77
—— Coal, 83, 95, 97
—— -east-trending belts of complex structure, 191, 194, 221–223, **227–229**
—— Haugh, St Andrews, 244, 247
Nydie Marine Band, 11, 37, **42–43**
—— Quarry, 42, 253
—— Wood, 42–43

Oil-shale, 29, 253
Old Red Sandstone: *see* Upper Old Red Sandstone

Olivine-analcimite, 207
—— -basalt, 6, 113, 137–140, 143–145, 148–158, 173, 186, 188–189, 201–202, 204–214, 217–218, 253; Craiglockhart type, 139, 158, 211; Dalmeny type, 113, 137–140, 143–145, 148–157, 205, 217, 253, chemistry, 148–154, mineralogy, 154–157, petrography, 148–157; Hillhouse type, 188–189, 202, 204–214; lavas, 112–113, 178–179; Markle type, 30
—— -dolerite, 1–2, 4–6, **136–161**, 177, 189, 202–203, 206, 217–218; non-ophitic, **137–142**, 148–150, 152, 154–157, 207, 217–218; ophitic, **137–141**, 148–150, 152, 154–157; sill-complex, 3–6, **136–161**, chemistry, 148–154, mineralogy, 154–158, petrography, 148–154; teschenitic, 137–140, 144–146, 148–150, 152, 154–156, 158, 203
—— -nephelinite, 207, 209
Orchard Beds, 104–105, **107**, 133
Orthoclase, xenocrysts of, 188
Orthopyroxene, xenocrysts of, 206, 213
Ovenstone, 55
—— *Lingula* Band, 11, **55**
—— Marine Band, 11, **55**

Palaeontology, Carboniferous, 122–135
Pans Fault, 28, 222, **225**
—— Marine Band, 11, **28**, 37, 125–126
—— section, 28
——, The, 225
Parrot Coal (of Capledrae), 104, 107
—— —— (of Denhead), 78
—— —— (of Denork), 101
—— —— (of Ladeddie), 98–99
—— —— (of Largoward), 92
—— —— (of Lochgelly), 83, 87–88, 93, 97
—— —— (of St Monance), 87
Passage Group, 2–3, 5, **111–116**, 119, 134, 163, 178, 183–184; bedded tuffs and lavas, 111, 178; palaeontology, 134; sedimentary rocks, 111
Pathhead, 14, 19, 26, 47, 52, 57, 66
—— -Anstruther coast section, 10, 13, **21–26**, 27–28, 34, 37–38, 44, 57
—— Beds, 11–13, 16–17, 19–20, **25–26**, 36, 41, 44–51, 54, 56–57, 123, 127–130
—— Lower Marine Band, 11, 20, **26**, 45–47, 128–129
—— Marine Bands, 26, 44, 51, 128–129
—— section, 20, **26**, 49

Pathhead Upper Marine Band, 11, 14, 20, **26**, 45, 47, 50, 128–129, 198
Peat, 3, 6, 241, 248–250
Peatlow Rock, 230–231
—— —— Fault, 27, 222, **225**
Pendleian Stage (E_1), 5, 82, 132; *see also* Limestone Coal Group
Pepper Knowe Anticline, 222–223, 226
—— —— neck, 181, 202–203
Peridotite blocks in neck, 213; xenoliths, 207, 213; *see also* lherzolite
Permian, 4, 6, 112, 162, 171, 177–178, 180
'Petrified forest', 27
Phreatomagmatic eruption, 175
Picroteschenite, 138, 144–146, 148–149, 151, 153–156, 158–159, 200
Pilkin Coal, 83, 88
Pilmour Links, 251
Pitcorthie, 29–30, 253
Pitcruvie, 109, 236
—— Den, 110
Pitmilly Burn, 35
—— Law, 242
Pitmullen, 198–199
Pitscottie, 44, 46, 166, 235–236, 250
—— Vale, 166, 236
Pittenweem bathing pool, 19, 24–26
—— Beds, 11–13, 15, 17, 20, **23–24**, 25, 33–40, 42–43, 51, 55–56, 124–127, 129–130, 253
—— East Harbour, 24, 198
—— Fault, 20, 23–24, 222, **225**
—— Harbour, 23–24, 198
—— —— *Lingula* Band, 11, 20, 24, 35
—— —— section, 20, **24–25**
—— Marine Band, 9, 11, 18, 20, **24**, 28, 37–38, 125–126
Plean Limestones, 104–106, 109, 134
—— No. 2 Limestone, 104–105, 109
—— No. 3 Limestone, 104–105, 109
Pleistocene and Recent, 3, 6, **233–251**
Post-Glacial marine and estuarine alluvium, 3, 238, 243, 246–247, **248–250**
—— —— marine transgression, 6, 237, 248
—— —— raised beach, 241, 248–250
—— —— sea, 248
—— —— sea-cliff, 248
Potassium–argon dating, 137, 175, 178–180
Priorletham, 250
Priormuir, 50, 235
Productus giganteus Limestone, 71
Pseudo-brecciated bed, 63, 66

Pyrope, xenocrysts of, 171, **215**, 255
Pyroxenite, 171, 207, 213

Quartz-dolerite, 1–2, 4–6, 136, **162–170**, 180, 202; blocks in necks, 162, 177, 180, 183; dykes, 4–6, 162, 164, **167–168**; petrography, 168–169; sills, 4–6, 162, **163–167**, 227, 229, 254
——, xenocrysts of, 216–217
Queenslie Marine Band, 117–119

Radernie, 60, 62, 65, 68, 70–71, 75
—— Brassie Coal, 61, 65, **67**, 70
—— Colliery, 62, 71
—— Duffie Coal, 62, 65, 67, **71**, 72–74, 142
—— Fault, 68, 141, 147, 222, **227**
—— Main Coal, 62, 65, 67, **71**, 72–73, 75
—— Marl Coal, 62, 65, 67, **71**, 72–73, 75
—— Quarry, 68, 70, 206, 212, 253
—— Syncline, 68, 71, **222–223**, 227
Radiometric dating, 137, 175, 177–180
Randerston Castle Fault, 20, 31, 222, **230–231**
—— Fault, 31, **222**
—— limestones, 20, 22, **31–32**, 124, 129–130
—— *Lingula* bands, 11
—— marine bands, 11, 20, **32**, 124
—— section, 14, 20, 22, **31–32**, 36, 124
—— Syncline, 31, 222, 231
Red facies (of Calciferous Sandstone Measures), 7, 13, 16–18, 21, 24–25, 54
—— Limestone, 80
Ribbonfield, 30
Ring-fractures, 176–177, 183, 189, 193, 196, 200
Rires Coalfield, 82, **89**
—— Coals, 93
—— Craigs, 173, 205, 208–209, 211
—— Main Coal, 89
—— neck, 112, 173, 181, **204–205**, 211
Roadstone, 253–254
Rock and Spindle, 199–200, 214
Roome Bay, 28–29, 231, 242
—— —— Fault, 28–29, 222
—— —— section, 28–29
—— Harbour, 29
—— Rocks, 28–29
Ruddons Point, 105, 173, 184

Ruddons Point neck, 178, 181, **184–186**, 211, 213, 215, 255
'Rums', 72–73, 77, 78

Salt Lake, 32
—— —— neck, 181, **201**, 210, 213, 216
Salt marsh deposits, 3, 241
Sand, 3, 6, 233, 235–250, 254
Sandy Craig, 25
—— —— Beds, 11, 13–17, 20, **24–25**, 36–42, 44, 49, 51, 54, 56, 123, 126–127, 253, 255
—— —— Fault, 20, 25, 222, **225**
—— —— section, 20, **25**
Sauchar Point, 190
Scarrat Loft Coal, 77
Scooniehill, 201
SCOTT, A., 250
——, T., 250
Scrumpie Coal (of Denhead), 101
Seafield clay pit, 247, 255
—— Marine Band, 61–62, 64–65, 67, **77**, 132
Seven Foot Coal, 83, 87–88, 98
Shear joints, 232
Shell Bay, 93, 103, 106, 133, 184, 186, 249
Shooters Point, 184
'Shrimp band', 57, 196
Sill-complex, olivine-dolerite, 136–161; *see also* olivine-dolerite sill-complex
Sills, quartz-dolerite, 4–6, 162, **163–167**, 227, 229, 254
Six Feet Coal (of Lathallan), 92
Six Foot Coal (of Ceres), 98–99
Skipsey's Marine Band, 117–118, 120
Small Coalfields, 94, 98–101
SMITH, D. E., 236, 239–240, 242
——, R. F., 253
Smithy Coal, 83, 87–88, 90, **91–92**, 95–96
SNELLING, N. J., iii, iv
South Baldutho, 68
—— Flisk, 43
—— Haugh, St Andrews, 245
—— Lambieletham, 201
Spalefield, 29, 136
—— sill, 138, **147–148**, 156, 160
Spinel, xenocrysts of, 184, 196, 213
Splint Coal (of Drumcarro), 100–101
—— —— (of Largoward), 61, 65, 67, 76, **77–78**, 88, 252
—— —— (of Lochgelly), 83, 87–88, 93, 97
—— —— (of St Monance), 87

Sprattyhall, 89
St Andrews, 1, 23, 122, 125, 127, 129, 242–249
—— —— Bay, 248
—— —— Castle Marine Band, 11, 13, 23, 25, 34, **37–38**, 39–40, 125–127
—— —— shore section, 37–39
—— —— Wells, 69, 166–167
St Monance, 1, 60–63, 67, 70, 76–78, 122, 240, 252
—— —— Brecciated Limestone, 9, 26, 44–45, 47, 50–52, 56–57, 60–65, **66–70**, 75, 109, 130–131, 164, 166–167, 195, 252–253
—— —— Burn, 198
—— —— Coalfield, 84, **86–87,** 93, 228
—— —— Harbour Coal, 76
—— —— intrusion, 207–209, 212
—— —— Little Limestone, 61, 64–65, 67, 70, **71**, 73–74, 131
—— —— neck, 163, 173, 175, 178–179, 181, 195, **197–198**
—— —— Station, 241
—— —— Syncline, 61–63, 66, 71, 75–79, 84, **222**
—— —— White Limestone, 11, 19–20, **26,** 45, 47–48, 50–52, 57, 63, 129–130, 193, 195, 252
St Nicholas, 251
—— —— Marine Band, 11, **40–41,** 127
Stephanian, 5–6, 113, 175, 177–180
Step Lake, 38
—— —— mudstones, 38–39
Strathairly, 112
—— Burn, 112, 115
—— coals, 112, 119
—— House, 119
Strathkinness, 42, 253–254
Strathtyrum, 42
—— House, 201, 215
Striations, 233–234
Sub-Carse Peat, 241, 248–250
Submerged forest peat, 249
Swallowdrum Coal, 87, 97
Sypsies Plantation, 30

TAIT, D., 14, 26, 57, 63, 66
Teasses, 60, 65, 70, 75, 163–164
—— Colliery, 62, 78
—— Den, 144–145
—— Fault, 222
—— House, 75, 164
—— Main Coal, 75
—— Mill, 69

Teasses Quarry (limestone), 70
—— —— (quartz-dolerite), 164
—— Under Coal, 75
TELFER, A., 100
Temple, 114–115, 224, 230
—— Fault, 112, 114, 222, **224**
—— Lower Marine Band, 111, **114–115**
—— Upper Marine Band, 111, **114–115**
Teschenite, 4, 139, **144–146,** 149–150, 152, 154–156, 158
Teschenitic essexite: *see* essexite
Teschenitic olivine-dolerite: *see* olivine-dolerite
Teuchats, 164
—— Quarries, 70
—— Toll, 235
Thick Coal (of Ceres), 83, **95, 97,** 100–101
—— —— (of Drumcarro), 100–101
—— —— (of Elie), 88
—— —— (of Largoward), 87–88, **90, 91,** 92, 96
—— sandstone facies (of Calciferous Sandstone Measures), 4, 13–14, 17–18, 21–26
Thirdpart, 29
Thirteenth Coal of St Monance, 87
Tholeiite, 5, 162–170
Tholeiitic dolerite, 165, 168
Thomsford Bridge, 109
Till, 6, 233–236, 241–246, 255; grey, 233–235; reddish-brown, 233–235; submarine ablation, 244
Toll Cottage, 236, 240
Tosh, 35, 236
Tournaisian, 130
Trace-fossils, 17, **19,** 22–23, 25–26, 32, 38, 62, 76–77, 79, 86, 91, 93–94, 96, 99, 105–108
TRAQUAIR, R. H., 29, 72, 122
Tuff, 2, 4, 85, 105–107, 112–113, 115, 117, 119, 165, 171–180, 182–206, 217
Tuffisite, 99, 139, 146, 173–174, 176–177, 183–184, 186, 188–200, 203
Tuffisitic breccia, 173–174, 186–189, 192–193, 195–200
Tuff-rings, 175, 177, 184, 186
Two Foot Coal, 83, 87–88, 90–91, **92–93**
—— —— —— (of Ceres), 83, 95, 97–98

Ultrabasic inclusions, 171, 184, 196; *see also* nodules, basic and ultrabasic, and xenoliths
Upper Ardross Limestone, 11, 20, 26, 45–48, 57, 195, 197

Upper Bollandian Stage (P_2), 4, 60, 130; see also Lower Limestone Group
—— Carboniferous, 3–4, **82–121,** 128, 132, 171, 221; see also Namurian, Stephanian, Westphalian
—— Coal, 91
—— —— Measures, 3, 5, 117–118, **120,** 134–135, 179; see also Stephanian, Westphalian C–D
—— Coxtool Coal, 118, 120
—— Dryas (pollen Zone III), 237, 248
—— Jersey Coal, 83, 87–88, 97
—— Kinniny Limestone, 60–61, 63–64, 67, **80,** 82–83, 87, 89–90, 94–95, 98, 132
—— Largo, 236, 240
—— Limestone Group, 2–3, 5, **103–110,** 112, 119, 133–134, 136, 162–163, 165, 178; lithology, 103; palaeontology, 133–134; stratigraphy, 103–110; thickness, 103; volcanism 105–106
—— Magus sill, 166
—— Old Red Sandstone, 2–4, **7–8,** 20

Vents, volcanic: see volcanic necks
Viewforth neck, 114–115, 117, 119, 162, 175, 177–179, 181, **182–184**
Viséan, 4, 9–81, 123–132; see also Calciferous Sandstone Measures, Lower Limestone Group
Volcanic detritus, 85, 105, 117, 119
—— necks, 1–2, 4–6, 105, 136, 140, 145, 167, **171–220;** age relations, 177–180; field relations, 180–206; intrusions in, 154, 173, 180–205, **206–218,** 254; mechanism of emplacement, 174–177; structure and lithology, 172–173; subsidence in, 175–177
Volcanism, 5, 30, 85, 105–107, 109, 112–113, 115, 117, 119, 171–220
——, basanitic, 179–180

Wadeslea neck, 66, 173–175, 178, 181, **191–193**
Wakefield Burn, 34–35
WALKER, F., 138–139, 141, 145–149, 154, 158–159, 162, 165, 168–169, 206
——, R., 29, 247, 249
WALLACE, MRS I. F., 162–163, 171, 177, 180, 183, 211–212
Wall Coal, 118, 120

Wanderer Coal, 87
WATTISON, A., 29, 122
Wehrlite, 196
West Braes Marine Band, 11, 19–20, **26,** 45–47, 49, 57, 127, 129–130
—— Cassingray, 226
—— Coates sill, 139, 158
Wester Balrymonth, 201
—— —— Hill, 201
—— Lathallan, 204, 211, 214
—— Pitscottie, 164, 166
—— Radernie, 147
Westland Skelly Marine Band, 11, **28,** 37, 126
Westphalian, 5–6, 117, 119, 136–137, 175, 177–179, 183; see also Coal Measures
—— A, 5, 117, 119, 137, 178–179, 183
—— B, 5, 177
—— C–D, 5–6, 175–177
West Pitcorthie, 29
—— Sands Marine Band, 11, 23, 33, **37–38,** 125–126
WHATLEY, R. C., iii, 247
Whinnyhall Quarry, 29, 147
White trap, 173
Wilkieston Burn, 202
—— Quarry, 68, 70, 138, 141, 155–156
WILSON, G. V., iii
——, J. S. G., iii
——, R. B., iii
Winchester, 236
Winthank, 72
—— Blackband Ironstone, 62, 72–73
—— Limeworks, 68, 70
—— Quarry, 70
Witch Lake, 19, 38, 230
—— —— Marine Band, 11, 28, 33, 35, **37–38,** 39–40, 125–126
WOOD, W., 68, 75–76, 194, 239–241, 249–250
Wood Haven, 68, 75, 139, 187, 189–190, 197, 228, 240–241, 248–249
Woodhaven Limestone, 68, 75
Woodtop, 70
—— sill, **144–145,** 154, 156, 158
Wormistone Fault, 12, 20–21, 31, 222, **230–231**
—— Hind, 230–231
—— Lower Marine Band, 11, **31**
—— Marine Bands, 11, 20, **31,** 124
—— section, 14, 20, **31,** 124
—— Syncline, 31
—— Upper Marine Band, 11
WRIGHT, J., 14, 26, 57, 63, 66, 76, 122

Xenocrysts: anorthoclase, 207, 214–215; apatite, 206, 214; biotite, 206, 214; chrome-diopside, 184; chrome-spinel, 184; garnet, 171, 190, 215, 255; kaersutitic hornblende, 206–207, **213–214;** orthoclase, 188; orthopyroxene, 206, 213; pyrope, 171, 190, 215, 255; quartz, 216–217; spinel, 206, **213;** zircon, 190, 214

Xenolithic basalt of Lundin Links, 207, 216–217

Xenolithic dolerite, 143
—— monchiquite, 203

Xenoliths: igneous, 207, **213–215;** of peridotite, 207, 213; of pyroxenite, 207, 213; sedimentary, 212–213, **215–216**

Zircon, xenocrysts of, 190, 214